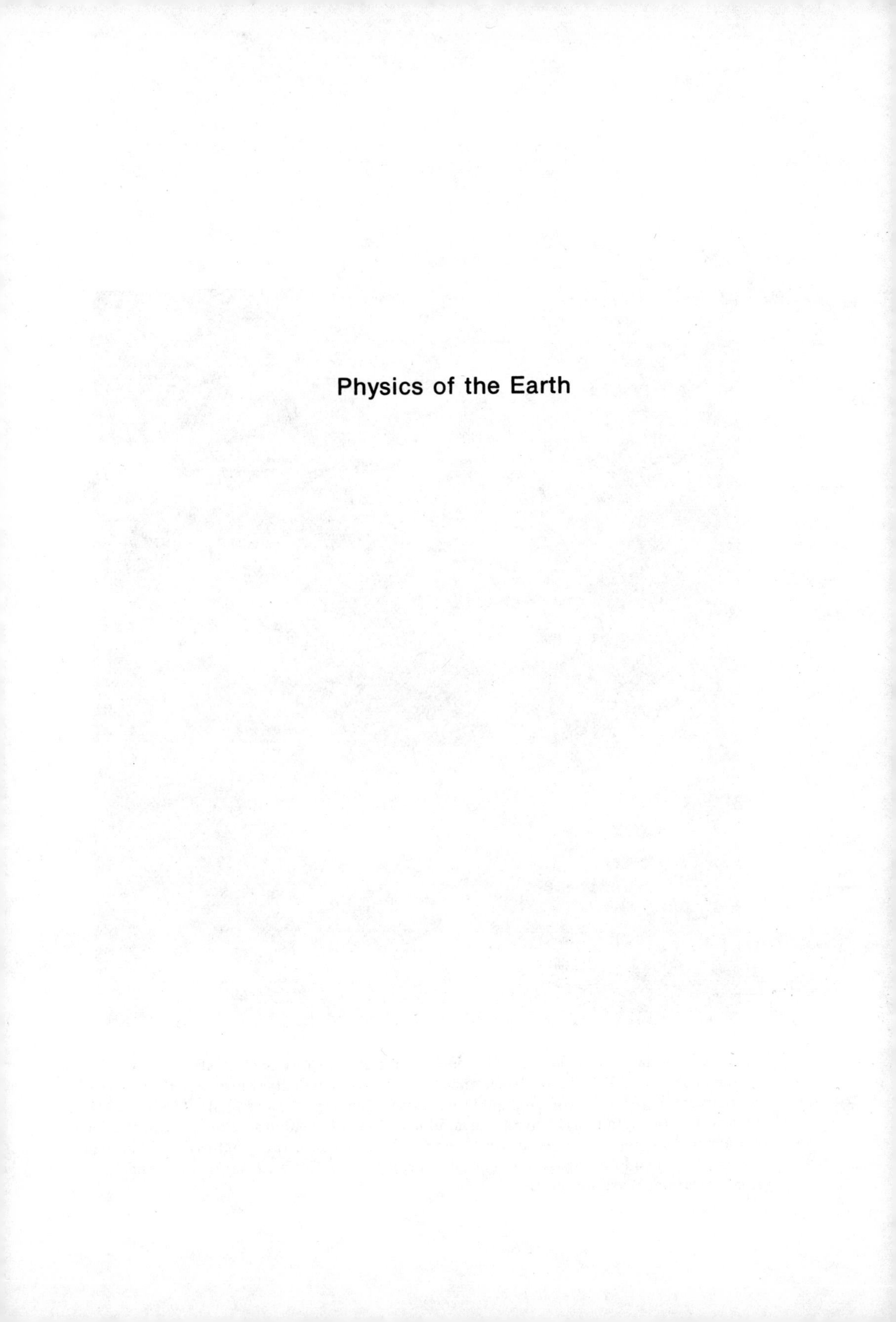

# Physics of the Earth

Gemini XI photograph of the Gulf of Aden and the Red Sea by NASA astronauts Charles Conrad and Richard F. Gordon. This is one of the areas of particular interest in the theory of sea-floor spreading. A line of earthquake epicenters (see Figs. 5.1 and 10.5) extends from the ridge system in the Indian Ocean up the middle of the Gulf of Aden and into the Red Sea and is presumed to mark the axis of a new ridge, along which mantle material is rising and pushing Africa and Asia apart. Photograph courtesy of the National Aeronautics and Space Administration, Washington, D.C.

# PHYSICS OF THE EARTH

## Second Edition

Frank D. Stacey

Professor of Applied Physics
University of Queensland
Australia

JOHN WILEY & SONS
New York Chichester
Brisbane Toronto
Singapore

**Library of Congress Cataloging in Publication Data:**

Stacey, Frank D.
Physics of the earth.

Bibliography: p.
Includes index.
1. Geophysics. I. Title.
QC806.S65 1977 551 76-41891
ISBN 0-471-81956-5

Printed in the United States of America

10 9 8 7 6

# Preface

". . . . . Obviously there are no well qualified students of the Earth, and all of us, in different degrees, dig our own small specialised holes and sit in them."

BULLARD, 1960, p. 92

The original purpose of this book was to bring the fundamental problems in solid-earth geophysics to the attention of graduate and advanced undergraduate students of physics. The study of the Earth has a natural fascination, but for the physicist who has no previous contact with the earth sciences there is a wide range of new concepts. To make the physical arguments clearer, mathematical developments have been simplified, relegated to appendixes or simply outlined, with references to the original literature. The intention is that this should make the text more easily understood also by students of geology and related sciences.

The range of topics is so wide that no single author can aim for comprehensive coverage, and a selection, with some bias of personal interest, is inevitable. My choice is based partly on an attempt to predict the topics that will increase in importance. Thus I have accepted the need for the presentation of problems that are subjects of current controversy. It is important, therefore, that the arguments include relevant citations, to which the reader can turn for more extensive coverage of any topic. Hopefully the bibliography, which represents a careful selection from the now vast literature, coupled with the author index, will become a useful starting point for literature searches.

Although this edition is a complete rewrite, necessitated by developments both of the subject and of my understanding of it, the style and philosophy are essentially unchanged from the first edition. Two minor changes, which represent responses to comments or requests, are the introduction (in Appendix J) of some student problems and conversion to S.I. units. The merit of S.I. units is least apparent in the treatment of magnetism; the geomagnetism chapters may be read in the alternative electromagnetic units by omitting bracketed factors in the equations.

I thank authors and publishers for their response to my requests for permission to use their figures, sources of which are acknowledged in captions. Colleagues who have scrutinized the draft manuscript of this edition, the first edition, or both are A. E. Beck, S. J. Broughton, J. Cleary, W. Compston, H. J. Dorman, H. Doyle, W. M. Elsasser, M. T. Gladwin, D. H. Green, R. A. Haddon, R. D. Irvine, E. Irving, B. Isacks, I. N. S. Jackson, W. H. K. Lee, R. C. Liebermann, A. T. Linde, J. F. Lovering, F. J. Lowes, M. W. McElhinny, A. C. McLaren, N. D. Opdyke, V. M. Oversby, R. L. Parker, W. D. Parkinson, M. S. Paterson, J. R. Richards, W. I. Riley, M. G. Rochester, T. J. Shankland, S. R. Taylor, G. J. Tuck, J. P. Webb, and D. J. Whitford, whose interest and help are greatly appreciated.

Brisbane, Australia FRANK D. STACEY

# Contents

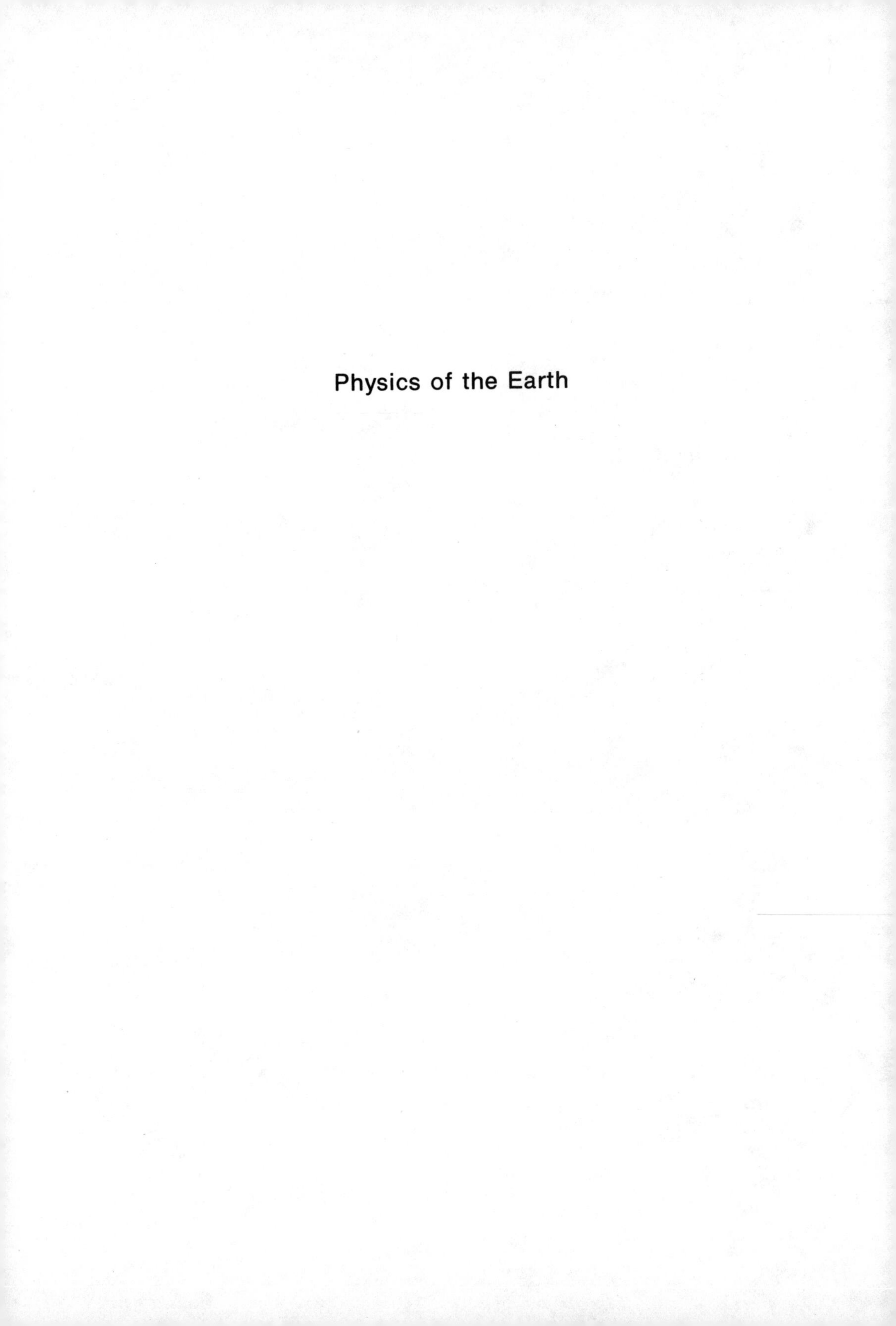

Physics of the Earth

# 1

# The Solar System

"... theoreticians have always succeeded in providing an understanding for all observed phenomena—even those which later proved to be incorrect."

ANON., *OBSERVATORY 94*, 6P (1974)

## 1.1 THE PLANETS

For many years theories of the origin of the solar system were developed from the observed regularity in planetary orbits. More recently the emphasis has been on chemical considerations, with appeals to the densities of the planets and isotopic abundances in the Earth and in meteorites. It is now apparent that the Earth had a common origin with the other planets, about $4.6 \times 10^9$ years ago, in a cloud of gas and dust surrounding the then youthful Sun. The abundances of the elements in the cloud were approximately what would be expected from theories of nuclear synthesis, so that the formation of the solar system required no special conditions; we suppose that there are millions of similar systems of planets even in our own galaxy.

Densities and orbital radii of the planets are listed in Table A.1 in Appendix A. Reliable values of density are much more recent than the orbital data, because precise earthbound measurements of planetary diameters are notoriously difficult to make and they enter as the third power in density estimates. However, the remaining uncertainties leave no doubt about the significant differences in composition, even between the basically similar inner four (terrestrial) planets. These differences are considered in Section 1.5.

The approximate geometrical progression of orbital radii of the planets is known as Bode's law (Roy, 1967) or, more correctly, as the Titius-Bode law. In its original form this gives the orbital radius $R_n$ of the $n$th planet (counted outward) as

$$R_n = a + b \,.\, 2^n \tag{1.1}$$

$a$ and $b$ being appropriate constants. In reviewing theories of the origin of the solar system, Ter Haar and Cameron (1963) suggested that a better fit was obtained with a simple geometrical progression

$$R_n = R_0 m^n \tag{1.2}$$

Including Pluto and counting the asteroids as one planet, the best fit is with $m = 1.71$ or, if the asteroids are counted as two planets, $m = 1.59$. Slightly larger values result from the exclusion of Pluto. We also note that orbital periods are related to orbital radii by Kepler's third law (Eqs. 4.41 or A.17) so that orbital periods follow a law similar to Equation 1.2, but with $m^* = m^{3/2}$, corresponding numerical values being $m^* = 2.235$ or 2.012. The possibility that orbital periods of adjacent planets favor 2:1 ratios as a result of resonant interactions has been suggested (Roy and Ovenden, 1954) but questioned (Dermott, 1968). Most recent discussions of Bode's law are influenced by von Weizsäcker's (1944) model of a turbulent solar nebula, in which the planets accumulated at the boundaries between vortices, whose sizes increased regularly outward from the Sun (with increase in the range over which self-gravitation of accreting material could compete with the gradient of solar gravity).

Historically, the significance of Bode's law arose from the recognition that the orbits of the inner four (terrestrial) planets fit an equation of the form (1.1) or (1.2) reasonably, as do the orbits of the major planets (Jupiter to Neptune), but the two groups can be fitted to a common law only by counting a "missing" planet between them. The search for a "missing" planet led to the discovery of the numerous asteroids or minor planets whose orbits are concentrated in the range between Mars and Jupiter. In Table A.1 the asteroids are counted together as planet number 5, to satisfy Bode's law. However, the approximate nature of Bode's law is indicated in Figure 1.1 by comparing alternative plots which assume either one or two missing planets, and the gap between Mars and Jupiter is so large that we should perhaps regard the terrestrial planets and the major planets as two distinct groups following two Bode's law relationships independently.

The orbits of all the planets are nearly coplanar and their orbital motions are all in the same sense, which also coincides with the axial rotation of the Sun, making it difficult to question the obvious conclusion that all of the planets were formed simultaneously from a disc of gas and dust corotating with the then youthful Sun. The most highly inclined

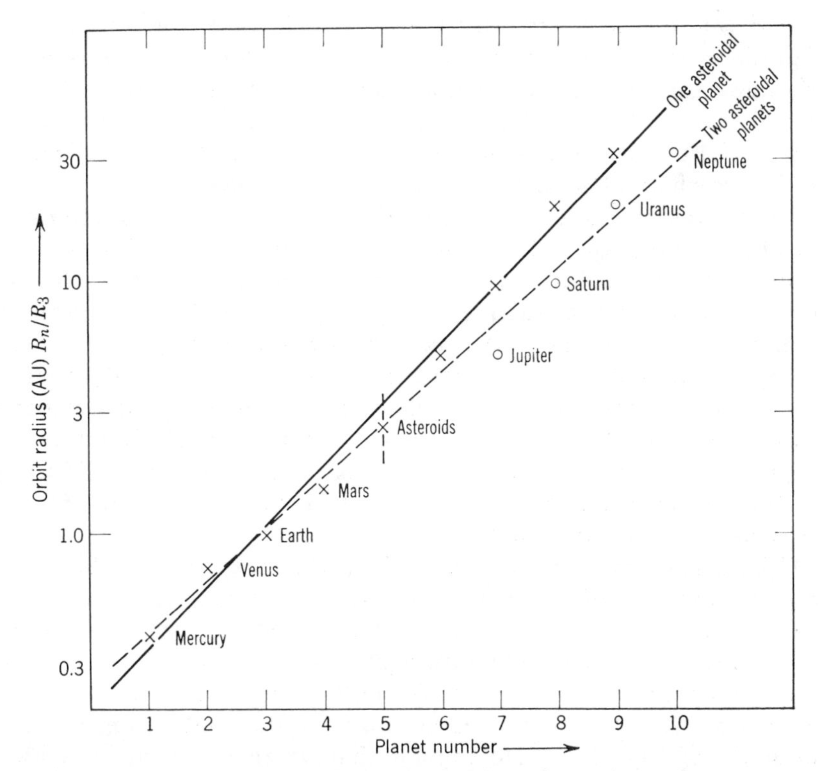


**Figure 1.1.** Radii of the planetary orbits, plotted to show the approximate geometrical progression (Bode's law). Crosses and the solid line apply to the assumption that there is one "missing" (asteroidal) planet; circles and the broken line are obtained by renumbering to allow for two "missing" planets between Mars and Jupiter.

(and most elliptical) orbit is that of Pluto, which appears to have been strongly perturbed by Neptune, whose orbit it crosses. There is even a suggestion that Pluto should be regarded as an escaped satellite of Neptune rather than an independent planet, and for this reason it is not counted with the numbered planets in Table A.1 (Appendix A) and Figure 1.1.

The axial rotations of the planets and the orbits of their satellites are less regular, although the rotations of most of the planets are in the sense of their orbital motions. But even in these cases the rotational axes are significantly misaligned relative to the normals to the orbital planes. In the case of the Earth the misalignment ($23\frac{1}{2}°$) is referred to as the inclination of the ecliptic (i.e., of the orbital plane, relative to the equatorial plane). The extreme cases are Uranus, with a rotational axis almost in the orbital plane, and Venus, which rotates very slowly in a retrograde sense. [Mercury is a special case in which rotation is controlled by tides; the 58.6-day rotation period is two thirds of the period of its elliptical orbit and is stabilized by a resonance of the orbit with the tide raised in the planet by the Sun (Colombo, 1965; Colombo and Shapiro, 1966; Goldreich and Peale, 1966).] The irregular orientations of planetary rotational axes indicate very late accretion of large fragments that plunged into the planets from various directions, imparting some randomness to their angular momenta. Impact cratering on the Moon, Mars, and Mercury shows that such late accreting fragments were common in the solar system for the first few hundred million years. The case of Mercury is particularly significant because its high density indicates that it has a proportionately very large iron core, whereas the surface craters appear in stony (silicate) material which must therefore have formed a stable outer shell before the final impacts occurred (Murray et al., 1974). The late impacts are important in interpreting the history of the Earth-Moon system (Section 4.5).

Although the Sun has nearly 99.9% of the mass of the solar system, its rotation carries only 2% of the angular momentum. If, as we suppose, the whole solar system condensed from a common, corotating cloud of gas and dust, then the distribution of angular momentum requires a special explanation. Alfvén (1954) emphasized the necessity for an outward transfer of angular momentum from the contracting Sun to the outer parts of the cloud. He argued that at an early stage, which is termed the T-Tauri stage by reference to a young star which is presently at that stage of evolution, the Sun had a very extensive magnetic field (drawn outward by an intense particle flux) which exerted a drag on the ionized gas of the surrounding nebula, accelerating the outer parts and slowing down the Sun's rotation. Some subsequent work (e.g., Hoyle, 1960; Sonett et al., 1970) has yielded estimates of the strength of the field in the range 0.1 to 1 G ($10^{-5}$ to $10^{-4}$ T) at the distance of the Earth or the asteroidal belt. This coincides reasonably well with estimates of the magnitude(s) of the field(s) which magnetized the chondritic meteorites, apparently at the time of their formation, about $4.6 \times 10^{9}$ years ago (Section 1.2). However, the probability that at its T-Tauri stage the Sun had closely neighboring stars, which have subsequently dispersed, allows an alternative explanation of the distribution of angular momentum.

Of the planetary satellites the Moon is by far the largest in relation to its primary, the Earth; so much so that we may be justified in regarding the Earth-Moon system as a double planet, rather than as a planet plus satellite. For comparison of density and composition with those of other planets it may therefore be more appropriate to consider values for the Earth and Moon together, rather than the Earth alone, and this possibility is allowed for in Table A.1. However, the Moon and smaller terrestrial planets are similar in both size and density to the inner satellites of Jupiter and Saturn. Io, a satellite of Jupiter, is very close in size to the Moon but has a mean density of about 4600 kg $m^{-3}$, which is appreciably greater than the mean lunar density, 3300 kg $m^{-3}$. In composition Io is almost certainly very similar to the terrestrial planets. A comparison of the masses and diameters of planetary satellites (which are conveniently tabulated by Blanco and McCuskey, 1961, Table VII, pp. 288–289) shows that the inner satellites of Jupiter and Saturn follow a pattern of decreasing density outward, similar to the outward decrease in uncompressed densities of the planets. It is probably that a similar process of segregation of the elements occurred in them.

The seven outermost, small satellites of Jupiter must be considered in a special category because they are almost certainly captured asteroids. Bailey (1971a,b) showed that the three satellites in prograde orbits close to $11.5 \times 10^6$ km were captured at the perihelion (closest point) of Jupiter's orbit about the Sun and the four satellites in retrograde orbits at about $23 \times 20^6$ km were captured when Jupiter was at aphelion (farthest from the Sun). The phenomenon of satellite capture was probably significant for other planets also. Of Neptune's satellites, Triton has a retrograde orbit and Nereid has a highly elliptical orbit, although the case for capture is confused by the possibility of interaction with Pluto. Singer (1970a) argued that the very slow retrograde rotation of Venus was best explained by the capture of a substantial satellite into a retrograde orbit and subsequent spiraling in by tidal friction (as discussed in Section 4.5); the final angular momentum of the (composite) planet, being the sum of its initial, prograde rotation and the captured, orbital retrograde angular momentum, happened to be close to zero. Capture of the Moon by the Earth has also been considered seriously (for reasons also discussed in Section 4.5), but cannot now be favored.

The obvious division of the planets into two groups can be made equally well on the basis of size or density. The four major planets, Jupiter to Neptune, are more remote from the Sun and larger but less dense than the four smaller terrestrial (Earthlike) planets, Mercury to Mars, with which must be grouped also the asteroids. The terrestrial planets and meteorites are composed mainly of nonvolatiles, especially iron and silicon in varying states of oxidation, whereas the major planets have densities so low that they must be composed largely of light, volatile materials, especially hydrogen, with the nonvolatiles in much smaller but unknown amounts. Presumably the light gases were ubiquitous in the solar nebula but did not accrete in significant quantities in the inner planets. Perhaps, being ionized, they were more effectively coupled to the T-Tauri magnetic field of the Sun and so were pulled outward into wider orbits than the nonvolatiles, which remained to accrete into the terrestrial planets. Woolfson (1969) has reviewed theories of the physical processes of planetary accretion.

## 1.2 METEORITES AND THEIR COMPOSITIONS

Meteorites are iron and stone bodies that fall on the Earth in small numbers, apparently on elliptical orbits extending from the asteroidal belt, beyond Mars. Observed arrivals are signaled by fiery trails through the atmosphere (fireballs or bolides), but they are distinct from the much fainter, and only briefly luminous meteor trails. Meteors are produced by small, friable particles of low density (meteoroids), many of which are identified from their orbits as fragments of comets, but which do not reach the ground. There is no association between the arrivals of meteorites and the occurrences of meteor showers, which are observed when the Earth passes through bands of orbiting comet debris, although it must be expected that a few of the sporadic meteors have the same origin as the meteorites (Jacchia, 1963). The total number of observed falls from which the recovery of meteorites has been documented now exceeds 1000 and the number of finds, which were not seen to fall but are certainly meteorites, approximately doubles the total number available for study. The meteorites contain clues to the early history of the solar system which make them scientific samples of the greatest interest. They are the subjects of a large and rapidly expanding literature; comprehensive reviews are by Mason (1962), Anders (1964), Wood (1968) and Wasson (1974).

The determination of preterrestrial orbits of meteorites is made difficult by the irregular and infrequent arrivals; only one fall has been observed with sufficient scientific control (by photographing the trail from several well-separated points) to allow a reliable calculation of its orbit. This occurred at Pribram, Czechoslovakia, in 1959 and the fall was one of the familiar type known as chondrites. The orbit, calculated by Z. Ceplecha, is reproduced as Figure 1.2. Substantially similar orbits have been deduced from photographic records of fireballs that did not yield recovered meteorites but were almost certainly meteoritic and from eyewitness reports of fireballs associated with recovered meteorites (Wood, 1968). These reports reinforce the conclusion that at least the majority of meteorites are asteroidal fragments with orbits that have apogees in the recognized asteroidal belt but are sufficiently elliptical to cross the Earth's orbit. Asteroids or asteroidal fragments smaller than

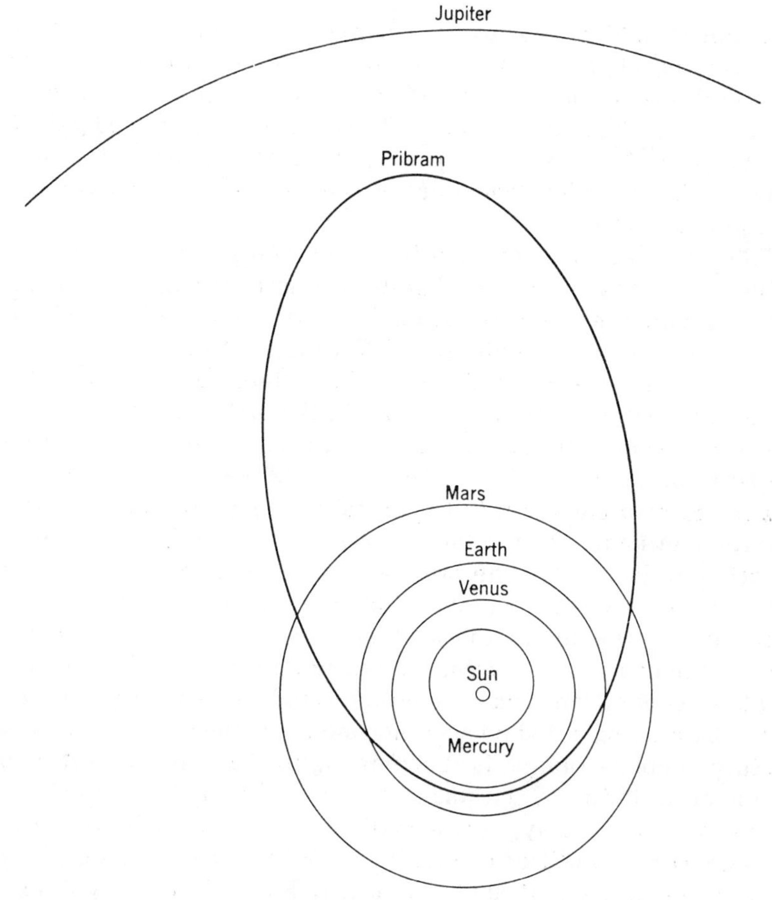


**Figure 1.2.** Orbit computed by Ceplecha (1961) for the meteorite that fell at Pribram, Czechoslovakia, in April 1959. Figure reproduced, by permission, from Mason (1962).

1 km across are not detectable astronomically but are presumed to be numerous from the fact that the distribution of sizes in the observable range shows a strong increase in numbers with decreasing size (Blanco and McCuskey, 1961, pp. 262–268).

Wetherill (1968) pointed out that afternoon falls of meteorites are twice as frequent as morning falls. This requires them to be orbiting substantially faster than the Earth when intercepted; not only must their orbit aphelia be near to Jupiter but their orbital motions must be in the same sense as those of the planets (whereas meteors and comets arrive from all orientations, not restricted to the plane of the solar system).

Although meteorites are, for convenience, classified into several structural and chemical types, there are no clear lines of demarcation and slightly different systems of classification have been in favor at different times. The four basic subdivisions generally used (e.g., Mason, 1962; Wood, 1968; Wasson, 1974) are chondrites, achondrites, stony-irons, and irons. A convenient checklist of their properties is given by Kaula (1968, pp. 380–382).

The irons are predominantly metallic iron, with nickel in solid solution averaging about 10%. Smaller amounts of sulphide and graphite are found and there are occasional inclusions of silicate. Two metal phases occur, the body-centered cubic $\alpha$ form (kamacite) with about 5.5% nickel, and the face-centered cubic $\gamma$ form (taenite) with variable nickel content, generally exceeding 27%. Normally both phases occur in close association, evidently as a phase separation from solid solution after solidification as a single phase from the melt. A third "phase"—plessite—is also common, but is in fact a very fine exsolution (phase separation) of kamacite and taenite. Common crystal orientations across meteorites indicate that the original metal crystals were very large, at least one meter across, which is indicative of extremely slow cooling. Quantitative evidence of the cooling rate has been obtained from the variations in composition across the kamacite-taenite phase boundaries. The phase separation forms a characteristic pattern, which is rendered obvious by etching polished sections and is known as the Widmanstätten structure. A good example is shown in Figure 1.3. This exsolution may be understood in terms of the phase diagram of nickel in iron (Fig. 1.4). A 10% nickel-in-iron alloy, corresponding to the average composition of iron meteorites, solidifies as taenite and remains as a single-phase solid as it cools down to about 690°C (point $A$ in Fig. 1.4). At this point it enters the two-phase region, and kamacite (with composition represented by point $B$) begins to exsolve along {111} planes in the taenite crystal lattice.* With further cooling to 500°C the taenite phase increases in nickel content, along path $A \rightarrow C$, so forming a decreasing proportion of the alloy, and the planes of kamacite grow in thickness and increase in nickel content, along path $B \rightarrow D$. The average nickel content is, of course, still 10% (represented by point $E$), so that at this temperature kamacite has become the major component. With further cooling the solubility of nickel in kamacite begins to decrease and nickel diffuses out of the boundaries of the kamacite zones into the taenite zones, but diffusion becomes too slow to maintain phase equilibrium and the inhomogeneities in composition shown in Figure 1.5 develop. Diffusion

*{111} planes form an octahedral pattern and iron meteorites with well developed Widmanstätten structures are termed octahedrites.

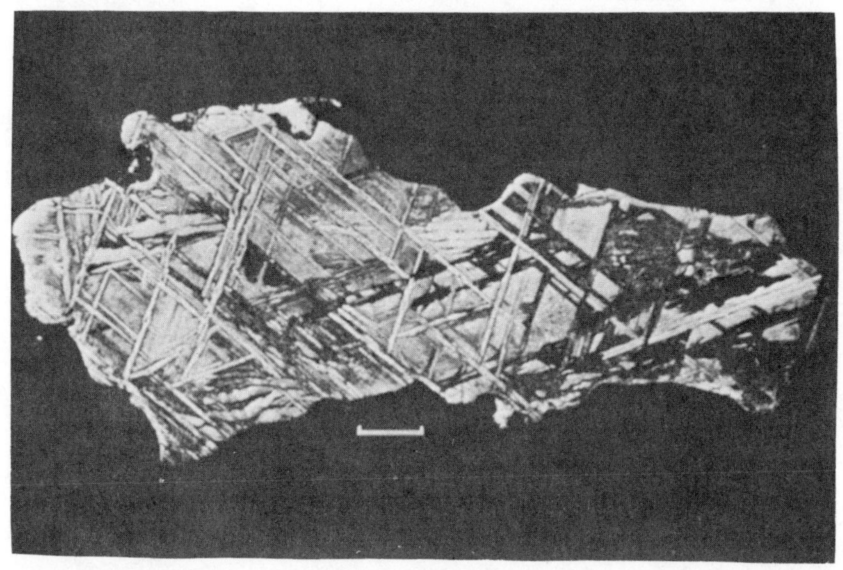

**Figure 1.3.** Widmanstätten structure of the metal phase in the Glorietta Mountain pallasite (stony-iron meteorite). The scale line is 1 cm. Photograph courtesy of J. F. Lovering.

of nickel is more effective in kamacite than in taenite, so that the margins of the kamacite are only slightly more depleted in nickel than the core regions, but nickel from the kamacite does not penetrate deeply into the taenite zones, remaining as a high nickel rim on the taenite in which the cores may exsolve as very finely divided plessite.

Widths of Widmanstätten diffusion zones have been used to estimate cooling rates of iron meteorites by Wood (1964), Short and Anderson (1965), and Goldstein and Short (1967), and of chondrites by Wood (1967). The estimated rates of cooling varied from 0.4°C to 40°C per $10^6$ years through the critical range from 650 to 350°C, in which the diffusion was effective. These slow rates are in keeping with the large crystal sizes and coarse Widmanstätten structures of both iron meteorites and the iron phases of ordinary chondrites, implying burial in parent bodies 100 km or more in radius. The wide range of cooling rates indicates that there were several parent bodies of different sizes.

Chondrites are the most common of the four types of meteorite, comprising more than 90% of observed falls, although iron meteorites are generally larger and the mass of meteoritic matter in space is probably less than 90% chondritic. The characteristic feature of chondrites is the occurrence of *chondrules*, partially devitrified glassy globules of silicate with diameters averaging about 1 mm, distributed through the matrix of

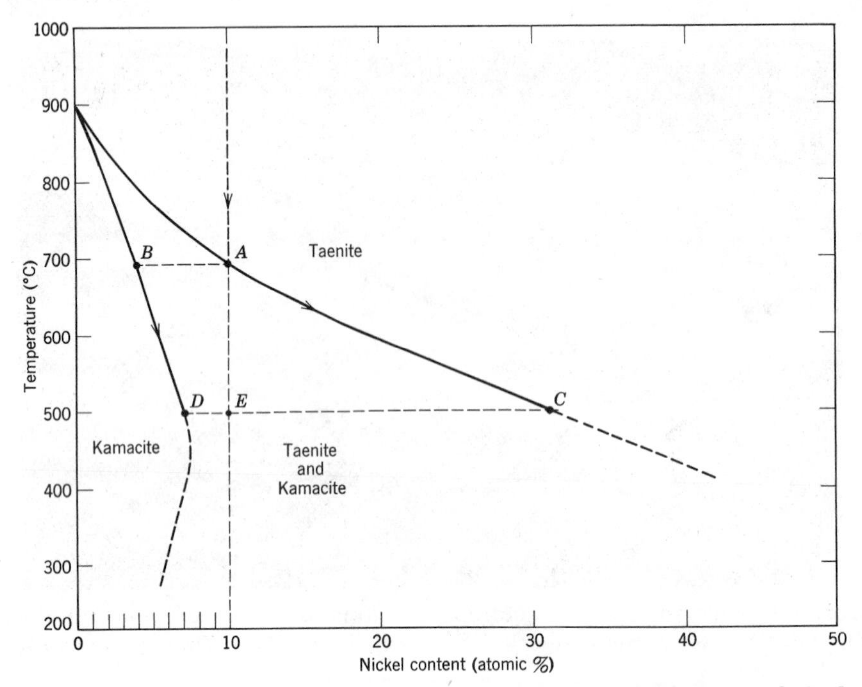


**Figure 1.4.** Phase diagram of nickel-iron according to Goldstein and Ogilvie (1965), with paths followed by exsolving kamacite and taenite in an alloy with 10% Ni, representative of the iron meteorites.

silicates and nickel-iron (Fig. 1.6). Chondrules do not occur in any of the terrestrial rocks accessible to us and they indicate that a large fraction of meteoritic material underwent a special process, of which there is no remaining evidence on the Earth. Their glassy nature and roughly spherical forms indicate that they were melted and rapidly cooled as independent objects in the solar nebula, being subsequently incorporated into larger bodies. The mechanism of transient heating is subject to conjecture. The explanation offered by Wood (1968) is that at its T-Tauri stage the Sun initiated violent shock waves through the surrounding nebula and that shock-wave heating of the dust produced the chondrules. This is a plausible mechanism in terms of the observed sudden variations in luminosity (flashes) of T-Tauri stars.

Much less frequent than chondrites are the achondrites, which are essentially similar to terrestrial rocks, being crystalline silicates with virtually no metal phase (and no chondrules). The merit of classifying separately the stony-iron meteorites appears doubtful, since they are merely about halfway through the continuous range between stones and

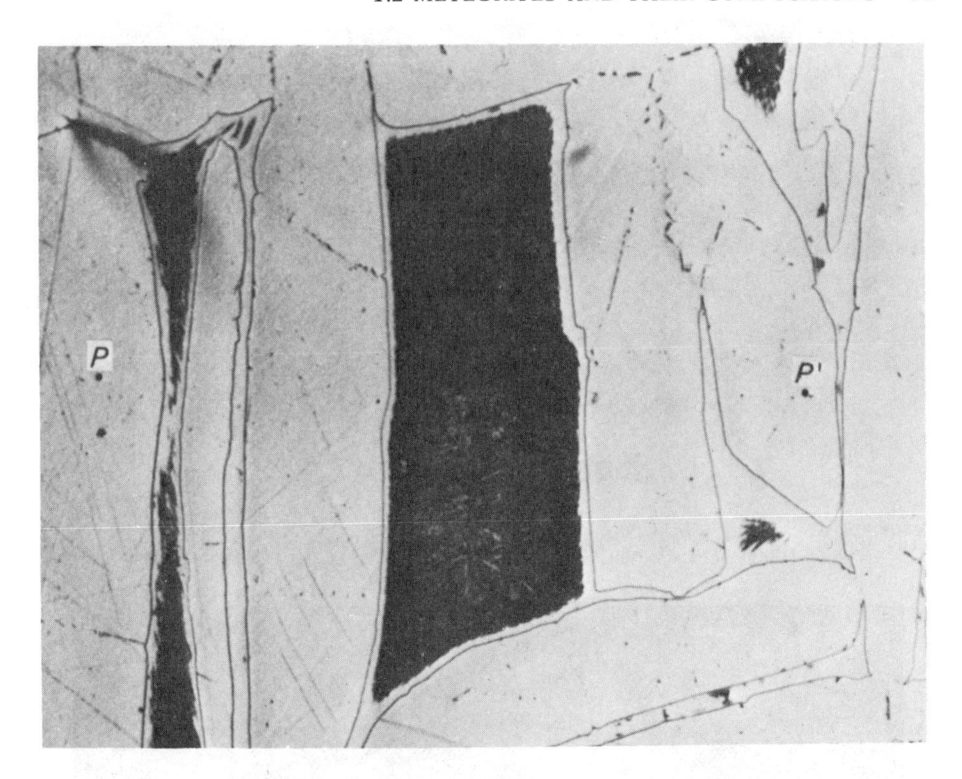


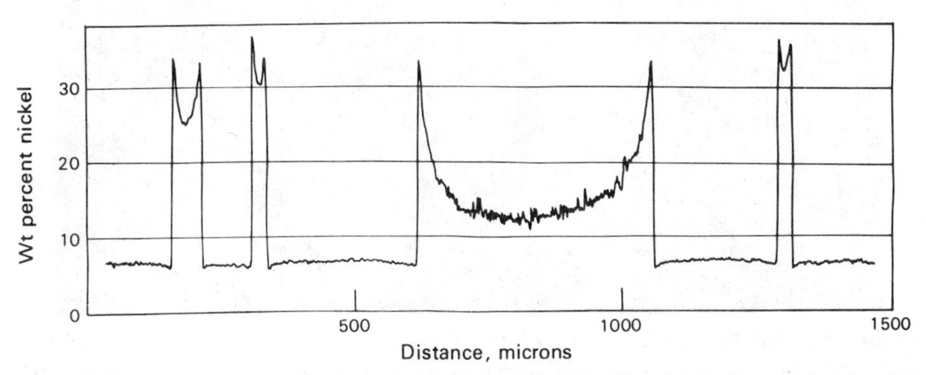


**Figure 1.5.** (Top) Microscopic picture of a polished and etched slice of the Anoka octahedrite (iron meteorite), showing details of the Widmanstätten pattern. The dark areas are plessite, rimmed by taenite, within the more uniform kamacite. (Below) Profile of the variation in nickel content along the line *PP'*, as determined by an electron microprobe. (This gives chemical analyses of small areas of surface in terms of the intensities of characteristic X rays excited by a sharply focused electron beam.) Figure reproduced by permission from Wood (1968).

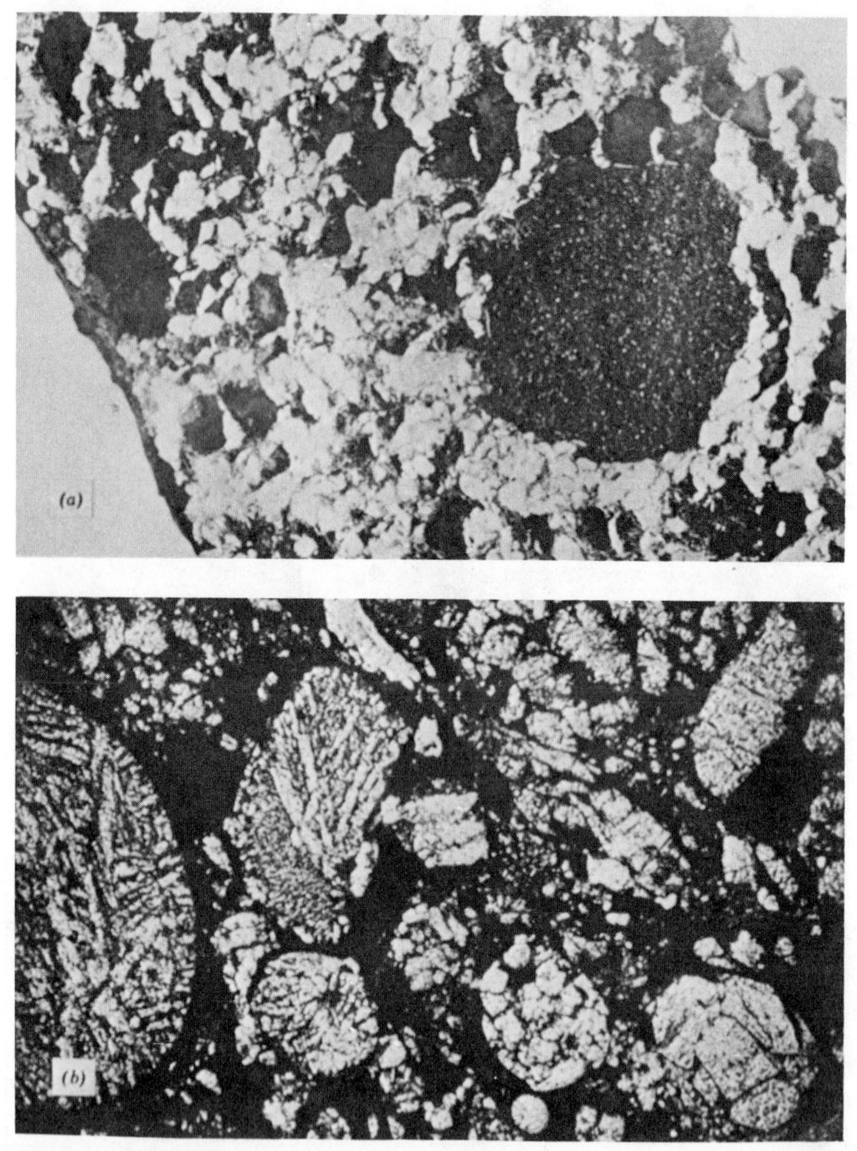


**Figure 1.6.** (*a*) Polished section of the Bencubbin meteorite. The bulk of the meteorite has a well-developed crystalline structure, but a large chondritic inclusion appears as the dark area on the right-hand side of this photograph. On the left-hand side is a carbonaceous chondritic inclusion. (*b*) Enlarged section of the chondrite fragment, showing the structure of the chondritic spherules, just apparent in (*a*). The significance of the occurrence of the three meteorite types in a single specimen is discussed by Lovering (1962). Photographs courtesy of J. F. Lovering.

irons. However, there is an important class of chondrites that deserves special mention—the carbonaceous chondrites which are relatively few in number and individually small in size but are of particular scientific interest (their properties are reviewed by DuFresne and Anders, 1963). Their rarity in meteorite collections is probably not due to an absolute rarity in space but to their extreme friability and consequent inability to withstand flight through the atmosphere. In fact, optical reflectivities of many asteroids indicate carbonaceous chondritic compositions, at least of the surface layers (Johnson and Fanale, 1973; McCord and Gaffey, 1974). As the name implies, carbonaceous chondrites contain several percent of carbon compounds. Metal phases are virtually absent, the iron occurring mainly in silicate, oxide, and sulfide phases. Carbonaceous chondrites have a dark amorphous appearance. Substantial amounts of volatiles—especially water—are present, so that these chondrites can never have been strongly heated. Of all the materials available for laboratory examination, the subclass of carbonaceous chondrites known as type I must be nearest to the primitive dust from which the terrestrial planets were formed. Types II and III are more like ordinary chondrites. The four known type I carbonaceous chondrites are homogeneous, finely particulate bodies (lacking even chondrules, in spite of their name), especially rich in volatiles. It is easy to see, at least in principle, how shock heating of this material, or perhaps high velocity impacts, would produce chondrules and probably reduce some iron to metal, so that subsequent progressive metamorphism produced first types II and III, then the ordinary chondrites and, finally, fully crystalline achondrites with metal separating into cores of meteoritic miniplanets and forming the iron meteorites.

Most meteorites are quite strongly magnetic by virtue of kamacite in them (or magnetite in the case of the carbonaceous chondrites). The techniques of paleomagnetism (Chapter 9) have been applied to them with the surprising but consistent result that all of the chondrites have remanent magnetism that cannot be attributed to their terrestrial histories but must have been induced in fields of 0.1 to 1 G ($10^{-5}$ to $10^{-4}$ T) at the time of formation or metamorphism (see, for example, the summary by Stacey, 1976). A convincing explanation has been elusive. The magnetic field of the youthful Sun was probably adequate if the magnetization could be accomplished quickly, but the magnetic remanences appear to be thermoremanent (or chemical-remanent) in nature; that is, the magnetizations were induced by cooling (or chemical formation or metamorphism) in the field and, as the Widmanstätten patterns show, many chondrites (as well as iron meteorites) cooled slowly, presumably by burial in a substantial parent body. The solar field could only be effective if the magnetization was induced sufficiently rapidly that no significant rotation occurred in the process. Therefore, it appears that strong but very brief reheating of the material may have occurred during breakup of the meteorite parent bodies and that the solar magnetic field was still effective at the time. Jain et al. (1972) estimated from the mechanical hardness of their metal phases that most meteorites were subjected to shock compressions exceeding 130 kbar ($1.3 \times 10^{10}$ Pa) by the asteroidal fragmentation events that produced them.

Many of the techniques for studying meteorites have been used also for tektites, which are rounded pieces of silica-rich glass that have been found in tens of thousands over limited areas of the Earth. Potassium-argon and fission track dating (Chapter 2) of the tektites reveals distinct formation ages (i.e., times since fusion) of the different geographical groups. In millions of years the age groupings are: 0.7 (Australia and S.E. Asia), 1.0 (West Africa), 4 (Australia), 14 (Central Europe), 35 (N. America), and possibly 26 (N. Africa) (Fleischer et al., 1970; Durrani, 1971). Shapes of the tektites clearly indicate rapid flight through the atmosphere, and the presence of fine metal grains leads to the widely accepted inference that tektites are fused splashes from massive meteorite impacts. The chemistry of tektites is consistent with their being products of impacts on terrestrial sedimentary rocks (Taylor, 1973) and now clearly not consistent with an origin in a lunar impact, as was commonly supposed before lunar rocks became available for comparison. However, the very low water contents of tektites (Gilchrist et al., 1969) still poses a problem.

## 1.3 COSMIC RAY EXPOSURES OF METEORITES

The "age" of a meteorite normally means the time that has elapsed since it was formed as solid material. More specifically, this is the *solidification age*, determined by the methods used for dating rocks and discussed in Chapter 2. The principal methods, based on the decay of uranium to lead, rubidium to strontium, and potassium to argon, are all applicable to stony meteorites, which contain the parent elements of these decay schemes. The meteorites have a common solidification age within narrow limits, $4.6 \times 10^9$ years (see Section 2.4), and this is identified with the formation of the solar system.

We are also interested in the *cosmic ray exposure ages* of meteorites, the intervals of time that have elapsed since they were broken down into meter-sized pieces and exposed to bombardment in space by cosmic radiation. It is only the outer 1 m or so of each independent body that is exposed to cosmic radiation, so that each fragmentation event exposes fresh material. From the residual abundances of certain short-lived cosmogenic (cosmic ray produced) nuclides ($Ar^{39}$, $C^{14}$, $Cl^{36}$) a *terrestrial age* can also be estimated. This is the time that has elapsed since a meteorite arrived on the Earth and cosmic radiation was effectively "switched off" and so ceased to maintain equilibrium concentrations of the short-lived nuclides. Note also that the Poynting-Robertson effect (Section 1.4) gives a limited life to small fragments in the solar system, so that cosmic ray exposure ages comparable to the age of the solar system are not possible.

Extremely energetic cosmic ray protons cause violent disruption (spallation) of the atomic nucleii in exposed meteorites. Anders (1962, 1963), who reviewed the whole subject of meteorite ages, used the following example to illustrate the spallation process:

$$Fe^{56} + H^1 \rightarrow Cl^{36} + H^3 + 2He^4 + He^3 + 3H^1 + 4n \tag{1.3}$$

Many products can arise from numerous similar reactions and, in principle, the total cosmic ray exposure can be estimated from the content of a particular product if its half-life exceeds the estimated exposure age and if its initial concentration in the unexposed meteorite can be neglected or reasonably estimated. Uncertainties in cosmic ray intensities and partial shielding by burial in a large meteorite can be minimized by comparing concentrations of two cosmogenic nuclides, one stable and the other radioactive and having a half-life short compared with the cosmic ray exposure times. Further selection to include only pairs of nuclides whose production cross sections have similar dependences upon cosmic ray energy, and which are sufficiently close in mass number to be produced from the same target nuclides, leaves two isobaric pairs ($H^3$—$He^3$ and $Cl^{36}$—$Ar^{36}$) and two isotopic pairs ($Ar^{38}$—$Ar^{39}$ and $K^{44}$—$K^{41}$) as species of greatest interest. The first three of these pairs, being gases, also avoid the problem of initial composition that arises in the case of nonvolatile spallation products. Then in terms of the measured concentrations $S$, $R$ of the stable and active nuclides and their production cross sections $\sigma_S$, $\sigma_R$ determined from laboratory data, the cosmic ray exposure age of a meteorite is given by

$$t = \frac{S}{R} \cdot \frac{\sigma_R}{\sigma_S} \cdot \frac{t_{1/2}}{\ln 2} \tag{1.4}$$

where $t_{1/2}$ is the half-life of the active nuclide which is assumed to be short compared with $t$ (Anders, 1962—see Section 2.2). In the cases of the isobaric pairs the stable nuclide is produced by decay of the active one as well as directly, so that the equation becomes

$$t = \frac{S}{R} \cdot \frac{\sigma_R}{\sigma_S + \sigma_R} \cdot \frac{t_{1/2}}{\ln 2} \tag{1.5}$$

The method of dating geological materials by counting nuclear fission tracks (Fleischer and Price, 1964; Fleischer, Price, and Walker, 1965; see also Section 2.2) has an interesting application to cosmic ray exposure ages. Fission tracks in meteorites are produced not only by the spontaneous fission of $U^{238}$, as in rocks, but also by neutron-induced fission of $U^{235}$. "Excess" tracks, due to $U^{235}$, can be estimated by comparing the solidification age with the expected number of tracks from $U^{238}$, and the excess represents the integrated exposure of the sample to cosmic ray neutrons or heavy cosmic ray primaries. In particular, Fleischer, Naesser, Price, and Walker (1965) found a complete absence of tracks attributable to cosmic rays in tektites, allowing them to conclude that the tektites cannot have existed independently in space for a period as long as 300 years.

If the meteorites all had the same cosmic ray exposure age, we would have to conclude that they were produced simultaneously by disruption of a single parent body or by the collision of two parents. However, the concentrations of spallation products are highly variable and indicate a more complex history. Most of the reliable exposure ages for stony meteorites are grouped around $4 \times 10^6$ and $23 \times 10^6$ years, but others cover the range from $2.8 \times 10^6$ to $100 \times 10^6$ years (Anders, 1964; Wänke, 1968) with obvious groupings of different types. Some of the lower estimates are probably invalidated by diffusion losses because the same meteorites have small potassium-argon solidification ages. Iron meteorites have generally had much greater exposures, up to a maximum of $2200 \times 10^6$ years with groupings at $630 \times 10^6$ and $900 \times 10^6$ years but not at $23 \times 10^6$ years (Anders, 1964). There is a wide scatter and, in many cases, very imperfect agreement between the alternative methods. Nevertheless, it is evident that the meteorites suffered a multiplicity of fragmentation events that were very much more recent than their solidifications.

The measurements may not preclude the possibility of a primary fragmentation of one or two parent bodies, because the large number of asteroids requires that we assume many more recent collisions of small bodies, although few major collisions of large ones. If collisions are occurring at a rate such that the interval between them, for any one fragment, is short compared with the time since the supposed primary fragmentation, then the cosmic ray exposures will indicate the more recent fragmentations. The fact that stony meteorites are more easily broken up, and are on average significantly smaller than the irons, is consistent with their shorter cosmic ray exposures. Furthermore, the exposed surfaces of stones are eroded more quickly by impact of small dust particles than are the surfaces of irons, an additional reason why stony material with very prolonged cosmic ray exposures may not be available for measurement (Fisher, 1966). A further bias to young exposure ages is imposed by the Poynting-Robertson effect (Section 1.4). Thus cosmic ray exposures cannot, in principle, give direct evidence of primary fragmentations. But the reasons for believing that the asteroids are collision fragments of at least four and probably many more planetismals that have never accreted to form a body as large as the Moon are strongly circumstantial:

1. It is difficult to devise a satisfactory mechanism for the breakup of a parent body of more than very limited size.
2. The compositions and oxidation states of the meteorites are very varied.
3. Cooling rates apparent from Widmanstätten patterns are quite different.
4. The grouping of exposure ages suggests that the irons and stones were produced by different events and therefore that the bodies from which they came were chemically different. However, measurements of isotopic ratios of lead and strontium, which have been used to date the meteorites (Section 2.4), show that the meteoritic parents differentiated from a common source and, if they are of asteroidal origin, then the asteroids had a history of chemical differentiation before the break-up of those sampled as meteorites. It is this differentiation process that is dated at $4.6 \times 10^9$ years ago.

## 1.4 THE POYNTING-ROBERTSON EFFECT

Solar radiation has an important influence on the orbits of small particles whose ratio of surface area to mass is large. Its effects on the meteor streams have been studied in detail and a historical and physical discussion is given by Lovell (1954). "Particles" up to about 10 cm diameter are affected on a time scale of $10^9$ years.

It is convenient to distinguish three effects of solar radiation pressure, although they are not really independent. First, there is a simple outward force from the Sun. For particles with diameters of a few thousand Angstroms or less this force may exceed the gravitational attraction of the Sun and blow them out of the solar system. This problem is complicated by the fact that the critical particle size is comparable to the wavelength of the radiation and the effective optical cross section is not the simple physical cross section, but we are concerned here with much larger particles. Second, the solar radiation received by a particle is Doppler shifted to cause an increase in radiation pressure if the particle is approaching the Sun and a decrease if it is receding; elliptical orbits are thus reduced to nearly circular orbits.

Third, the angular momentum of an orbiting particle is progressively destroyed by the fact that it receives solar radiation, which has only a radial momentum from the Sun (neglecting the solar rotation), and reradiates this energy with a forward momentum corresponding to its own motion about the Sun. This is the essential feature of the Poynting-Robertson effect, which is most conveniently analyzed as a problem in relativity.

We consider the special case of a spherical particle of mass $m$ and diameter $d$ in a circular orbit at radius $r$. Its orbital velocity is

$$v = \left(\frac{GM}{r}\right)^{1/2} \tag{1.6}$$

$M$ is the mass of the Sun and $G$ the gravitational constant, so that the total orbital energy is

$$E = -\frac{GMm}{r} + \frac{1}{2}mv^2 = -\frac{GMm}{2r} \tag{1.7}$$

It is convenient to consider separately the processes of absorption and reradiation of the energy.

In time $dt$ the particle receives energy $d\epsilon$ as solar radiation, and this causes an increase in mass

$$dm = \frac{d\epsilon}{c^2} \tag{1.8}$$

$c$ being the velocity of light. But since this radiation traveled radially from the Sun it carried no orbital angular momentum and the total angular momentum of the particle is conserved, that is, $d(mvr) = 0$, so that

$$md(vr) = -vr\,dm = -\frac{v}{c^2}r\,d\epsilon \tag{1.9}$$

The particle then reradiates the energy $d\epsilon$, but it does so isotropically in its own frame of reference and this process involves no reaction on the particle. The orbital velocity is therefore conserved in the radiation process and since the mass $dm$ is lost, a net loss of angular momentum by $vrdm$ occurs. This angular momentum is carried away by the radiation, which, when viewed in the stationary reference frame of the Sun, is seen to be Doppler shifted; the energy and momentum projected forward from the particle exceed the energy and momentum radiated backward.

The rate of loss of orbital angular momentum may be equated to a retarding torque $L$:

$$L = m\frac{d(vr)}{dt} = -\frac{v}{c^2}r\frac{d\epsilon}{dt} \tag{1.10}$$

so that

$$\frac{dE}{dt} = L\frac{v}{r} = -\frac{v^2}{c^2}\frac{d\epsilon}{dt} \tag{1.11}$$

Now $d\epsilon/dt$ is the rate at which the particle receives solar radiation and is given by

$$\frac{d\epsilon}{dt} = S\left(\frac{r_E}{r}\right)^2 A \tag{1.12}$$

where $S = 1360 \text{ W m}^{-2}$ is the solar constant, the energy flux through unit area at a distance equal to $r_E$, the radius of the Earth's orbit, and $A = (\pi/4)d^2$ is the cross-sectional area of the particle. Thus, by differentiating Equation 1.7 and equating to Equation 1.11 with the substitution of Equation 1.12, we obtain

$$\frac{GMm}{2r^2}\frac{dr}{dt} = -\frac{v^2}{c^2}S\left(\frac{r_E}{r}\right)^2 A \tag{1.13}$$

and since $v$ is given in terms of $r$ by Eq. (1.6), we obtain the differential equation for $r$:

$$r\frac{dr}{dt} = -\frac{2Sr_E^2A}{mc^2} \tag{1.14}$$

Integrating from the initial condition, $r = r_0$ at $t = 0$:

$$\frac{r_0^2 - r^2}{r_E^2} = \frac{4SA}{mc^2}t \tag{1.15}$$

which, for a spherical particle of density $\rho$, becomes

$$\left(\frac{r_0}{r_E}\right)^2 - \left(\frac{r}{r_E}\right)^2 = \frac{6S}{d\rho c^2}t \tag{1.16}$$

where $(r/r_E)$ is the radius of a particle orbit, expressed in astronomical units (AU).

We are interested in the time taken by particles, of diameter $d$, originating in the asteroidal belt at $2.7r_E$, to reach the Earth's orbit, $r_E$. Assuming a particle density of $4000 \text{ kg m}^{-3}$ and $d$ in meters, this is

$$t = 2.77 \times 10^{17}\, d \text{ seconds} = 8.8 \times 10^9\, d \text{ years} \tag{1.17}$$

A more complete analysis (Lovell, 1954, pp. 402–409) shows that a particle in an elliptical orbit is first reduced to a nearly circular orbit, just inside its initial perihelion distance. Since this process also depends on the Doppler shift of radiation due to motion of the particle relative to the Sun, the time required is similar to that for the spiraling effect.

The Poynting-Robertson effect thus ensures that any small particles in the common meteoroid range, less than 1 cm ($10^{-2}$ m) in diameter, which originated in the asteroidal belt about $10^8$ years ago, would have passed the Earth's orbit and spiraled into the Sun. McKinley (1961, pp. 169–171) pointed out that very few meteors appear to be due to particles having densities of stone or iron. They are envisaged as loose, dusty aggregates bound with volatile "ices," similar to the supposed structures of comets and quite different from the meteorites. The relative rarity of very small meteorites is consistent with the conclusion that they must be products of recent asteroidal collisions. Furthermore, we can see that if a primary asteroidal fragmentation had occurred very early in the history of the solar system, say $4 \times 10^9$ years ago, then all primary fragments smaller than 50 cm would have spiraled into the Sun and the terrestrial collection would be strongly biased toward the shorter cosmic ray exposures of more recent, secondary fragmentations.

## 1.5 COMPOSITIONS OF THE TERRESTRIAL PLANETS

Meteorites appear to be far more representative of the overall compositions of the terrestrial planets than are the rocks to which we have access near the surface of the Earth. Chemical considerations, based largely on meteorite observations, have dominated recent discussions of the origin of the Earth, Moon, and terrestrial planets as well as the meteorites themselves (Ringwood, 1966a,b,c, 1970a, 1971; Reynolds and Summers, 1969; Clark et al., 1972; Gast, 1972; Larimer, 1973; Grossman and Larimer, 1974).

In spite of doubts about the validity of spectroscopic determinations of elemental abundances in the Sun and stars (Worrall and Wilson, 1972), estimates of solar abundances of nonvolatile elements correspond reasonably with relative abundances found in meteorites and believed to represent the Earth also to a good approximation. The dominance of Si, Mg, and Fe supports the general conclusion that all of the terrestrial planets and the satellites of the major planets, as well as the meteorites are composed essentially of these elements, their oxides, and mutual compounds. An approximate composition for the Earth is given in Table 1.1. This has been deduced from the structure and density of the Earth (Chapter 6) and from geochemical arguments and was shown by Ringwood (1966a) to correspond very closely indeed to the composition which would result from appropriate reduction of type I carbonaceous chondrites, with removal of most of the volatile materials. The relative elemental abundances indicated by this composition are listed in Table 1.2 for comparison with solar abundances.

The agreement between the solar and terrestrial abundances in Table 1.2 is now almost suspiciously good. Differences in density of the

**Table 1.1** Chemical Composition of the Earth[a]

| | | |
|---|---|---|
| Mantle | $SiO_2$ | 32 |
| | MgO | 23 |
| | $FeO/Fe_2O_3$ | 7.5 |
| | $Al_2O_3$ | 2 |
| | CaO | 2 |
| | $Na_2O$ | 0.5 |
| | Others | 1 |
| Core | Fe | 24 |
| | Ni | 3 |
| | S | 5 |

[a]Approximate percentages by mass of major constituents, being rounded averages of estimates by Mason (1966) and Ringwood (1966a), with adjustment to include the probable sulfur content of the core (Murthy and Hall, 1970). Cameron (1973) has given a comprehensive review of solar system abundance data.

terrestrial planets (Table A.1 in Appendix A) preclude interpretation of their structures in terms of identical compositions, and perfect agreement of any of them with the solar composition cannot be expected. Venus is very similar to the Earth in both size and density and is presumed to be essentially similar internally, but if we treat Mars in the same way and assign to it an iron core of radius calculated to give the observed mean density with a silicate mantle of $\rho_0 = 3300\ \mathrm{kg\ m^{-3}}$, then we face a difficulty. The moment of inertia of Mars, as deduced from its precessional period of $1.73 \times 10^5$ years and the ellipticity deduced from the paths of space probes (Lorell and Shapiro, 1973; Wier, 1975) is $0.37Ma^2$ ($M$ = mass, $a$ = equatorial radius), which is too close to the value for a uniform sphere ($0.4Ma^2$) to admit a structure that is simply a scaled-down version of the Earth (moment of inertia $0.33078Ma^2$—see Section 3.1). It is necessary to suppose that Mars has a much smaller core, in proportion to its size, but also a denser mantle. This supports Ringwood's (1966a, 1966c, 1971) theory of planetary evolution, which leads to different oxidation states for the terrestrial planets, so that in Mars virtually all of the iron has remained oxidized and therefore has not separated from the silicates. According to Ringwood the overall Fe:Si ratio in Mars is approximately the same as that in the Earth, the

**Table 1.2** Elemental Abundances in the Sun and Earth[a]

| Element | Sun | Earth |
|---|---|---|
| H | $2.2 \times 10^4$ | |
| He | $1.4 \times 10^3$ | |
| O | 15 | 3.5 |
| C | 9.3 | |
| Si | 1.00 | 1.00 |
| Mg | 0.89 | 1.1 |
| Fe | 0.71 | 1.0 |
| S | 0.35 | 0.3 |
| Al | 0.074 | 0.07 |
| Ni | 0.043 | 0.10 |
| Ca | 0.050 | 0.06 |
| Na | 0.043 | 0.03 |

[a]Spectroscopic estimates of numbers of atoms of selected elements in the solar atmosphere, relative to silicon, compared with values for the Earth deduced from the composition in Table 1.1. Source of solar data is Ross and Aller (1976) and solar system estimates for non volatiles are given by Cameron (1973).

iron occurring as oxide, which has a density of 5200 kg m$^{-3}$. When added to silicate of $\rho_0 = 3300$ kg m$^{-3}$, it brings the uncompressed silicate density up to $\rho_0 = 3700$ kg m$^{-3}$, to which only a small core need be added to give the observed Martian density. If the mantle and crust of Mars does indeed have a high iron-oxide content it is appropriate that the surface of the "red planet" should appear rusty.

However, no plausible difference in oxidation state can account for the density of Mercury in terms of the composition of the Earth; it must have a proportionately larger iron core. Conversely the density of the Moon admits of no significant enrichment in iron relative to the mantle of the Earth and its moment of inertia ($0.392Ma^2$—Gapcynski et al., 1975) disallows the hypothesis of a significant core.* Thus fractionation of iron relative to other nonvolatiles occurred during the accretion of the terrestrial planets. This is consistent with the argument (Gast, 1972) that

*Seismological studies of lunar structure are reviewed by Lammlein et al. (1974) and Toksöz et al. (1974).

the Earth accreted from inhomogeneous materials which had undergone substantial preterrestrial processing.

A suggestion that the solar system is not isotopically homogeneous raises a completely new range of possibilities. A slight but measurable isotopic fractionation occurs in ordinary chemical processes; these are recognized in the relative abundances of isotopes of H, C, O, Si, S in geological materials and in meteorites, and can be used to deduce temperatures of formation (see, for example, Section 9.5). However, Clayton et al. (1973, 1975) report oxygen isotope variations in meteorites that do not appear to be explicable in terms of chemical or physical processing but point to the incomplete homogenization of isotopically different sources. Radioastronomical measurements of molecular spectra indicate quite significant isotopic variations in different parts of the galaxy (Bertojo et al., 1974), presumably arising from different conditions of nuclear synthesis. If these are indeed represented within the solar system then a new dimension is added to conjectures about planetary formation.

# 2

# Radioactivity and the Age of the Earth

"In terms of observational facts alone the behaviour of the inner earth now appears even more fantastic than when we knew much less about it."

HOLMES, 1965, p. vii

## 2.1 THE PRERADIOACTIVITY AGE PROBLEM

Until the discovery of radioactivity, the age of the Earth was a matter of heated debate. Geologists required times of hundreds of millions of years for the accumulation of known sedimentary strata, but Kelvin (1899) argued that the heat flow through the crust was incompatible with an age greater than about 25 million years. It is of interest to reconsider Kelvin's argument, which is not based on a limitation in the thermal capacity of the Earth but on the ineffectiveness of thermal conduction through the crust. The total heat capacity of the Earth ($2.0 \times 10^{26}$ atomic moles), neglecting latent heat of solidification, is about $5 \times 10^{27}$ J deg$^{-1}$ ($3R$/mole), which could maintain the present heat flux of $3 \times 10^{13}$ W for $4.5 \times 10^{9}$ years with an average temperature drop of 900°C. If Kelvin had supposed that heat from the entire Earth was available to maintain the geothermal flux, he could not have discounted the geological evidence. However, he considered a model of the Earth, which was more plausible in terms of the physics known at the time, and assumed an initially molten crust that solidified progressively inward from the outside. The thermal gradient in the crust was determined by the fixed difference in temperature across it (i.e., melting point of rock minus surface temperature) and decreased as the crust thickened. Thus the heat flux to the surface and the consequent rate of solidification were determined by the thermal conductivity of the crust; the heat flux would have fallen to the present value after only 25 million years.

Kelvin's model could have been modified to agree with geological evidence if the recognition of convection within the Earth (Chapters 7 and 10) had preceded the discovery of heat generation by radioactivity. However, an even greater difficulty arose in the case of the Sun, whose prodigious heat output can be estimated from the solar constant $S$ (Section 1.4), the heat received per unit area at the radius of the Earth's orbit, which is

$$S = 1360 \text{ W m}^{-2} \tag{2.1}$$

from which the total heat output of the sun is found to be $3.8 \times 10^{26}$ W. The two sources of heat to which preradioactivity physicists could appeal were gravitational energy, due to shrinking of the Sun, and the simple heat capacity of an initially very hot Sun. An approximate value of the gravitational energy $E_G$ released by shrinking of mass $M$ to the present size is obtained by assuming the Sun to be a sphere of uniform density and radius $R$, in which case

$$E_G = \frac{3}{5}\frac{GM^2}{R} = 2.4 \times 10^{41} \text{ J} \tag{2.2}$$

In fact, we can expect the mass to be more concentrated toward the center, so that $E_G$ is somewhat higher. At the present rate of dissipation this energy would last for 20 million years, which is, by coincidence, in agreement with Kelvin's estimate of the age of the Earth and thus supports the estimate. Alternatively, we may assume the specific heat of the Sun to have the classical value, applied to ionized hydrogen,

$$C = 3R = 25 \text{ J mole}^{-1} \text{ deg}^{-1} = 2.5 \times 10^4 \text{ J kg}^{-1} \text{ deg}^{-1} \tag{2.3}$$

which gives a total heat capacity of $4.8 \times 10^{34}$ J deg$^{-1}$. On this basis the average, unmaintained temperature of the Sun would decrease by 0.25 deg year$^{-1}$. Temperatures of hundreds of millions of degrees were not seriously contemplated in Kelvin's time and since, from the fossil record, the Earth was apparently warmed by the Sun for hundreds of millions of years, the heat supply of the Sun would have remained a problem, even if Kelvin's thermal conductivity argument had been circumvented in estimating the age of the Earth. As is well known, the discovery of nuclear energy offered solutions to both problems.

The uneven distribution of radioactive sources within the Earth was recognized in an early paper by Strutt (1906) (later Lord Rayleigh), who found that the concentration of radioactivity in igneous rocks greatly exceeded what was required in the Earth as a whole to produce the observed geothermal flux. He made the suggestion that the radioactivity of the Earth is confined to a crust a few tens of kilometers thick, which was, by that time, recognized as distinct from the deeper material or mantle. This shallow distribution of heat sources neatly removed the thermal conductivity problem that had led to Kelvin's erroneous conclusion.

Strutt (1906) also noted that basic igneous rocks (basalt, gabbro) were much less radioactive than the lighter acidic (silica rich) or granitic rocks, which are characteristic of much of the accessible part of the crust in continental areas. This observation is basic to the theory of the formation of the Earth's crust by differentiation from the mantle (Section 2.3) and to the thermal model of the Earth (Chapter 7).

## 2.2 RADIOACTIVE ELEMENTS AND THE PRINCIPLES OF RADIOMETRIC DATING

Independently of the thermal problem, which is considered further in Chapter 7, radioactivity provides the means for precise dating of geological events. Excellent summaries of radioactive dating methods are given by Faul (1966) and York and Farquhar (1972). Monographic discussions with comprehensive references are by Kanasewich (1968), Dalrymple and Lanphere (1969), and Faure and Powell (1972). The important isotopes, listed in Appendix H1, are those that have half-lives comparable to the age of the Earth and are widely distributed in sufficient quantities to be measurable in many rock types. A more complete list is given by Faul (1966). Potassium and rubidium decays are simple but the uranium and thorium decay schemes have intermediate daughter products with measured half-lives, in one case ($U^{234}$, great-granddaughter of $U^{238}$) exceeding $10^5$ years, so that decay to lead is not immediate. However, the half-lives of all intermediate products are very short by comparison with those of the parent isotopes. Provided the ages to be measured are also much longer than $10^5$ years, which is always the case with rocks examined by uranium-lead methods, the estimation of elapsed time from the ratios of parent and final daughter isotopes requires only that the material under consideration be a closed system, that is, there must be no fresh introduction to it or escape from it of any component, including all intermediate products. Fortunately, the gaseous intermediate product, radon, of which there are isotopes in the decay series of $Th^{232}$ as well as both $U^{238}$ and $U^{235}$, is short-lived and appears not to diffuse out of rocks appreciably, although the steady outgassing of soil produces a measurable radon content of the atmosphere. The shorter-lived isotopes that are generated by cosmic rays in the upper atmosphere, or in interplanetary dust collected by the Earth, are not considered here, although they can be useful in dating sediments. The best known is $C^{14}$, which has become very important in the dating of archeological remains.

Almost all measurements of the ages of rocks and minerals use one or more of the four decay schemes of $U^{238}$, $U^{235}$, $Rb^{87}$, and $K^{40}$. $Th^{232}$ is less useful than $U^{238}$ and $U^{235}$, which, being chemically identical, exist in a ratio that is a function of time only and so, when considered together, either provide an internal check on the data or else allow ages to be deduced from lead isotopes alone (Eq. 2.17).

Radioactive decay is described in terms of the probability that a constituent particle in a nucleus will escape through the potential barrier binding it to the nucleus. The energies are so large (and the nuclear size is so small) that no ordinary physical conditions influence the probability that a particular nucleus will decay.* "Ordinary physical conditions" here include all pressures and temperatures in the interior of the Earth. It follows that the rate of decay of $N$ nuclei of a particular species is directly proportional to $N$:

$$\frac{dN}{dt} = -\lambda N \tag{2.4}$$

where $\lambda$ is the decay constant given in Appendix H. Integrating from an initial number $N_0$ at time $t = 0$ we obtain the basic equation of all radioactive age work:

$$N = N_0 e^{-\lambda t} \tag{2.5}$$

The half-life $\tau_{1/2}$ of the isotope is obtained by substituting $N = N_0/2$ at $t = \tau_{1/2}$ in Equation 2.5:

$$\tau_{1/2} = \frac{ln\ 2}{\lambda} = \frac{0.69315}{\lambda} \tag{2.6}$$

If the initial concentration of an isotope is known, then the age of the body in which its present concentration is measured can be found directly from Equation 2.5. However, this is only possible in $C^{14}$ dating†.

*The decay of $K^{40}$ to $Ar^{40}$ may be slightly influenced by chemical bonding of the potassium and by pressure, because the decay in this case occurs by capture of an inner, orbital electron and thus depends on the local electron density at the nucleus. This effect has been observed in the decay of $Be^7$ (Hensley et al., 1973).

†In principle, $C^{14}$ dating assumes a constant proportion of $C^{14}$ in $CO_2$, from which organic carbon is derived; in fact atmospheric $C^{14}$ has varied due to variable geomagnetic shielding of the atmosphere from cosmic ray bombardment, but the $C^{14}$ clock has been calibrated in terms of samples of independently determined ages.

For the other decay schemes the concentration $D^*$ of the daughter isotope is measured; then

$$D^* = N_0 - N = N_0(1 - e^{-\lambda t}) \tag{2.7}$$

so that, combining (2.5) and (2.7), we can eliminate the unknown $N_0$:

$$\frac{D^*}{N} = e^{\lambda t} - 1 \tag{2.8}$$

$D^*$ is used to designate the number of *radiogenic* daughter nuclei, which are produced by decay, as, in general, the same isotope may occur independently of the decay and the nonradiogenic or *initial* contribution must be allowed for.

The simplest of the so-called accumulation clocks, which are based on measurements of the concentrations of daughter isotopes, is $K^{40}$—$Ar^{40}$, because when an extrusive igneous rock is formed, its initial concentration of $Ar^{40}$ is usually negligible, the previously generated argon having been lost by the melt to the atmosphere. Correction for contamination by atmospheric argon may be necessary and can be made on the basis of the $Ar^{36}$ and $Ar^{38}$ contents, which are present in the atmosphere to the extent of 0.337% and 0.063%, but are not produced as decay products. This correction is very conveniently applied routinely, since the measurement of $Ar^{40}$ is made by driving the argon out of a sample by melting it in a vacuum and mixing it in a mass spectrometer with a known quantity of isotopically separated $Ar^{38}$ (the spike). (Details of the method may be found in Dalrymple and Lanphere, 1969.) A measurement is simply a determination of the ratios $Ar^{40}:Ar^{38}:Ar^{36}$, which allow the concentrations of both radiogenic and atmospheric argon to be estimated in terms of the spike concentration. Potassium is commonly determined by a flame photometer comparison with a known standard and $K^{40}$ is estimated by assuming that it is always a constant fraction (0.01167%) of the total potassium. An alternative method, first used by T. Sigurgeirsson, is to estimate potassium from the $Ar^{39}$, which is produced from $K^{39}$ by $(n, p)$ reactions in a nuclear reactor. Since the measurement is a comparison of $Ar^{39}$ and $Ar^{40}$, argon and potassium are measured in the same sample; the necessity for a spike is removed and extremely small samples can be examined. Mitchell (1968) and Dalrymple and Lanphere (1971) have reported detailed measurements.

In the case of $K^{40}$—$Ar^{40}$ decay, Equation 2.8 must be modified to determine ages because $Ar^{40}$ is produced only by one of the two competing decay processes. However, the branching ratio* $\lambda_{Ar}/\lambda_{Ca}$ is

*The branching ratio is the ratio in which the two daughters are produced by the competing decay mechanisms.

well known, so that age $t$ can be determined from the radiogenic (i.e., corrected) $Ar^{40}$ and total $K^{40}$ contents:

$$Ar^{40} = \frac{\lambda_{Ar}}{\lambda} K^{40}(e^{\lambda t} - 1)$$

where $$\lambda = \lambda_{Ar} + \lambda_{Ca} \tag{2.9}$$

and $$\lambda_{Ar}/\lambda = 0.105$$

The most important limitation of the K—Ar method of dating is the loss of argon by diffusion, which has been studied as a function of temperature in a wide range of minerals (Fechtig and Kalbitzer, 1960). The argon loss is strongly dependent on temperature, as expected for a rate process, but several activation energies are involved and argon appears to be held in minerals in several types of site, from some of which it diffuses out more readily than others. We can, for example, imagine that an argon atom occupying a lattice vacancy will be held much more tightly than an interstitial atom. Certain minerals, notably hornblende, resist argon diffusion better than others. The only test for diffusion losses in nature is to compare the apparent K—Ar ages of different minerals and also compare these with other age measurements. If agreement is found the ages are termed *concordant*. Lower K—Ar ages normally indicate argon loss, which may be due to metamorphic heating or to very slow initial cooling and consequent delay in the start of argon retention. Comparison between minerals also serves as a check on the possible assimilation by some minerals of original argon in magma which was incompletely outgassed when it solidified. The potassium-argon method is particularly suitable for the dating of relatively young igneous rocks; rocks as young as $10^5$ years or even less may be dated if they are potassium rich.

Another simple accumulation clock of a different kind is based upon the spontaneous fission of $U^{238}$. Although spontaneous fission is a very rare process, recoiling fission fragments are very energetic and cause intense radiation damage in mineral crystals. Individual tracks can be made visible for counting under a microscope by an etching process developed by Fleischer et al. (1964, 1965, 1967). The number of fission tracks $T$ can be treated as an accumulated daughter product of the spontaneous fission of $U^{238}$:

$$\frac{\lambda}{\lambda_F} \frac{T}{U^{238}} = e^{\lambda t} - 1 \tag{2.10}$$

where $\lambda_F$ is the decay constant for fission and $\lambda$ is the total decay constant of $U^{238}$. It is not possible to count the total number of tracks in any sample because only one surface is etched, but the count is calibrated by comparison with the number of additional tracks produced by neutron-induced fission of $U^{235}$ during a controlled neutron irradiation in a nuclear reactor.

The neutron-fission cross section of $U^{235}$ is known ($\sigma_F = 582 \times 10^{-24}$ cm$^2$) and the present-day ratio $U^{238}/U^{235} = 137.88$ is constant in virtually all geological materials.* Then the number of neutron-induced fission tracks is

$$T_N = \phi \sigma U^{235} \quad (2.11)$$

where $\phi$ is the total neutron flux, in particles per unit area, and, combining Equations 2.10 and 2.11,

$$\frac{T}{T_N} = (2) \frac{\lambda_F}{\lambda} \cdot \frac{U^{238}}{U^{235}} \frac{(e^{\lambda t} - 1)}{\phi \sigma} \quad (2.12)\dagger$$

Regarding the tracks as a daughter product, in the general sense, the method is an accumulation clock with the special merit of having no nonradiogenic daughter (i.e., no nonradiogenic tracks) and depending only on the measurement of a ratio, $T/T_N$, and not upon a count of an absolute number of tracks. The production of neutrons in the Earth is insignificant, so that $U^{235}$ fission in rocks can be neglected, but in work on meteorites the neutrons produced by cosmic rays lead to additional fission tracks, which must be allowed for. Fission tracks are annealed out at temperatures from about 50 to 600°C, depending on the mineral, so that differences between fission track ages for different minerals in a rock may allow the dating of a mild metamorphism or possibly indicate uranium diffusion.

The other important clocks, based on Rb—Sr and U—Pb decays, are complicated by the occurrence of original as well as radiogenic daughter nuclides. In these cases the estimation of the proportion of a daughter isotope, which is of radiogenic origin, is made possible by two essential features in the occurrence of the elements concerned:

1. Because of their chemical differences the various minerals in a rock have quite different initial parent/daughter ratios.
2. At least one entirely nonradiogenic isotope of the daughter element is also present, and the initial isotopic ratio of the daughter element is homogeneous throughout the rock.

Considering the case of Rb—Sr, we can rewrite Equation 2.7 with the

*An exception, apparently resulting from a natural nuclear chain reaction, has been reported by Bodu et al. (1972)—see also *New Scientist*, *56*, 6 (1972).

†Equation 2.12 applies without the factor 2 if $T_N$ is measured in a surface which is freshly cut after irradiation. If the same surface is used as for the count of $T$ then the factor 2 is included because the surface was irradiated from both sides by the spontaneous fissions, but only from the one remaining side by neutron-induced fissions.

addition of an original amount $Sr^{87}{}_o$ of the daughter isotope $Sr^{87}$:

$$Sr^{87} = Sr^{87}{}_o + Sr^{87*} = Sr^{87}{}_o + Rb^{87}(e^{\lambda t} - 1) \tag{2.13}$$

Measurements of relative abundances of isotopes are made by mass spectrometer comparisons using spiked samples although the equality of atomic masses of $Rb^{87}$ and $Sr^{87}$ necessitates a chemical separation of the two elements before measurement. All values are referred to the abundance of the nonradiogenic isotope $Sr^{86}$, so that

$$\frac{Sr^{87}}{Sr^{86}} = \frac{Sr^{87}{}_o}{Sr^{86}} + \frac{Rb^{87}}{Sr^{86}}(e^{\lambda t} - 1) \tag{2.14}$$

The unknowns in this equation are $Sr^{87}{}_o/Sr^{86}$ and $(e^{\lambda t} - 1)$, but both factors are constant for all of the minerals in a rock with a simple igneous history, so that there is a direct linear relationship between the measured ratios $Sr^{87}/Sr^{86}$ and $Rb^{87}/Sr^{86}$ for the several minerals. From this relationship both $Sr^{87}{}_o/Sr^{86}$ and $(e^{\lambda t} - 1)$ are determined. Since $\lambda t < 0.1$ even for $t = 5 \times 10^9$ yr, $(e^{\lambda t} - 1) = \lambda t$ is normally a sufficient approximation. The determination of $t$ dates the time when strontium was last isotopically homogeneous.

The Rb—Sr technique is useful in examining the metamorphic histories of rocks and several graphical methods of representing the data have been important to the development of the subject (Compston et al., 1960; Nicolaysen, 1961; Lanphere, et al., 1964). The principle is illustrated in Figure 2.1. Suppose that three rocks $A$, $B$, $C$, differentiated from a common magma $T$ years ago, with a common initial ratio $Sr^{87}{}_o/Sr^{86}$, but different ratios $Rb^{87}/Sr^{86}$, by virtue of their chemical differences. In the course of the time $T$, the $Rb^{87}$ was depleted by decay and $Sr^{87}$ was enriched correspondingly, so that the isotopic compositions of the rocks as a whole migrated from $A$, $B$, $C$ to $A'$, $B'$, $C'$ in Figure 2.1. The line through $A'$, $B'$, $C'$ is known as the *T isochron*; its gradient is, by Equation 2.14, $(e^{\lambda T} - 1) \approx \lambda T$, from which $T$ is determined, and the intercept gives the initial isotopic composition of Sr $T$ years ago. However, if the rocks were all reheated during a metamorphic event $t$ years ago, much more recently than the original formation of the rocks, and during this event the minerals were changed somewhat and within each rock the strontium isotopes were rehomogenized, this would have given a new, different starting point for strontium evolution in each of the rocks. The isotopes did not migrate appreciable distances so that the whole rock compositions and the determination of the $T$ isochron were unaffected. However, isotopic measurements on individual minerals, indicated by the solid circles in Figure 2.1, give isochrons, with gradients $(e^{\lambda t} - 1)$, for the time $t$ when the rocks were individually rehomogenized with respect to strontium isotopes. Thus the times $t$ and $T$ and the

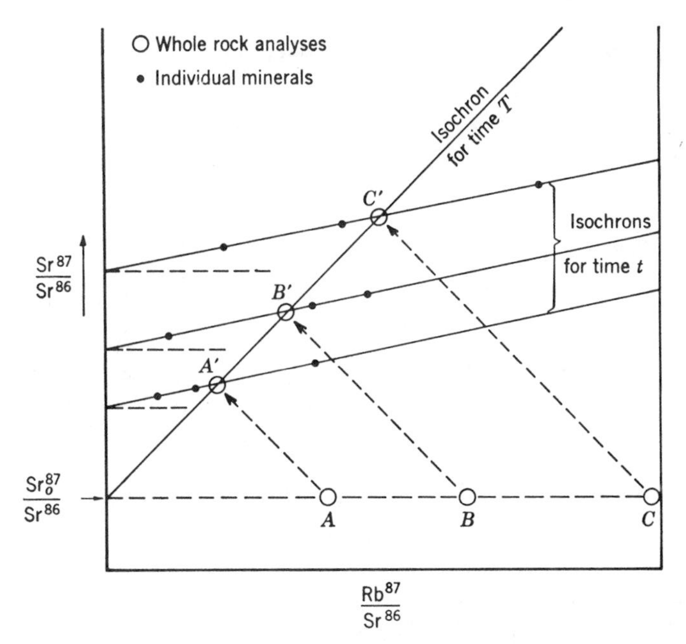


**Figure 2.1.** Rb–Sr evolution of three hypothetical rocks originating $T$ years ago from a common source and undergoing simultaneous metamorphosis $t(\ll T)$ years ago. Original whole-rock isotopic ratios are represented by $A$, $B$, $C$ and present ratios by $A'$, $B'$, $C'$. Isochrons through individual mineral analyses for each rock date the metamorphic event and the isochron through the whole rock analyses dates the original magma differentiation.

original isotopic ratio of the parent magma, $Sr^{87}/Sr^{86}$, are all determined.

Lead-uranium evolution follows equations, which are exact analogies of Equation 2.14 except that there are two parallel decay schemes $U^{238}$—$Pb^{206}$ and $U^{235}$—$Pb^{207}$, each of which is referred to the abundance of the nonradiogenic isotope $Pb^{204}$:

$$\begin{aligned}\frac{Pb^{206}}{Pb^{204}} &= \frac{Pb^{206}{}_o}{Pb_{204}} + \frac{U^{238}}{Pb^{204}}(e^{\lambda_{238}t} - 1)\\ \frac{Pb^{207}}{Pb^{204}} &= \frac{Pb^{207}{}_o}{Pb^{204}} + \frac{U^{235}}{Pb^{204}}(e^{\lambda_{235}t} - 1)\end{aligned} \tag{2.15}$$

The simultaneous equations (2.15) give two semi-independent estimates of the age of a rock, in which lead and uranium isotopes can be measured for several minerals, or of a suite of related rocks. There is, therefore, a test for concordance solely from measurements on lead and uranium. But it is normally more convenient to make measurements of lead isotopes only; Equations 2.15 can be combined to give a linear relationship between $Pb^{207}/Pb^{204}$ and $Pb^{206}/Pb^{204}$ by first writing

$$\frac{\dfrac{Pb^{206}}{Pb^{204}} - \dfrac{Pb^{206}_o}{Pb^{204}}}{\dfrac{Pb^{207}}{Pb^{204}} - \dfrac{Pb^{207}_o}{Pb^{204}}} = \frac{U^{238}(e^{\lambda_{238}t} - 1)}{U^{235}(e^{\lambda_{235}t} - 1)} \tag{2.16}$$

whence

$$\frac{Pb^{207}}{Pb^{204}} = \left[\frac{U^{235}}{U^{238}} \cdot \frac{(e^{\lambda_{235}t} - 1)}{(e^{\lambda_{238}t} - 1)}\right] \cdot \frac{Pb^{206}}{Pb^{204}} + \left[\frac{Pb^{207}_o}{Pb^{204}} - \frac{U^{235}}{U^{238}} \frac{(e^{\lambda_{235}t} - 1)}{(e^{\lambda_{238}t} - 1)} \cdot \frac{Pb^{206}_o}{Pb_{204}}\right] \tag{2.17}$$

Equation 2.17 introduces $U^{235}$, $U^{238}$ only as a ratio which is constant ($U^{235}/U^{238} = 0.007253$) so that the two bracketed quantities are constants for any set of samples that were isotopically homogeneous $t$ years ago and have remained isolated (closed) systems since then. Equation 2.17 is therefore a linear relationship between $Pb^{207}/Pb^{204}$ and $Pb^{206}/Pb^{204}$ and is known as a lead-lead isochron; the age $t$ may be calculated from its gradient.

Uranium-lead ages, and the thorium-lead ages with which they can be compared, are discordant sufficiently often to demonstrate that lead and uranium in the Earth have had complicated histories. However, the occurrence of the parallel decay schemes of $U^{238}$ and $U^{235}$, with chemically identical parents and final products, provides a powerful tool for the study of this history and, therefore, of the chemical evolution of the Earth (Kanasewich, 1968; Russell, 1972). The relevance to the age of the Earth as a whole is considered in Section 2.4.

## 2.3 GROWTH OF THE CONTINENTS AND OF ATMOSPHERIC ARGON

Before the advent of dating methods based on the radioactive decay schemes, the ages of geological formations in different parts of the world were correlated using the fossil record, which is essentially similar everywhere. Sedimentary strata back to the early Cambrian period, about 600 million years ago, were placed in chronological order and an approximate time scale was obtained from the estimation of sedimentation rates (see Hudson, 1964). Ages of igneous rocks were fitted into the time scale by relating them to sediments and the whole scale was subsequently made quantitative by dating igneous rocks radiometrically. Although there are still discrepancies between estimates of the ages of boundaries between geological periods the general validity of the geological time scale has been amply demonstrated. A table of dates is given in Appendix I.

Although pre-Cambrian paleontology is developing (Glaessner, 1966; Walter, 1972) the fossil record becomes very sketchy more than 500 or 600 million years ago. Thus detailed dating by fossil identification is restricted to little more than 10% of what we now understand to be the life span of the Earth. It is in the dating of pre-Cambrian events that radiometric methods have been most essential. However a basic difference between the two approaches must be recognized. The methods of fossil dating use sedimentary rocks and, since erosion and sedimentation are continuous processes, by comparing sedimentary strata in different parts of the Earth we have an essentially continuous sedimentary record. Radiometric methods date events, igneous and metamorphic, which are discontinuous in both place and time. Dates can be obtained only for times and localities of igneous activity and, if the present is typical of the past, then volcanism is quite limited in extent at any one time.

The basement rocks of the continents, the solid igneous rocks that underlie the veneer of sediments, must have been produced by processes we can describe, in general terms, as igneous (although many may be better regarded as high-grade metamorphics). Thus a geographical plot of radiometric ages of basement rocks is a plot of the migration of zones of tectonic activity. It has been known for a long time that the continents have "cores" of great ages, the pre-Cambrian "shield" areas, and that the surrounding basement rocks are younger. This observation has developed into the theory of continental growth by lateral accretion; the continental cores, now the shields, were formed about $3 \times 10^9$ years ago, and the continents have grown, and are still growing, by progressive additions of younger material (see Figure 2.2).

The idea that the material of the continental crust has originated by progressive differentiation from the mantle was given a quantitative basis by the work of Hurley et al. (1962) on Sr isotopes. They pointed out that, in differentiating from basic (low silica) sources, sialic (continental) material becomes enriched in Rb relative to Sr. They estimated that the weighted average Rb/Sr ratio in continental material was about 0.25 and that this was much greater than in the mantle. Using abundance ratios $Rb^{87}/Rb = 0.28$ and $Sr^{86}/Sr = 0.10$, the continental estimate corresponds to $Rb^{87}/Sr^{86} = 0.70$. Differentiating Equation 2.14, and noting that $d(Rb^{87})/dt = -d(Sr^{87})/dt$, we obtain as an average for continental material

$$\frac{d}{dt}\left(\frac{Sr^{87}}{Sr^{86}}\right) = \lambda\left(\frac{Rb^{87}}{Sr^{86}}\right) = 0.01 \times 10^{-9}\ \text{year}^{-1*} \qquad (2.18)$$

*The total range of $Sr^{87}/Sr^{86}$ values is very small, but measurements have become exceedingly sensitive to small differences.

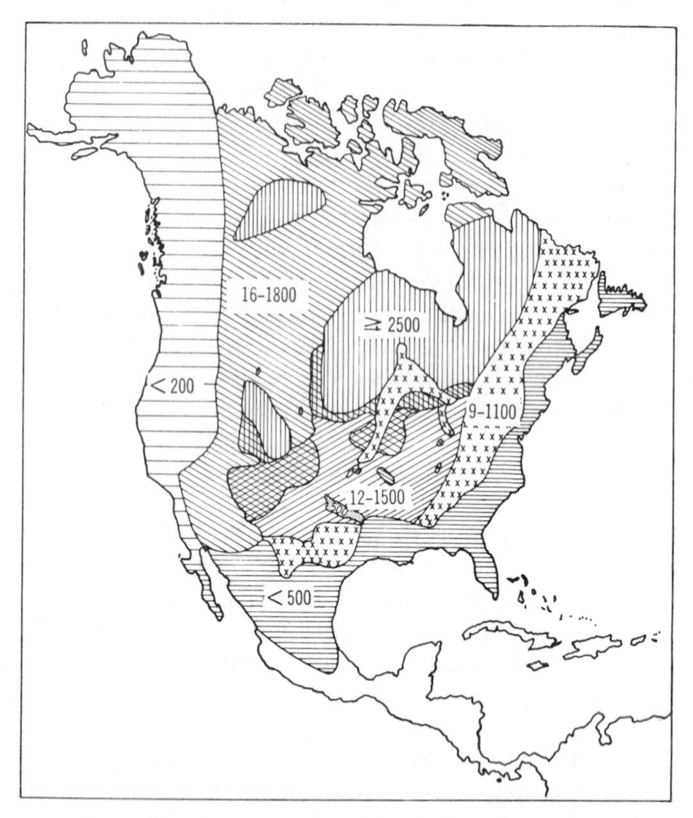


**Figure 2.2.** Age zones for basement rocks of North America. Numbers give ages in millions of years. Figure based on data by Hurley et al. (1962) and Goldich et al. (1966). (See also Muelberger et al., 1967.)

so that Hurley et al. (1962) suggested that the ratio $Sr^{87}/Sr^{86}$ increased faster in sialic material than in the mantle, from which it was ultimately derived, by nearly 0.01 per $10^9$ years, several times the uncertainty of measurement. Thus primary sial may be distinguished from material of the same geological (emplacement) age, which is merely reworked sedimentary or continental igneous material, by the characteristic value of $(Sr^{87}_o/Sr^{86})$, measured for whole rock samples. Hurley et al. (1962) concluded that the $Sr^{87}/Sr^{86}$ values of continental rocks corresponded to their ages well enough to demonstrate that the proportion of older, reworked material incorporated in them could only be large if it was itself of recent primary subsialic origin; in other words, the continents must have developed by progressive differentiation (chemical separation) from the mantle.

A closer look at the Sr isotopic ratios of mantle-derived volcanics indicates a significant heterogeneity of the sources of basaltic lava. The principal sources are of two kinds, one being the ocean rises, or lines of convective upwelling where new ocean floor is formed (Chapter 10), and the other being "point" sources, not necessarily coinciding with ocean rises, Hawaii being the best known example. Metz (1974) reviewed the evidence that the "point" sources are geographically fixed with respect to one another and therefore by implication to the lower part of the Earth's solid mantle (producing chains of islands as the crust moves across), while the ocean rises and corresponding sinks (subduction zones) are continuously changing features in the pattern of upper mantle convection. Atlantic and Pacific ocean ridge basalts have characteristic $Sr^{87}/Sr^{86}$ ratios of 0.7023 to 0.7027, while the island chain basalts have a ratio of 0.7030 plus (Hart et al. 1973) and have therefore been derived from sources richer in Rb. If, as appears plausible, the ridge basalts are the light differentiates that have "floated" to the top in the process of upper mantle convection, whereas the island chains are products of deeper convective "plumes" (Morgan, 1971) based in the lower mantle, then the Sr isotopes indicate a significant difference in composition between the lower and upper parts of the mantle. Possibly the Rb in the upper mantle has largely been swept into the crust by numerous convective cycles over several billion years but the lower mantle has been much less affected by convection and more nearly retains its original Rb/Sr ratio. In view of the chemical similarity of Rb and K and the importance of K as a heat source, this consideration has a direct bearing on the thermal state of the Earth (Chapter 7) as well as the argon in the atmosphere. But superimposed on these apparently consistent differences are differences between oceans. Subbarao and Hedge (1973) and Hedge et al. (1973) found that Indian Ocean islands were similarly richer in radiogenic Sr than neighboring ridge basalts, but both were consistently richer than the corresponding materials from the Atlantic and Pacific Oceans. Evidently the mantle has suffered less differentiation under the Indian Ocean.

The atmosphere contains $6.5 \times 10^{16}$ kg of argon, which is dominated by the radiogenic isotope $Ar^{40}$. The other argon isotopes, as well as other inert gases, notably neon and krypton, are very much rarer, relative to $Ar^{40}$, than would be expected from their cosmic abundances.* The atmospheric argon is a decay product of $K^{40}$, outgassed from the crust and mantle in the course of volcanic activity. This is entirely in keeping with the evidence that the atmosphere and oceans, as well as the

*The ratio $Ar^{40}/Ar^{36}$ is 296 in atmospheric argon but Kirsten et al. (1971) found that the ratio was 0.15 for solar wind argon trapped by grains of lunar surface material.

continental crust have evolved as volcanic products (Rubey, 1951). A basic assumption of the K-Ar dating method is that any pre-existing Ar is completely outgassed when an igneous rock is first formed; this is effectively true for extrusive rocks but incompletely so for intrusives, or for rapidly cooled submarine basalts. However retention of initial argon is probably sufficiently slight to be neglected in a calculation of the gradual accumulation of argon in the atmosphere. Thus the $6.5 \times 10^{16}$ kg of $Ar^{40}$ must be related to the available $K^{40}$ in the Earth in terms of its igneous history. A calculation on these lines by Shillibeer and Russell (1955) first established the significance of argon outgassing.

We obtain an idea of the numbers involved by calculating from Eq. 2.9 the residual mass of the $K^{40}$ which would have produced $6.5 \times 10^{16}$ kg of $Ar^{40}$ in the life of the Earth ($4.5_5 \times 10^9$ years-section 2.4). This gives $5.6 \times 10^{16}$ kg and since $K^{40}$ is 1.18 parts in $10^4$ of total K the corresponding total mass of potassium is $4.7 \times 10^{20}$ kg. Within the uncertainties of the estimates this happens to coincide with the potassium content of the crust. However, since parts of the crust are of great ages and have retained the argon produced in them since their formation, partial outgassing of the mantle must be invoked to make up the observed atmospheric argon. The obvious assumption is that from any volume of mantle which differentiates to produce crust all of the argon then present is carried up and lost to the atmosphere. So if we consider a particular volume of material in which the present mass of radioactive potassium is represented by $K^{40}$, and which was last outgassed $t$ years ago then its contribution to the atmospheric argon is

$$Ar^{40} = K^{40} \frac{\lambda_{Ar}}{\lambda} (e^{\lambda T} - e^{\lambda t}) \tag{2.19}$$

where $T$ is the age of the Earth, or the time since argon accumulation began, which could be somewhat later. This expression can be integrated over the total estimated potassium in the Earth for any assumed igneous history, with the constraint that it must yield the measured atmospheric argon.

If we make the simple assumption that the fraction of the Earth that has been outgassed has grown progressively and at a uniform rate since the origin of the Earth (at $T = 4.5 \times 10^9$ years) and that no part has been outgassed twice, then the total potassium content of that fraction is related to the atmospheric argon by a straightforward integral:

$$\begin{aligned} Ar^{40} &= K^{40} \frac{\lambda_{Ar}}{\lambda} \int_o^T (e^{\lambda T} - e^{\lambda t}) \frac{dt}{T} \\ &= K^{40} \frac{\lambda_{Ar}}{\lambda} \left[ e^{\lambda T}\left(1 - \frac{1}{\lambda T}\right) + \frac{1}{\lambda T} \right] \end{aligned} \tag{2.20}$$

which gives $K^{40} = 8.8 \times 10^{16}$ kg and the corresponding total $K = 7.4 \times 10^{20}$ kg. But the total potassium content of the Earth must be at least twice this value if we accept the average crustal K/U ratio as representative of the whole Earth. Since the total content of the important heat-producing elements, K, U, Th, whose heat productions are listed in Table H3 (Appendix H), is presumed to match the total geothermal flux ($3.1_4 \times 10^{13}$ W) and the ratio Th/U $\sim 3.5$ is roughly constant in most geological materials, the total K is uniquely related to the K/U ratio of the Earth, as considered in Section 7.1. Assuming the crustal value, $K/U \approx 10^4$, gives $K \approx 1.4 \times 10^{21}$ kg.

On this basis we must assume that a substantial fraction of the interior of the Earth, representing not less than 50% of the total potassium content, has not participated in the outgassing, in spite of its substantial heat source. This assumption is difficult to accept without further strong evidence. Ozima (1975) noted the problem and placed bounds on possible outgassing models, favoring an early "catastrophic" event, such as formation of the core (with probable substantial loss of the original atmosphere), followed by progressive partial outgassing. The difficulty is reduced but not avoided by postponing as long as possible the postulated catastrophic event, to dispose of early radiogenic argon. Alternatively Fisher (1975) suggested that the mantle K/U ratio was much lower than the crustal value, perhaps even lower than that of the Moon ($\sim 2000$) and this suggestion is supported by the generally low K/U ratios of ultrabasic rocks. His primary argument was based on the high $He^4/Ar^{40}$ ratios (10 to 20) reported in rapidly chilled, glassy submarine basalts, which are presumed to have retained ratios of these gases characteristic of the mantle regions from which they were derived. Such high values are incompatible with a crustal K/U ratio and even more so with a meteoritic ratio and suggest a value nearer to 1500.

We must also recognize the need for a core heat source of perhaps $5 \times 10^{12}$ W (section 7.3), which could be provided by $1.5 \times 10^{21}$ kg of potassium, but then it is necessary to suppose that very little of the argon from the core has escaped into the atmosphere. However, the segregation of potassium from large regions of the lower mantle into the core, as considered in Section 7.1 for geothermal reasons, would provide the potassium-depleted regions of the mantle needed to satisfy Fisher's (1975) $He^4/Ar^{40}$ argument. In Chapter 7 numerical values are given for a thermal model of the Earth in which an overall carbonaceous chondritic K/U ratio $\sim 2 \times 10^4$ is assumed, partly because the heat output of an Earth of overall carbonaceous chondritic composition coincides with the geothermal flux (Table 7.2). This is probably still as good an assumption as we can make at present, but further work on the outgassing problem may cause it to be drastically revised.

Gastil (1960) examined the distribution in time of radiometric dates and concluded that they were grouped in a manner that implied that igneous activity was a spasmodic process and occurred for periods of about 200 million years at intervals of 350 to 500 million years, being essentially simultaneous on all continents. However Hurley and Rand (1970) suggested that the accumulation of 10 years' more data smeared Gastil's peaks. They suggested that the rate of production of continental crust has progressively increased with time, but it appears more likely that more of the older rocks have been reworked or buried and that the rate of production has remained constant (or slowly decreased with the diminishing radioactive heat in the Earth). In Chapters 7 and 10 steady rates of convection and differentiation are assumed, but such calculations should be asterisked with the comment that if Gastil's peaks are real and not an artifact of specimen selection they modify the conclusions.

## 2.4 AGE OF THE EARTH AND OF METEORITES

We have, at least at present, no means of determining the age of the Earth independently of data on meteorites. The oldest well-dated rocks, which are found in Greenland, Rhodesia, Minnesota and the Kola Peninsula of northern Russia, have ages of about $3.7 \times 10^9$ years (see, for example, Moorbath et al., 1972) and there is no geological record at all of the first 800 million years of the Earth's existence as a planet. This is not altogether surprising, when we realize that the internal heat generated by radioactive elements, especially $U^{235}$, was very much greater $4 \times 10^9$ years ago than now and that there had presumably been less differentiation of the crust from the mantle, so that the radioactivity was, on average, distributed more deeply. In these circumstances igneous activity would have been more intense, so that the material of any very early rocks has been completely "reworked" into later rocks. It is also probable that bombardment of the Earth by substantial planetismals persisted for several hundred million years, as for the Moon.

A simple model of the Earth proposed independently by A. Holmes and F. G. Houtermans in 1946 has been widely used as a basis for discussions of its isotopic history, especially the uranium-lead evolution. The supposition is that at or very soon after its accretion, the Earth separated into a number of subsystems (e.g., core and mantle), with different U/Pb ratios, and that each subsystem remained completely closed thereafter. The lead isotopes would then have developed differently from a common primordial lead. With radiogenic lead distributed according to the U/Pb and Th/Pb ratios in the subsystems, the primary differentiation can be dated by the methods of Section 2.2. However, the mantle and crust have had too complex a history to allow us to find independent subsystems in the regions accessible to sampling. Geological processes have led to partial rehomogenization of lead isotopes and perhaps more importantly, uranium apparently diffuses quite readily in rocks, as is evident from the difficulty experienced in obtaining concordant lead ages. Thus we cannot date the formation of the Earth from terrestrial samples alone.

A primary differentiation of the kind considered by Holmes and Houtermans evidently occurred in the material that formed the meteorites, many of which do appear to have remained closed subsystems and can therefore be dated. The methods used for rocks are all available (Anders, 1962). Until recently the lead isotope method received most attention, being, in principle, the most accurate, because the decay constants of uranium are the best known and because the range of variation in isotopic abundances can be as great as 10:1, but improvements in Rb/Sr techniques have led to elucidation of some of the finer details of the history of meteorite formation. A lead-lead isochron (Eq. 2.17) for the meteorites is plotted as Figure 2.3, using data selected from

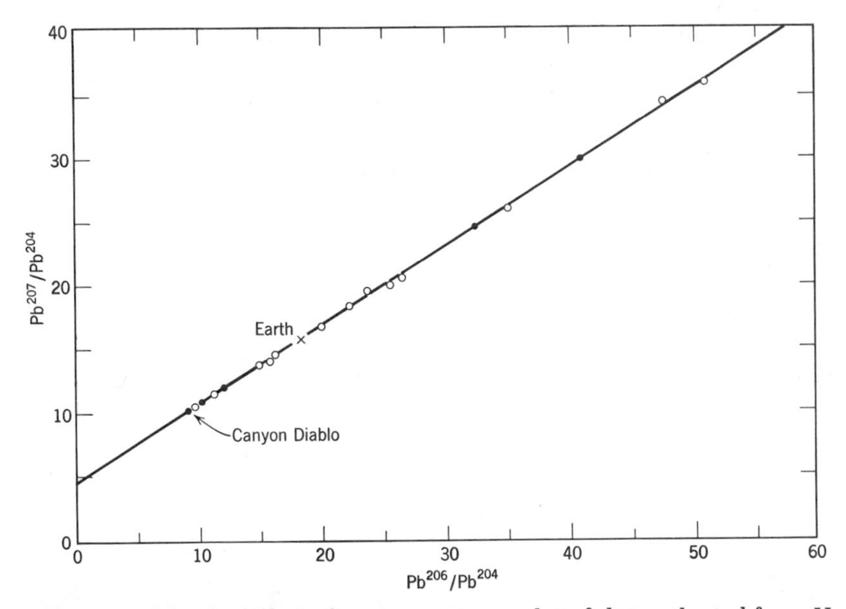


**Figure 2.3.** Lead-lead isochron for meteorites; a plot of data selected from Huey and Kohman (1973) (open circles) and Tatsumoto et al. (1973) (solid circles). The average terrestrial (marine sediment) data point of Chow and Patterson (1962) is shown as a cross. Some overlapping data points are omitted and values for highly radiogenic samples are well off the range plotted here, but these values constrain the least-squares fitted line (Eq. 2.21) which is also shown.

Huey and Kohman (1973) and Tatsumoto et al. (1973) which give an equal-weight, least-squares-fitted relationship

$$\frac{Pb^{207}}{Pb^{204}} = (0.6128 \pm 0.014)\frac{Pb^{206}}{Pb^{204}} + (4.46 \pm 0.10) \tag{2.21}$$

With the decay constants listed in Appendix H and taking $U^{238}/U^{235} = 137.88$, this gives an age of $4.54 \times 10^9$ years, which happens to coincide with the now-classical early estimate, $4.55 \times 10^9$ years, by Patterson (1956).

By comparison with values for chondrites, most iron meteorites have closely similar lead isotope ratios, since their uranium and thorium contents are very small. The lead in the troilite (iron sulfide) of the Canyon Diablo iron meteorite is the least radiogenic of all meteorite leads; it is referred to as "primordial" lead, from which other samples have evolved by additions of various proportions of radiogenic lead since the solar system formed. The primordial lead composition is usually considered more secure than most other individual estimates and is sometimes used as a constraint in isochron determinations, but this has not been done in Equation 2.21. Comparison of primordial lead with any other single sample allows an estimate of age. In Table 2.1 primordial lead is compared with the average lead composition of the Earth's crust, as estimated by Chow and Patterson (1962) from marine sediments, and with an ancient ($2.7 \times 10^9$ year) galena from Manitouwadge, Ontario, which Tilton and Steiger (1965) favored for estimating the age of the Earth.

Applying Equation 2.17 to both the primordial lead (represented by

**Table 2.1** Lead Isotope Ratios[a]

| | $Pb^{206}/Pb^{204}$ | $Pb^{207}/Pb^{204}$ | $Pb^{208}/Pb^{204}$ |
|---|---|---|---|
| Primordial (Canyon Diablo troilite) | 9.307 | 10.294 | 29.476 |
| Crustal average (marine sediments) | $18.5_8$ | $15.7_7$ | $38.8_7$ |
| Ancient galena (Manitouwadge, Canada) | 13.30 | 14.52 | 33.58 |

[a]A comparison of data by Tatsumoto et al. (1973) on troilite from the Canyon Diablo iron meteorite, which is widely accepted as "primordial," that is, the least radiogenic of any natural lead, with the average crustal lead estimated by Chow and Patterson (1962) from marine sediments and an ancient galena ($\sim 2.7 \times 10^9$ years) favored by Tilton and Steiger (1965) for estimating the age of the Earth.

subscript $o$) and crustal average (subscript $a$) we obtain the relationship for the age of the Earth, $T$:

$$\frac{e^{\lambda_{235}T}-1}{e^{\lambda_{238}T}-1}=\frac{U^{238}}{U^{235}}\cdot\frac{(Pb^{207}/Pb^{204})_a-(Pb^{207}/Pb^{204})_o}{(Pb^{206}/Pb^{204})_a-(Pb^{206}/Pb^{204})_o}=81.42 \qquad (2.22)$$

and hence $T = 4.48 \times 10^9$ years, but with an uncertainty of about $0.1 \times 10^9$ years, within which range it coincides with the estimate from the meteorite isochron. Tilton and Steiger (1965) noted the variability of terrestrial leads, which have clearly not been confined to closed systems, and favored an estimate based on a 2700-million-year-old lead ore, for which they obtained isotope ratios listed in Table 2.1. With the presumption that this lead was completely isolated from uranium at $t = 2.7 \times 10^9$ years ago, but that before that time it was part of a closed system, the time interval for accumulation of its radiogenic lead $(T - t)$ can be determined from the gradient of an isochron that would have been obtained $t$ years ago by using lead ratios from the then-new galena and the iron meteorites.

$$\frac{e^{\lambda_{235}(T-t)}-1}{e^{\lambda_{238}(T-t)}-1}=\left(\frac{U^{238}}{U^{235}}\right)_t\cdot\frac{(Pb^{207}/Pb^{204})_g-(Pb^{207}/Pb^{204})_o}{(Pb^{206}/Pb^{204})_g-(Pb^{206}/Pb^{204})_o}=15.53 \qquad (2.23)$$

where subscript $g$ represents the galena values and

$$\left(\frac{U^{238}}{U^{235}}\right)_t=\left(\frac{U^{238}}{U^{235}}\right)_{\text{Present}}\cdot\frac{e^{\lambda_{238}t}}{e^{\lambda_{235}t}}=14.6_{76} \qquad (2.24)$$

for $t = 2.7 \times 10^9$ years. This is the value of the uranium isotope ratio when the galena was formed, that is, the value that a geochronologist working at that time would have used in Equations 2.17 or 2.22. With the values in Table 2.1 we obtain $(T - t) = 1.8_4 \times 10^9$ years and thus $T = 4.5_4 \times 10^9$ years in fortuitously precise agreement with the value from the meteorite isochron.

The essential point made by Tilton and Steiger (1965) can be examined in terms of the data in Table 2.2. This gives values of the ratio $[(Pb^{207}/Pb^{204})_s - (Pb^{207}/Pb^{204})_o]/[(Pb^{206}/Pb^{204})_s - (Pb^{206}/Pb^{204})_o]$ for several samples ($s$) referred to primordial lead ($o$). This ratio must have the same constant value (0.613) as that obtained from the meteorite isochron (Eq. 2.21) for any specimens that represent closed systems originating from the same isotopic source as the meteorites. The extreme departure is represented by the Manitouwadge galena, which was isolated from uranium 2700 million years ago, and so has a composition strongly biased toward *early* radiogenic lead, since the proportionately greater amount of $U^{235}$ then present produced a greater relative enrichment in $Pb^{207}$. Conversely, all of the other terrestrial samples are biased the other way, that is, they are relatively depleted in early radiogenic lead. This

**Table 2.2** Radiogenic Lead[a]

| | |
|---|---|
| Meteorites (Eq. 2.21) | 0.613 |
| Marine sediments (Table 2.1) | 0.591 |
| Manitouwadge galena (Table 2.1) | 1.058 |
| Easter Island[b] | 0.532 |
| Guadalupe Island[b] | 0.478 |
| East Pacific Rise[b] | 0.572 |
| Mid-Atlantic Ridge[b] | 0.584 |

[a]Values of the ratio $\frac{(\mathrm{Pb}^{207}/\mathrm{Pb}^{204})_s - (\mathrm{Pb}^{207}/\mathrm{Pb}^{204})_o}{(\mathrm{Pb}^{206}/\mathrm{Pb}^{204})_s - (\mathrm{Pb}^{206}/\mathrm{Pb}^{204})_o}$ for various samples ($s$) compared with primordial lead ($o$).

[b]Data from Tatsumoto (1966).

depletion is much more pronounced for the island basalts than for the ocean ridge basalts and may be used to infer different lead isotope histories for the upper and lower mantles. In principle the preferential depletion of the lower mantle could be explained by supposing the "missing" early lead to be hidden in lead ore bodies, such as that at Manitouwadge, but it is more natural to suppose that lead was removed into the core at some stage or stages after the development of appreciable radiogenic lead in the mantle. Oversby and Ringwood (1971) showed that lead is quite strongly partitioned into the metal phase of a silicate-iron melt and argued that the close fit of the marine sediment lead (taken to be an effective mantle average) to the meteorite lead isochron (Fig. 2.3) implied an insignificant and probably nonexistent time lag between the accretion of the Earth and the formation and isotopic isolation of the core. However, the island basalt data suggest a continuing exchange with the core much later than Oversby and Ringwood (1971) infer and the variability of this data suggests that the gleaning of early radiogenic lead from the mantle into the core may have persisted to different times in different regions.

Rubidium and strontium measurements on stony meteorites also give a convincing isochron (e.g., Gast, 1962). Members of the class known as achondrites are all very low in Rb, which conveniently allows their strontium isotope ratio, $\mathrm{Sr}^{87}/\mathrm{Sr}^{86} = 0.6989$, to be identified as that of "primordial" strontium. The chondrites are distributed along an isochron through the "primordial" point and having a gradient

$$\frac{d(\mathrm{Sr}^{87}/\mathrm{Sr}^{86})}{d(\mathrm{Rb}^{87}/\mathrm{Sr}^{86})} = 0.0664 \qquad (2.25)$$

The exact estimate of an age from this value was, until recently, subject to uncertainty in the decay constant of $Rb^{87}$. Using $\lambda = 1.42 \times 10^{-11}$ years$^{-1}$, as in Appendix H, the calculated age by Equation 2.14 is $4.53 \times 10^9$ years, in agreement with the lead value. Burnett and Wasserburg (1967a) first reported that silicate intrusions in iron meteorites give the same age and Sanz et al. (1970) made a very detailed study of Sr and Rb isotopes in individual minerals of the silicate intrusions of the iron meteorite Colomera. The mineral isochron from their data, with the revised $Rb^{87}$ decay constant, gives an age of $4.51 \times 10^9$ years. Sanz et al. (1970) noted that the interval between formation and stabilization of Sr and Rb was evidently some tens of millions of years, so that this figure is in remarkable accord with the "whole rock" Sr isochron for meteorites. However, the ages are not all precisely coincident and at least some meteorites have undergone more recent isotopic homogenization and give mineral ages much younger than the accepted age of the solar system (Compston et al., 1965; Burnett and Wasserburg, 1967b; Wasserburg and Burnett, 1969; Wetherill et al., 1973). Clues to the very early development of the solar system are obtained from initial $Sr^{87}/Sr^{86}$ ratios, very low values indicating isolation of material from a mix that was richer overall in Rb relative to Sr. The lowest recorded ratio $Sr^{87}/Sr^{86} = 0.69877$, was reported by Gray et al. (1973) for Rb-deficient chondrules in the Allende carbonaceous chondrite, which are presumed to be the earliest known condensate from the solar nebula. It is evident that this subject will receive much more attention in the next few years. But, in the present context, it is important to note that the evidence for the nearly simultaneous formation of meteorites and the solid materials that produced the Earth, the Moon and, by implication, at least the other terrestrial planets, approximately $4.53 \times 10^9$ years ago, is completely convincing.

## 2.5 DATING THE NUCLEAR SYNTHESIS

That radioactive species such as $U^{235}$, with half-lives of $10^9$ years or less, have survived to the present time is evidence that the process or processes of nuclear synthesis that formed them must have occurred only a few billion years ago, that is, not more than about $10^{10}$ years ago and probably less. If it is assumed that the uranium and lead isotopic ratios in the Earth's crust and in meteorites are representative of the solar system, then a more precise upper limit to the age of the elements can be imposed. The first approach was to adopt from theories of nuclear synthesis an original ratio of uranium isotopes $(U^{235}/U^{238})_0$. This is uncertain, since it depends on the processes of nuclear synthesis; the heavier elements are presumed to have been built up by neutron bombardment, but this would rapidly destroy $U^{235}$ by fission, so that the $U^{235}$ with which the solar system started was presumably itself produced by decay of now-extinct heavier parents. An early guess by Lord Rutherford was $(U^{235}/U^{238})_o = 0.8$. Then, using Equation 2.5 for the $U^{235}$ and $U^{238}$ decays, the present and original isotopic ratios are related by the equation

$$\left(\frac{U^{235}}{U^{238}}\right) = \left(\frac{U^{235}}{U^{238}}\right)_o \exp\left[-(\lambda_{235} - \lambda_{238})\tau_o\right] \tag{2.26}$$

which gives $\tau_o = 5.7 \times 10^9$ years as the time since nuclear synthesis.

A more satisfying method is to assume that all of the $Pb^{207}$ has been derived by decay of $U^{235}$, which is unlikely to be the case and therefore gives a very clear upper limit to the age of the elements. The relevant terrestrial ratio of U/Pb is obtained from the data of Section 2.4, since

$$\left(\frac{Pb^{207}}{Pb^{204}}\right)_E - \left(\frac{Pb^{207}}{Pb^{204}}\right)_o = \left(\frac{U^{235}}{Pb^{204}}\right)_E \left[\exp(\lambda_{235}\tau) - 1\right] \tag{2.27}$$

where subscripts $E$ and $o$ refer to Earth (crustal average) and primordial values, as in Table 2.1, and $\tau = 4.5 \times 10^9$ years is the age of the Earth, as determined from Equation 2.21. This gives $(U^{235}/Pb^{204})_E = 0.06335$. Then

$$\left(\frac{Pb^{207}}{Pb^{204}}\right)_E \geqslant \left(\frac{U^{235}}{Pb^{204}}\right)_E \left[\exp(\lambda_{235}\tau_o) - 1\right] \tag{2.28}$$

which gives $\tau_0 \leqslant 5.6 \times 10^9$ years.

This indicates that the interval between nuclear synthesis and the accretion of the Earth was no more than $10^9$ years and could have been much shorter. We might therefore expect to find in the Earth, or in meteorites, evidence of orphaned daughter isotopes, whose short-lived parents no longer exist in measurable quantities. Study of this problem—*cosmochronology*—has been reviewed by Schramm (1973). Reynolds (1960) first identified radiogenic $Xe^{129}$ from extinct $I^{129}$ in meteorites. There are nine stable xenon isotopes and they occur in meteorites in proportions which indicate at least three distinct sources; it is not only $Xe^{129}$ that is variable with respect to the others. But the relative enrichment of $Xe^{129}$ has been positively identified with iodine-bearing minerals, so that the amount of enrichment allows the synthesis-accretion interval to be estimated in terms of the half-life of $I^{129}$, $1.64 \times 10^7$ years. By assuming that the nonradioactive isotope $I^{127}$ and the extinct isotope $I^{129}$ were produced with equal abundances in the nuclear synthesis, Reynolds (1960) estimated from the abundances of trapped $Xe^{129}$ that the $I^{129}$ abundance had fallen by a factor $10^5$ between the synthesis and the accretion of the chondrites, giving an interval of $2.9 \times 10^8$ years. On the assumption of a very prolonged process of synthesis, the "initial" $I^{129}$ is less abundant and the estimated interval was reduced to 1.2 to $1.4 \times 10^8$ years. An interval estimated in this way is the time between nuclear synthesis and the solidification and cooling of the meteorite parent body to the point at which it could retain xenon. The intervals estimated for various meteorites vary, as does the initial strontium ratio, $(Sr^{87}/Sr^{86})_o$ (Papanastassiou and Wasserburg, 1969), indicating significant differences in the dates of formation of the meteorites. But Manuel et al. (1968) found that $Xe^{129}$ was held in different crystalline sites characterized by both low and high xenon retentivities and that the cooling rates of individual meteorite bodies could be estimated. They obtained the estimate 5 to 9 deg/$10^6$ years, which is within the wider range of cooling rates deduced from Widmanstätten patterns in meteoritic iron (Section 1.2). Podosek (1970) has reviewed the general field of $Xe^{129}$ chronology.

The other xenon nuclides found in meteorites are substantially, or possibly entirely, of *fissiogenic* origin, that is, they are stable products of spontaneous fission of extinct transuranic elements that were present in significant quantities 4.5 billion years ago. The plutonium isotope $Pu^{244}$ (see Appendix H2) is clearly established as the most important source of this xenon (Alexander et al., 1971) and Reynolds et al. (1969) reported that concordant values of about 200 million years were obtained from the $I^{129}$ and $Pu^{244}$ methods of estimating the interval between nuclear synthesis and xenon retention in meteorites. However, $I^{129}$ and $Pu^{244}$ do not account for all of the variability in xenon isotope abundances. The existence of a second fissiogenic component in volatile-rich (especially carbonaceous) chondrites has been suggested (Anders and Larimer, 1972) and Rao and Gopalan (1973) favoured curium, $Cm^{248}$, which has a normal $\alpha$ decay to $Pu^{244}$ but a very high (7%) spontaneous fission yield. However, since this has a half-life shorter than $10^6$ years it appears incompatible with the $I^{129}$ and $Pu^{244}$ estimates of the synthesis-accretion interval. Anders et al. (1975) strongly favored a volatile element of atomic number about 115, but the necessity for a second fissiogenic xenon component is doubted by other authors (Kuroda et al., 1974). Another possibility, suggested by the variability of isotopic abundances of light elements in meteorites (Bernas et al., 1969), is that some neutron flux persisted even after accretion and that neutron-induced fission (as of $U^{235}$, $Pu^{239}$) may account for some fissiogenic xenon.

Theories of nuclear synthesis are reviewed by Truran (1973). The processes of greatest interest in the present context are referred to as *r* (rapid) processes*, that is, successive neutron captures that occurred too

*Also important to the buildup of the less-heavy elements are the *s* (slow time scale) neutron capture processes, which progress along the main nuclide series and, to a lesser extent, *p* (proton capture) processes.

rapidly to allow normal $\beta$-decay to the most stable nuclear series. Neutron-rich heavy nucleii were built up and subsequently $\beta$-decayed to the more familiar species, bypassing isotopes such as $U^{235}$, $Pu^{239}$ which are strongly neutron fissionable. The event believed to have been responsible for these *r* processes is a supernova explosion. It is, therefore, a fertile source of conjecture that accretion of the solar system followed a supernova so rapidly. But whatever the details of the preaccretion solar system, which are undoubtedly more complicated than any current hypotheses and cannot reasonably be the result of a single stage of nucleosynthesis, it is inevitable that the early planetary and meteorite parent bodies were rich in relatively short-lived nucleii which would have caused strong internal heating and probably large-scale melting in most cases, including the Earth. An important representative of these short lived nuclides is $Al^{26}$ (listed in Table H2).

# 3

# Rotation of the Earth

"... consider the wobble induced by a hula dancer of mass *m* on the geographic north pole ..."

MUNK AND MacDONALD, 1960b, p. 55

## 3.1 FIGURE OF THE EARTH

The centrifugal effect of the Earth's rotation causes an equatorial bulge, which is the principal departure of the Earth from spherical shape. If the whole Earth were covered with a shallow sea then, apart from minor disturbances due to wind and internal currents, the surface would assume the shape determined by hydrodynamic equilibrium of the water subjected to gravitation and rotation; the sea level equipotential surface is the geoid or figure of the Earth.* Tidal effects are superimposed on the mean geoid by gravitational gradients of the Moon and Sun, but are very small in comparison with the rotational ellipticity with which we are concerned here. Crustal features, continents and mountain ranges, are significant departures of the actual surface of the Earth from the geoid, but mass compensation at depth (the principle of isostasy discussed in Section 4.3) minimizes the influence of surface features on the geoid.

The form of the geoid has been determined from astrogeodetic surveys over several extended continental survey arcs by measuring the direction of the vertical, or local gravity vector, relative to stars. The process is described in detail by Bomford (1971) and is indicated in

*For a physical idea of the geoid in continental areas, picture canals cut through the continents and connected with the oceans, so that the water level is the geoidal surface.

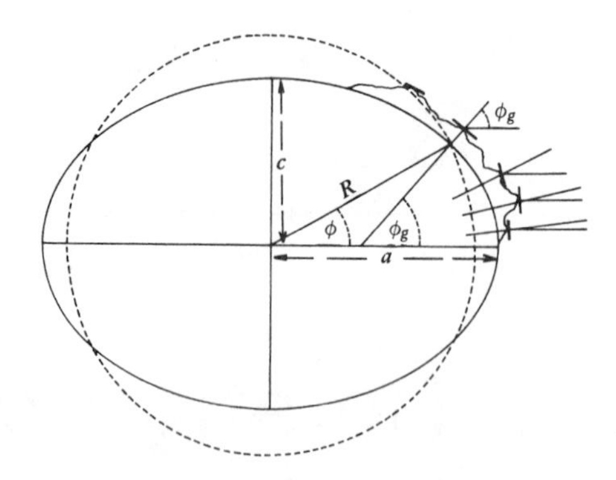


**Figure 3.1.** A comparison of the equilibrium geoid (solid line) with a sphere of equal volume (dashed line). The ellipticity of the geoid is exaggerated by a factor of about 50. The radius of the sphere is $R = (a^2c)^{1/3}$, $a$ and $c$ being the major and minor semiaxes of the geoidal ellipsoid. $\phi$ is the geocentric latitude at any point and $\phi_g$ is the geographic latitude or angle between the normal to the geoidal equipotential surface at the point of observation and the equatorial plane. This angle is determined by reference to stars at each point around an astrogeodetic survey arc, so that the local orientation of the geoidal surface is measured.

Figure 3.1. Values of the flattening* of the geoid deduced in this way from surveys completed between 1900 and 1960 are within the range 1/297 to 1/298.3 and deductions from gravity observations (Section 4.1) cover the same range. These estimates were derived from measurements limited to the continental areas of the Earth's surface. A more accurate value, which properly represents the Earth as a whole, is now available from analyses of satellite orbits (Section 4.2), which give $1/298.25_6$ with considerable confidence, so that this figure is now preferred. It is, however, important to note that the geodetic estimates are not in conflict with the satellite value, although they are biased to one side of it.

The total geopotential $U$ at any external point rotating with the Earth is a sum of gravitational potential $V$ and rotational potential:

$$U = V - \tfrac{1}{2}\omega^2(x^2 + y^2) = V - \tfrac{1}{2}\omega^2 r^2 \sin^2\theta = V - \tfrac{1}{2}\omega^2 r^2 \cos^2\phi \qquad (3.1)$$

where $\omega$ is the angular velocity of the Earth's rotation about the $z$ axis and $(x, y)$ or $(r, \theta)$ are the coordinates of the point, referred to the centre of the Earth. $\theta$ is the colatitude or angular distance from the pole of rotation; it is often convenient to refer to latitude, $\phi = [(\pi/2) - \theta]$. An additional potential term due to pressure must be added for Equation 3.1 to apply to points in the interior of the Earth. The geoid is a particular

*See Equation 3.21.

surface of constant geopotential $U_0$ and gravitational acceleration at the surface is normal to the geoid, being given by

$$g = -\text{grad}\, U \tag{3.2}$$

The problem of calculating the form of the geoid is, in principle, a matter of obtaining an expression for $V$. If the distribution of mass within the earth were completely known, $V$ could be obtained by direct integration, but we must proceed in the reverse direction and derive information about the Earth's interior from the form of the geoid.

We may approach the problem in a general way through Laplace's equation, which $V$ must satisfy at all points external to the Earth and therefore, in the limit, on the surface itself:

$$\nabla^2 V = \frac{1}{r^2}\frac{\partial}{\partial r}\left(r^2 \frac{\partial V}{\partial r}\right) + \frac{1}{r^2 \sin\theta}\frac{\partial}{\partial \theta}\left(\sin\theta \frac{\partial V}{\partial \theta}\right) + \frac{1}{r^2 \sin^2\theta}\frac{\partial^2 V}{\partial \lambda^2} = 0 \tag{3.3}$$

where $\lambda$ is longitude. $V$ can be expressed as a polynomial in $1/r$ with coefficients that are spherical harmonic functions of $\theta$ and $\lambda$. Our present interest is limited to rotational symmetry about $z$, so that variation with $\lambda$ is discounted and the coefficients reduce to zonal harmonics or Legendre polynomials, $P_0, P_1, \ldots$ (see Appendix C):

$$V = -\frac{GM}{r}\left(J_0 P_0 - J_1 \frac{a}{r} P_1(\theta) - J_2 \left(\frac{a}{r}\right)^2 P_2(\theta) \ldots\right) \tag{3.4}$$

in which $G$ is the absolute gravitational constant, $J_0, J_1, \ldots$ are dimensionless coefficients, which we wish to determine because they represent the distribution of mass within the Earth, and $a$ is the equatorial radius.

$J_0$ is known to be unity from the fact that at great distances all other terms become insignificant and we are, in effect, considering the potential due to a point mass $M$:

$$V = -\frac{GM}{r} \tag{3.5}$$

By taking the origin of the coordinate system to be the center of mass of the Earth we make $J_1$ identically zero. Our particular interest is in the $J_2$ term, which is the principal one required to give the observed oblate ellipsoidal form of the geoid. In the first-order theory of the ellipticity presented here this term suffices and the associated even harmonics, $J_4, J_6, \ldots$ are neglected. These and other higher harmonics are included in the discussion of the departure of the geoid from an ellipsoid (section 4.2) but they are smaller by factors of order 1000.

We now have

$$V = -\frac{GM}{r} + \frac{GMa^2J_2}{2r^3}(3\sin^2\phi - 1) \tag{3.6}$$

This is the starting point for two approaches to the problem of the geoid. In Section 4.1 we consider the gravity on the equilibrium geoid as the gradient of the geopotential. Here we are concerned with the internal distribution of mass.

Having accepted the two terms in Equation 3.6 as sufficient for the present purpose, we can express $J_2$ in terms of the principal moments of inertia of the Earth. The geometry of the problem is given in Figure 3.2. The gravitational potential at $P$ due to the mass element is

$$dV = -G\frac{dM}{q} = -\frac{G\,dM}{r[1+s^2/r^2 - 2(s/r)\cos\psi]^{1/2}} \tag{3.7}$$

This may be expanded in powers of $1/r$ to $1/r^3$ by noting that

$$\left(1+\frac{s^2}{r^2} - 2\frac{s}{r}\cos\psi\right)^{-1/2} = \left(1+\frac{s}{r}\cos\psi - \frac{1}{2}\frac{s^2}{r^2} + \frac{3}{2}\frac{s^2}{r^2}\cos^2\psi + \cdots\right)$$

$$= \left(1+\frac{s}{r}\cos\psi + \frac{s^2}{r^2} - \frac{3}{2}\frac{s^2}{r^2}\sin^2\psi + \cdots\right) \tag{3.8}$$

To this order the total potential can be expressed as a series of integrals obtained by substituting Equation 3.8 in Equation 3.7:

$$V = -\frac{G}{r}\int dM - \frac{G}{r^2}\int s\cos\psi\,dM - \frac{G}{r^3}\int s^2\,dM$$

$$+\frac{3}{2}\frac{G}{r^3}\int s^2\sin^2\psi\,dM \tag{3.9}$$

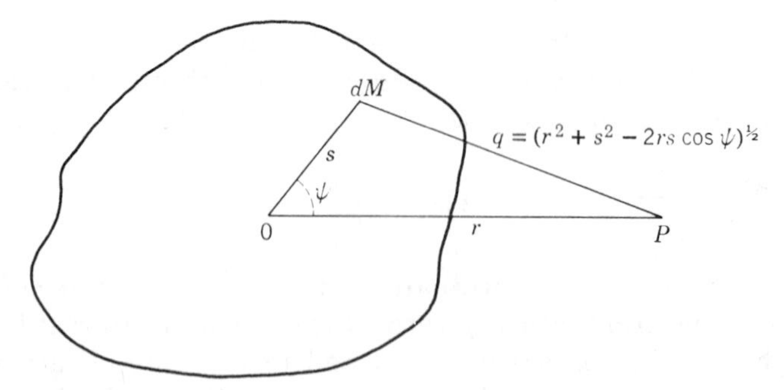


**Figure 3.2.** Geometry for the integration of gravitational potential to obtain MacCullagh's formula. The potential is calculated at a point $P$ external to the mass $M$ and distant $r$ from its center of mass, $O$; $r$ is a constant in the integration, the variables being $s$ and $\psi$, the coordinates of the mass element with respect to $O$ and the line $OP$.

The first integral is the potential of the centered mass, $-GM/r$. The second is identically zero because the center of mass was chosen as the origin. By assigning to the elementary mass coordinates $(x, y, z)$, the axes being arbitrary, we can write the third integral as

$$-\frac{G}{r^3}\int (x^2+y^2+z^2)\,dM = -\frac{G}{2r^3}\left[\int (y^2+z^2)\,dM + \int (x^2+z^2)\,dM + \int (x^2+y^2)\,dM\right]$$
$$= -\frac{G}{2r^3}(A+B+C) \qquad (3.10)$$

where $A$, $B$, $C$ are moments of inertia about the $x$, $y$, $z$ axes. The fourth integral in Equation 3.9 is the moment of inertia $I$ of the mass $M$ about $OP$, so that

$$V = -\frac{GM}{r} - \frac{G}{2r^3}(A+B+C-3I) - \cdots \qquad (3.11)$$

This is MacCullagh's formula. It is very useful for application to nearly spherical bodies, including the Earth, and is a valid approximation at large distances from any body. The limitation implicit in the truncation of the expansion of Equation 3.8 is that higher-order terms are required for accurate representation of the gravitational field, but in the case of the Earth the second term of Equation 3.11, representing the ellipticity, is of order 1000 times the magnitudes of higher-order terms, or of the fourth-order ($J_4$) term in the polynomial expansion of the ellipticity (see Appendix C).

The moment of inertia $I$ may be represented in terms of $A$, $B$, $C$:

$$I = Al^2 + Bm^2 + Cn^2 \qquad (3.12)$$

where $l$, $m$, $n$ are direction cosines of $OP$ with respect to $x$, $y$, $z$. Then with rotational symmetry of the Earth about $z$,

$$A = B \qquad (3.13)$$

$$n^2 = \sin^2\phi = 1 - l^2 - m^2 \qquad (3.14)$$

so that

$$V = -\frac{GM}{r} + \frac{G}{2r^3}(C-A)(3\sin^2\phi - 1) \qquad (3.15)$$

Thus the coefficient $J_2$ of the gravitational potential (Eq 3.6) is

$$J_2 = \frac{C-A}{Ma^2} \qquad (3.16)$$

The total geopotential is, therefore,

$$U = -\frac{GM}{r} + \frac{G}{2r^3}(C-A)(3\sin^2\phi - 1) - \tfrac{1}{2}r^2\omega^2\cos^2\phi \tag{3.17}$$

The geoid is defined as the surface of constant potential $U_0$. At the equator ($r = a$, $\phi = 0$) and poles ($r = c$, $\phi = \pi/2$) we obtain

$$U_0 = -\frac{GM}{a} - \frac{G}{2a^3}(C-A) - \tfrac{1}{2}a^2\omega^2 \tag{3.18}$$

$$U_0 = -\frac{GM}{c} + \frac{G}{c^3}(C-A) \tag{3.19}$$

from which

$$a - c = \frac{C-A}{M}\left(\frac{a}{c^2} + \frac{c}{2a^2}\right) + \frac{1}{2}\frac{ca^3\omega^2}{GM} \tag{3.20}$$

and since $a \simeq c$, the flattening $f$* is slight:

$$f = \frac{a-c}{a} = \frac{3}{2}\frac{C-A}{Ma^2} + \frac{1}{2}\frac{\omega^2a^3}{GM} \tag{3.21}$$

This gives the flattening correct to first order and, since $f \simeq 10^{-3}$, the error in neglecting terms of order $f^2$ is comparable in magnitude to the neglected higher harmonics in the gravitational potential. To this accuracy the equation of the surface is

$$r = a(1 - f\sin^2\phi) \tag{3.22}$$

Equation 3.21 allowed the absolute difference $(C - A)$ between moments of inertia of the Earth about polar and equatorial axes to be calculated in terms of the geodetically determined flattening $f$. However, since $(C - A)/Ma^2$ has been determined from satellite orbits with precision more than an order of magnitude greater than that obtained goedetically, the flattening of the geoid estimated from satellite data is now used in geodesy. The second term in Equation 3.21 is well deter-

*The term *flattening* is used for the fractional difference in radii, to avoid confusion with the conventionally defined ellipticity, $e = [1-(c^2/a^2)]^{1/2}$. For small ellipticities $e^2 \approx 2f$. As considered in Appendix A, the equation of an ellipse referred to a coordinate origin at one focus is

$$r = a(1-e^2)(1+e\cos\phi)^{-1}$$

The equation of an ellipse referred to a coordinate origin at its geometrical center, from which the foci are separated by a distance $ae$, is

$$r = a\left[1 + \left(\frac{e^2}{1-e^2}\right)\sin^2\phi\right]^{-1/2}$$

to which Equation 3.22 is an approximation for small $e$.

mined from the ratio of centrifugal acceleration to gravitational acceleration at the equator (see Eq. 4.8):

$$\frac{\omega^2 a^3}{GM} \approx \frac{\omega^2 a}{g_e} = m = 3.46775 \times 10^{-3} \tag{3.23}$$

Using the satellite value of $J_2$, now adopted as the standard for geodetic reference (International Union of Geodesy and Geophysics, 1967),

$$J_2 = 1.08270 \times 10^{-3} \tag{3.24}$$

we obtain for the flattening by Equation 3.21

$$f = \tfrac{3}{2}J_2 + \tfrac{1}{2}m = 3.3579 \times 10^{-3} \tag{3.25}$$

Retaining second-order terms in the theory,* we obtain a better value (Rapp, 1974):

$$f = 3.35282 \times 10^{-3} = \frac{1}{298.256} \tag{3.25a}$$

It is useful at this point to anticipate a result obtained in Section 3.2. The gravitational attractions of the Sun and Moon, acting on the equatorial bulge of the Earth, exert torques which cause a precession of the axis of rotation and from the rate of precession the dynamical ellipticity, $H$, or fractional difference in principal moments of inertia of the Earth, is estimated:

$$H = \frac{C - A}{C} = 3.2732 \times 10^{-3} = \frac{1}{305.51} \tag{3.26}$$

From Equations 3.16, 3.24 and 3.26 we obtain the axial moment of inertia of the Earth:

$$C = \frac{J_2}{H} Ma^2 = 0.33078 Ma^2 \tag{3.27}$$

The value of $C$ thus obtained is a vital boundary condition on calculations of the radial density profile within the Earth (Chapter 6). The moment of inertia of a uniform sphere is $0.4Ma^2$, so that Equation 3.27 indicates a strong concentration of mass toward the center of the Earth.

Assuming a knowledge of the density profile from seismology (Chapter 6) and hydrostatic equilibrium at all levels, we can bring the problem of the figure of the Earth back to its starting point and calculate the equilibrium figure. Comparison of the geoid and dynamical ellip-

*Second-order equations relating $f$, $J_2$, $J_4$, and $m$ are

$$J_2 = \tfrac{2}{3}f\left(1 - \frac{f}{2}\right) - \frac{m}{3}\left(1 - \tfrac{3}{2}m - \tfrac{2}{7}f\right)$$

$$J_4 = -\tfrac{4}{35}f(7f - 5m)$$

ticities (Eqs. 3.25 and 3.26) shows that the inner layers are less elliptical than the surface, because if the ellipticities of all layers of constant density were equal, we would have

$$\frac{C-A}{C}=\frac{a^2-(1/2)(a^2+c^2)}{a^2}\approx\frac{a-c}{a}=f \tag{3.28}$$

The observed difference is expected because the internal surfaces of constant potential enclose material of higher average density than the Earth as a whole, so that the rotational contribution to the potential is smaller, relative to the gravitational contribution.

The hydrostatic theory is given to first order by Jeffreys (1970), following an elegant approximation procedure due to R. Radau, which shows that, to this order, the surface flattening can be expressed in terms of the moment of inertia:

$$f_H=\frac{(5/2)m}{1+(25/4)[1-(3/2)(C/Ma^2)]^2} \tag{3.29}$$

in which $m$ is given by Equation 3.23 and $C/Ma^2$ by 3.27. However, the difference between $f_H$ and $f$ is small and first-order theory is insufficient. Higher-order treatment, or a numerical correction to Equation 3.29 (Jeffreys, 1963; Khan, 1969), using the details of the Earth's density profile, which are reasonably well known, gives $f_H=1/299.75$. The observed flattening is about 0.5% greater than that which would result from hydrostatic conditions in the interior.

As Goldreich and Toomre (1969) pointed out, the excess ellipticity is not remarkable. When the second-order harmonics of gravitational potential are considered alone, and the equilibrium ellipticity is subtracted, the Earth appears as a triaxial ellipsoid with principal moments of inertia $A'<B'<C'$, such that $(C'-A')/(B'-A')\approx 2$. This is the axial ratio to be expected with highest probability in a spheroid that evolved randomly from a large number of masses of differing densities. By causing such a body to rotate, a rotational bulge is superimposed. If the body is imperfectly elastic, the bulge assumes the equilibrium (hydrostatic) value, superimposed on the intrinsic ellipticity due to its internal mass distribution, and the body turns so that its axis of greatest intrinsic (nonequilibrium) moment of inertia coincides with the axis of rotation.

It is, of course, possible that part of the excess ellipticity represents a delay in the response of the equatorial bulge to slowing of the Earth's rotation, as suggested by Munk and MacDonald (1960a), but it appears to be no more than a minor part. An analysis of the stresses associated with the mass "anomalies" responsible for higher harmonics of the geoid (Section 4.2) compels us to assign them to the upper mantle, where they are probably dynamically maintained. The stress due to the excess ellipticity is slightly greater than for the other harmonics, but this may be explained also as an excess ellipticity due to incomplete recovery from former more massive loads of polar ice (McConnell, 1968; O'Connell, 1971).

## 3.2 PRECESSION OF THE EQUINOXES

Equation 3.15 shows that, in addition to the principal term in $r^{-1}$, the gravitational potential of the Earth has a smaller term in $r^{-3}$, which arises from the ellipticity and has a latitude dependence. Thus there is, in addition to the central gravitational force $-m(\partial V/\partial r)$ exerted on mass $m$ at $(r, \phi)$, a torque $-m(\partial V/\partial \phi)$. The reaction is therefore an equal and opposite torque exerted by the mass on the Earth. The torques exerted by the Moon and the Sun on the Earth's equatorial bulge are responsible for the precession of the equinox.

The magnitude of the precession deserves some emphasis because it is commonly supposed that the axis of the Earth's rotation has a fixed direction in space. It is inclined at about 23.5° to the pole of the ecliptic (normal to the Earth's orbital plane) as in Figure 3.3 and precesses slowly

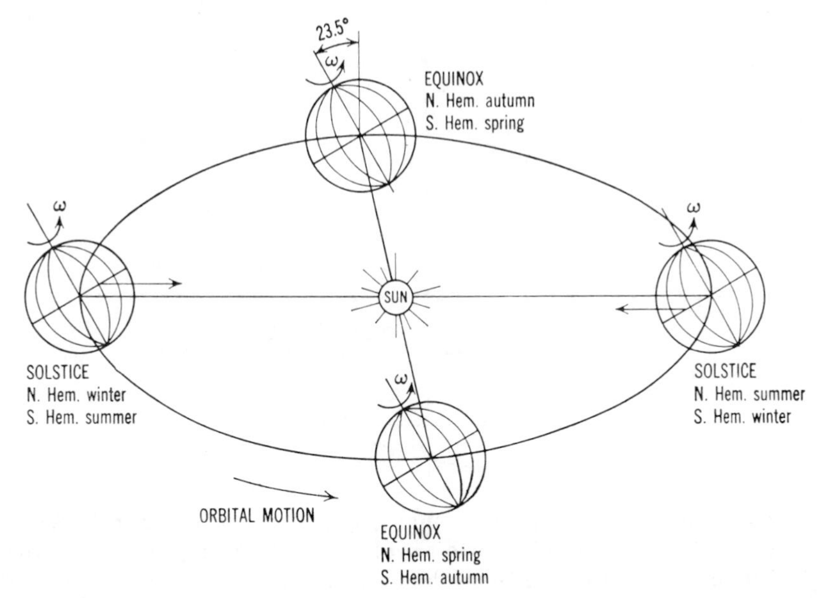


**Figure 3.3.** Cause of the precession of the equinox. Gravitational action of the Sun on the equatorial bulge causes a torque in the same sense at both solstices but no torque at the equinoxes.

about it, so that the orientation of the north pole will diverge from the pole star, reaching a maximum deviation of 47° before returning 25,730 years later. Navigation by the stars would be much more complex if the precession were not so slow. Nevertheless, it is sufficiently rapid to be measured astronomically with considerable precision, the mean rate being about 50″.37 per year. ($7.74 \times 10^{-12}$ rad sec$^{-1}$), corresponding to a mean angular movement of the pole of $3.08 \times 10^{-12}$ rad sec$^{-1}$.

A rigorous solution to the problem of precession requires the application of Euler's equations (see, for example, Fowles, 1970) but since the angular rate of precession is very small compared with the rate of rotation (by a factor of nearly $10^7$) a simpler approach, in which precession is treated as a perturbation of the rotation, is sufficient for the present discussion. We consider the Sun, of Mass $M_\odot$ at $(R, \phi)$ with respect to the center of the Earth, as in Figure 3.4, so that the gravitational potential at the center of the Sun due to the Earth is given by Equation 3.15 and the torque exerted on the Earth is therefore

$$L = M_\odot \frac{\partial V}{\partial \phi} = \frac{3GM_\odot}{R^3}(C - A)\sin\phi\cos\phi \qquad (3.30)$$

Now this torque acts about an axis in the equatorial plane normal to the Earth-Sun line, that is, it tends to pull the equatorial bulge into line with the instantaneous Earth-Sun axis. We can resolve the torque into

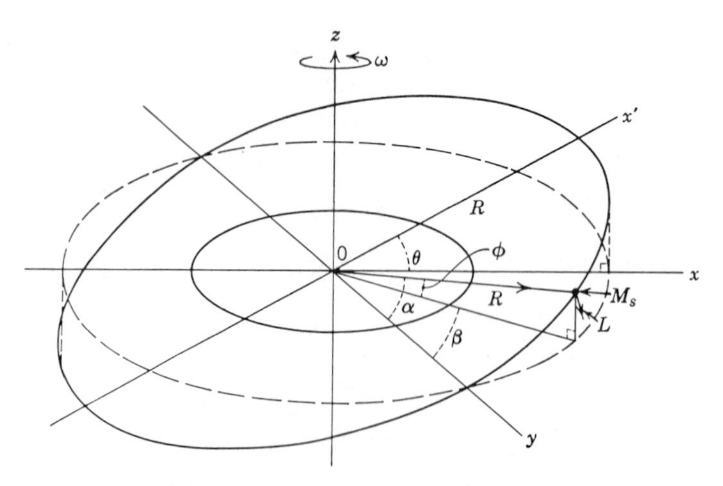


**Figure 3.4.** Geometry of the precessional torque. $O$ is the center of the Earth and $Oxy$ the equatorial plane. The Earth's orbit about the Sun is equivalent for this purpose to rotation of the Sun about the Earth at radius $R$ in a plane $Ox'y$ which makes an angle $\theta$ with the equatorial plane. The projection of the "solar orbit" on the equatorial plane is shown as a dashed ellipse. $\phi$ is the instantaneous geocentric latitude of the Sun.

components, $L_x$ about the axis at the solstice ($Ox$ in Fig. 3.4) and $L_y$ about the axis at the equinox ($Oy$). Our principal interest is in $L_y$, which causes the precession; $L_x$ integrates to zero over one solar orbit (one year), but causes a nutation or "nodding" of the pole of rotation toward and away from the ecliptic pole. Then

$$L_y = \frac{3GM_\odot}{R^3}(C - A)\sin\phi\cos\phi\sin\beta \qquad (3.31)$$

where $\beta$ is the longitude of the sun relative to its longitude at the equinox and is related to the corresponding angle $\alpha$ of the Sun in its orbital plane by the trigonometric identities

$$\sin\phi = \sin\theta\sin\alpha$$
$$\tan\phi = \tan\theta\sin\beta$$

from which

$$L_y = \frac{3GM_\odot}{R^3}(C - A)\sin\theta\cos\theta\sin^2\alpha \qquad (3.32)$$

Since $\overline{\sin^2\alpha} = \frac{1}{2}$, the mean torque over one year ($\alpha = 0 \rightarrow 2\pi$) is

$$\overline{L_y} = \frac{3}{2}\frac{GM_\odot}{R^3}(C - A)\sin\theta\cos\theta \qquad (3.33)$$

The component of the Earth's rotational angular momentum in the ecliptic plane is $C\omega\sin\theta$, so that the angular rate of precession is

$$\omega_{PS} = -\frac{\overline{L_y}}{C\omega\sin\theta} = -\frac{3}{2}\frac{G}{\omega}\cdot\frac{C - A}{C}\cdot\frac{M_\odot}{R^3}\cos\theta$$

$$= -\frac{3}{2}\frac{\omega_\odot^2}{\omega}\cdot\frac{C - A}{C}\cos\theta \qquad (3.34)$$

since

$$\frac{GM_\odot}{R^3} = \omega_\odot^2$$

by Kepler's third law (as in Eq. 4.41). The motion is retrograde, that is, opposite to the sense of $\omega$ and $\omega_\odot$.

Equation 3.34 gives the contribution to the rate of precession by the solar torque only. The lunar contribution is obtained similarly:

$$\omega_{PL} = -\frac{3}{2}\frac{G}{\omega}\frac{C - A}{C}\frac{M_☽}{r^3}\cos\theta' \qquad (3.35)$$

Since the plane of the lunar orbit is close to the ecliptic, $\theta'$ is not very different from $\theta(23.5°)$. The proximity of the Moon more than compensates for its smaller mass, the effect being that of a gravitational gradient inversely proportional to the cube of distance, as in the case of tides (Section 4.4), and $\omega_{PL}$ is slightly more than twice as great as $\omega_{PS}$. The total mean rate of precession is the observed quantity

$$\omega_P = \omega_{PS} + \omega_{PL} = 50''.37 \text{ year}^{-1}$$

Since all of the quantities in Equations 3.34 and 3.35, except the moments of inertia, are observable, the dynamical ellipticity

$$H = \frac{C - A}{C} = 0.0032732 = \frac{1}{305.51} \tag{3.36}$$

is determined.

By averaging the torque $L_y$, as in Equation 3.33, we have obtained the average rate of precession, but the precession is not a steady process. Each time the Sun (or Moon) crosses the equatorial plane, the solar (or lunar) torque is switched off. Thus the two (solar and lunar) components of precession proceed in semiannual and semimonthly "bursts." Similarly, the perpendicular torque component, $L_x$, causes semiannual and semimonthly nutations ("nodding" of the pole toward and away from the ecliptic pole), so that the path of the pole of rotation due to the combined effects of precession and nutation is an elliptical motion about the mean precessional path.

The preceding analysis applies to a simplified situation, in which the Sun and Moon are assumed to be in circular orbits in the plane of the ecliptic. In fact the Moon's orbit is both eccentric and inclined to the ecliptic plane and the eccentricity and inclination both vary by virtue of the three-body Sun-Earth-Moon interactions. The long-term trends in the lunar orbit are considered in Section 4.5. The effect on the Earth's motion is to produce several superimposed periodicities with periods in the range $10^4$ to $10^5$ years, which are of interest in connection with climatic variations (Section 9.5).

As is apparent from Equation 3.34, the rate of precession is proportional to the dynamical ellipticity, but this is itself the result of a balance between self-gravitation, which tends to pull the Earth into spherical form, and rotation. The ellipticities of the surfaces of constant density within the Earth decrease inwards, as is apparent from the fact that the surface flattening is greater than the dynamical ellipticity (see Eq. 3.28), and in particular the core flattening ($2.54 \times 10^{-3}$—Bullen and Haddon, 1973) is only 75% of that of the Earth as a whole. Thus the lunar and solar torques exerted on the core suffice to make it precess at only three quarters of the rate of precession of the Earth. As Bullard (1949) noted, the effects of allowing the rotational axis of the core to lag the mantle axis by more than a very small angle are so violent as to be inadmissible; 25% of the torque required to maintain precession of the core must be provided by its coupling to the mantle.

The dominant mechanism of precessional coupling of the core has been termed *inertial coupling*, the theory of which is associated particularly with the French mathematician H. Poincaré. A qualitative insight is given by a simple mechanical analogue due to Toomre (1966), who considered a frictionless particle sliding initially in the equatorial plane of an oblate ellipsoidal cavity free from gravity. When the axis of the cavity is turned through a small angle, the particle continues to orbit in the original plane, now inclined to the equator of the cavity, but the orbit has become slightly elliptical, and the force exerted on the particle by the cavity wall, being frictionless, is normal to the wall and no longer precisely in the plane of the orbit. There is then a precessional torque which causes the plane of the orbit to precess in a retrograde sense about the equator of the cavity. In the same way, the angular momentum vector of the core tends to precess about the axis of its cavity (the mantle), so that instead of being left behind by the precession of the mantle it swings out into a slightly wider precessional angle. The angular difference between core and mantle axes is of order $10^{-5}$ rad. But the internal motion within the core due to the inertial coupling ("Poincaré flow") dramatically reduces the energy dissipation that this angular difference would suggest, and Rochester (1974), Rochester et al. (1975), and Loper (1975) reported that the power available for dynamo action from this coupling is less than $10^{8}$ W, which is not a significant contribution to the total power required (Section 8.4).

## 3.3 THE CHANDLER WOBBLE

Independently of its gravitational interaction with external bodies* the Earth undergoes a free, Eulerian precession, commonly called the free nutation in geophysical literature—although it is not correctly termed a nutation. To distinguish this from the forced motions due to interactions with other bodies, the term wobble or, in the name of its discoverer, Chandler wobble is preferred. The wobble results from rotation of the Earth about an axis that departs slightly from its axis of greatest moment of inertia. The total angular momentum remains constant in magnitude and direction, but the Earth wobbles so that, relative to the surface features, the pole of rotation describes a circle about the

*Except for Mach's principle, whereby the stationary axes to which absolute rotation refers are determined by the whole mass of the universe.

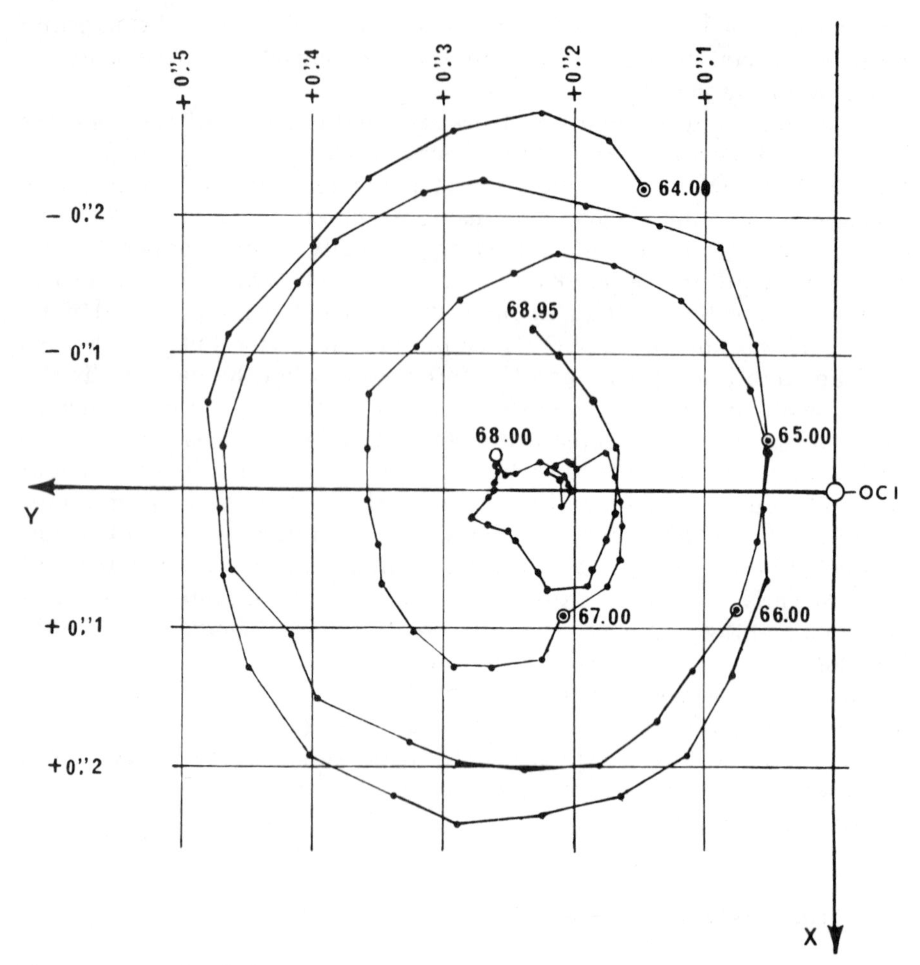


**Figure 3.5.** Path of the pole from 1964.00 to 1968.95 (unsmoothed averages for 0.05-year intervals). Reproduced by permission from Guinot (1970). Note that beating of the 12-month and 14-month periods causes the circular pole path to collapse periodically.

geometrical axis of greatest moment of inertia (Fig. 3.5). The spin axis is almost fixed in absolute orientation, so that the wobble is apparent as a cyclic variation of latitude with a period of 430 to 435 days (1.18 years) and variable amplitude of average (r.m.s.) value 0.14 second of arc ($6.8 \times 10^{-7}$ rad). Recent spectral analyses of the wobble are by Pedersen and Rochester (1972), Currie (1974), and Graber (1976). It is superimposed on a 12-month (seasonal) variation of similar amplitude. The free precession of rigid bodies is given in terms of Euler's equations by standard texts on dynamics (see, for example, Fowles, 1970, Chapter 9). A detailed discussion of the wobble of the Earth has been given by Munk and MacDonald (1960b) and a very readable and up-to-date review is by Rochester (1973); a simplified treatment follows here.

We consider initially a rigid Earth or any similar body with principal moments of inertia $C$, $A$, $A$, where $C > A$, rotating with angular velocity $\omega$ about an axis at a very small angle $\alpha$ to the $C$ axis (Fig. 3.6). The total energy of rotation is the sum of energies of the components of rotation about the three principal axes:

$$E_T = \tfrac{1}{2}(Cm_3^2 + Am_2^2 + Am_1^2)\omega^2 \tag{3.37}$$

where $m_1$, $m_2$, $m_3$ are direction cosines of the rotational axis with respect to principal axes, and

$$m_1^2 + m_2^2 + m_3^2 = 1 \tag{3.38}$$

$$m_1^2 + m_2^2 = \alpha^2 \tag{3.39}$$

The energy of rotation with the same angular momentum about the $C$ axis, that is, with $\alpha = 0$ and no wobble, is

$$E_0 = \tfrac{1}{2}C\omega_0^2 \tag{3.40}$$

so that we can write the wobble energy as

$$\begin{aligned} E_w = E_T - E_0 &= \tfrac{1}{2}C\omega^2\left[m_3^2 + \frac{A}{C}(m_1^2 + m_2^2)\right] - \tfrac{1}{2}C\omega_0^2 \\ &= \tfrac{1}{2}C\omega^2\left[1 - \left(\frac{C-A}{C}\right)(m_1^2 + m_2^2)\right] - \tfrac{1}{2}C\omega_0^2 \end{aligned} \tag{3.41}$$

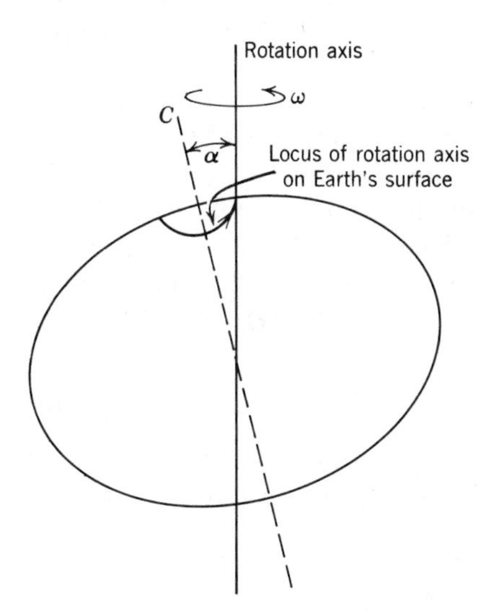


**Figure 3.6.** The Chandler wobble. A cyclic variation of latitude is observed as the Earth moves bodily so that the axis of rotation, which is almost constant in space, describes a cone of semiangle $\alpha$ about the axis of maximum moment of inertia.

Since the components of angular momentum add vectorially we can equate the angular momenta in the two states:

$$C\omega_0 = (C^2 m_3^2 + A^2 m_2^2 + A^2 m_1^2)^{1/2}\omega$$
$$= \left[1 - \left(\frac{C^2 - A^2}{C^2}\right)(m_1^2 + m_2^2)\right]^{1/2} C\omega \tag{3.42}$$

We thus substitute for $\omega_0$ from Equation 3.42 in 3.41 and use also Equation 3.39 to obtain the wobble energy in terms of its amplitude $\alpha$:

$$E_w = \tfrac{1}{2}A\left(\frac{C - A}{C}\right)\omega^2\alpha^2 = \tfrac{1}{2}AH\omega^2\alpha^2 \tag{3.43}$$

The total energy is greater than for symmetrical rotation about the $C$ axis and, as the Earth tends toward its state of lowest energy (rotation about the axis of maximum moment of inertia, with $\alpha = 0$), the excess energy exerts, in effect, a gyroscopic torque:

$$L = -\frac{dE_w}{d\alpha} = -AH\omega^2\alpha \tag{3.44}$$

Treating it in the same way as the external torque in forced precession, we obtain the angular velocity of the free precessional motion at the Chandler frequency:

$$\omega_c = -\frac{L}{A\omega\alpha} = H\omega \tag{3.45}$$

The period of the wobble is $C/(C - A)$ days, viewed in stationary axes, or $A/(C - A)$ days when referred to Earth axes. The precessional rotation is prograde, that is, in the same sense as $\omega$, but the period calculated for a rigid Earth is 305 days compared with the observed period of about 430 days. A 305-day variation of latitude was sought for many years before the discovery in 1891 by S. C. Chandler of the 430-day period, which was subsequently shown to be the free Eulerian period. The difference is due to elastic yielding of the Earth, which results in a partial accommodation of the equatorial bulge to the instantaneous axis of rotation and thus reduces the gyroscopic torque.* The increase in Chandler period by the factor $f = 1.42$ means that the gyroscopic torque is reduced to

$$L = -AH\omega^2\alpha/f \tag{3.46}$$

implying elastic deformation of the Earth by an angle $\epsilon = H[\alpha - (\alpha/f)] = 1.0 \times 10^{-3}\,\alpha$ ($H$ being here identified with the flattening). It is of some

*Damping lengthens the wobble period by the factor $[1 - (1/4Q^2)]^{-1/2}$ which, for $Q$ in the range 72 to 600, as considered in Section 3.4, is not significant.

interest to compare the resulting elastic energy of the Earth

$$E_{el} = \tfrac{1}{2}\bar{\mu}\epsilon^2 V = \tfrac{1}{2}\bar{\mu}\left[H\alpha\left(1-\frac{1}{f}\right)\right]^2 V \approx 7.6 \times 10^{25}\, \alpha^2 \,\mathrm{J} \qquad (3.47)$$

where $\bar{\mu} = 1.45 \times 10^{11}$ Pa ($1.45 \times 10^{12}$ dyn cm$^{-2}$) is the effective rigidity of the Earth (which may be obtained as the weighted mean for an earth model, as in Appendix G) and $V$ is the volume of the Earth, with the wobble energy by Equation 3.43:

$$E_W = 7.0 \times 10^{26}\, \alpha^2 \,\mathrm{J} \qquad (3.48)$$

Thus the elastic energy is only about 10% of the wobble energy. This is important in considering the extent to which anelasticity of the Earth may damp the wobble (Section 10.4).

Geophysical interest in the wobble centers on the possible mechanisms for excitation and damping, which have been the subject of strenuous debate for the past decade. This problem is considered in Sections 3.4 and 10.4.

## 3.4 FLUCTUATIONS IN ROTATION AND THE EXCITATION OF THE CHANDLER WOBBLE

There is a slow decrease in the angular velocity of the Earth's rotation which is due to tidal friction (Section 4.5) but, in short-term astronomical observations, this effect is obscured by irregular fluctuations amounting to a few parts in $10^8$. These are often referred to as length-of-day (lod) fluctuations. There are no plausible torques of external origin to which such variations can be attributed and an explanation must be sought in terms of internal redistribution of angular momentum. In principle, relative motions within or between any components of the Earth, (core, mantle, atmosphere) may be invoked and it now appears that there is no single cause. Irregularities in core motions may cause changes over a few years; very rapid changes are more reasonably attributed to atmospheric circulation. Also, the mechanism of excitation of the Chandler wobble may not be an entirely separate effect but arise at least partly from interaction with the atmosphere as well as mass redistribution in the mantle associated with earthquakes.

Bullard et al. (1950) recognised the possibility of the coupling of the mantle to irregular motions within the fluid core, but doubted its effectiveness. Electrical conduction in the lower mantle, due to the semiconducting properties of silicates at high temperatures (Section 8.3), couples the mantle to the geomagnetic field, which is generated in the core (Section 8.4), and the adequacy of this coupling is the subject of papers by Rochester (1960) and Roden (1963). Relative motions within the fluid core are essential to the operation of the geomagnetic dynamo (Section 8.4) and the secular variation of the field (Section 8.2) is evidence that the motion is variable, so that a partial transmission of these irregularities to the mantle provides a natural explanation for the slower lod variations.

It is of interest to relate the relaxation time for restoration of equilibrium, following a disturbance in the relative rotations of the core and mantle, to an empirical coupling constant. Suppose, for simplicity, that the core is rotating coherently at a rate $(\omega + \Delta\omega_c)$ and that the mantle is rotating at a rate $(\omega + \Delta\omega_m)$ where $\omega$ is the common rate at equilibrium. Then by conservation of angular momentum

$$I_m \, \Delta\omega_m + I_c \, \Delta\omega_c = 0 \tag{3.49}$$

where $I_m$, $I_c$ are the moments of inertia of the core and mantle. The relative angular velocity is

$$\Delta\omega = \Delta\omega_m - \Delta\omega_c = \Delta\omega_m\left(1 + \frac{I_m}{I_c}\right) = 8.74 \, \Delta\omega_m \tag{3.50}$$

If the coupling is linear, in the sense that the mutual torque, $L$, between the core and mantle is proportional to $\Delta\omega$, then we can define a coupling coefficient, $K_R$, such that

$$L = -I_m \frac{d}{dt}(\Delta\omega_m) = K_R \, \Delta\omega = K_R\left(1 + \frac{I_m}{I_c}\right)\Delta\omega_m \tag{3.51}$$

which integrates to

$$\Delta\omega_m = (\Delta\omega_m)_o \exp(-t/\tau) \tag{3.52}$$

where the relaxation time is

$$\tau = \left[K_R\left(\frac{1}{I_m} + \frac{1}{I_c}\right)\right]^{-1} \tag{3.53}$$

If we suppose $\tau \approx 10$ years, then $K_R \approx 2.6 \times 10^{28}$ J sec. This is a reasonable value in terms of the strength of the geomagnetic field and our estimates of the electrical conductivity at the base of the mantle (Section 8.3).

Evidence of angular momentum exchange with the core was reported by Vestine (1953), Vestine and Kahle (1968), and Kahle et al. (1969) who found a direct correspondence between lod variations and the rate of westward drift of features of the geomagnetic field (Section 8.2) which are presumed to indicate the rotation of the core, but Yukutake (1972, 1973) claimed a more significant correlation between rotation rate and the strength of the field. In any case the effect of the atmosphere must also be taken into account, especially with respect to rapid fluctuations (Lambeck and Cazenave, 1973, 1974). The moment of inertia of the atmosphere ($1.39 \times 10^{32}$ kg m$^2$) is only $1.7 \times 10^{-6}$ of that of the Earth as a whole, so that a change in the rate of rotation the Earth by 3 parts in $10^8$ due to angular momentum exchange with the atmosphere requires a 2% change in mean angular velocity or moment of inertia of the atmosphere, or a combination of the two. At first sight this appears to be an implausibly large effect, corresponding to a net average equatorial wind of about 9 m sec$^{-1}$, but Lambeck and Cazenave (1973, 1974) reported that strong, world-encircling upper atmosphere winds do in fact suffice to explain the rapid lod variations. Whether or not the atmosphere contributes to longer term changes, they can be adequately explained by electromagnetic core-mantle coupling.

The Chandler wobble (Section 3.3) is also a fluctuation in rotation, in the broadest sense, but arising from movements in the rotational axis (or of the Earth with respect to its axis) instead of changes in angular velocity. It is therefore natural to consider the possibility that it has a common cause with the lod variations. The amplitude (and frequency) of the 14-month wobble are variable within limited ranges over a time scale of a few years, so that spectral analysis of the latitude variation yields a spectral line at the Chandler frequency whose width has usually been interpreted in terms of the damping of the motion according to Equation 10.26. However, both analyses and interpretations have yielded widely disparate estimates of the wobble $Q$, $Q_W$*. The corrected value of Currie (1974) is $Q_W = 72 \pm 20$ (corresponding to natural damping with a time constant of 27 years) but Graber (1976) finds $Q_W = 600$ (corresponding to 225 years). Several authors have noted that variability of the wobble suggests that the excitation is nonstationary in the statistical sense (i.e., nonrandom), in which case the estimation of $Q_W$ from the spectral line width yields too low a value. Estimates of $Q_W$ less than 100 have been conventional, but it should be admitted that Graber may be nearer correct. It is important to work toward an agreement on this matter because the wobble excitation is a geophysical problem of considerable interest, being possibly connected with earthquakes and related tectonic movements as well as geomagnetism (Press and Briggs, 1975).

Excitation of the wobble by core-mantle coupling can be examined in terms of a coupling coefficient, $K_w$, which must be supposed comparable to $K_R$, as estimated from Equation 3.53. The relative angular velocity of the mantle over the core due to wobble of angular amplitude $\alpha$, in which the core does not significantly participate, is

$$\Delta\omega_W = \omega_C\alpha \tag{3.54}$$

*See Section 10.4 for a discussion of $Q$.

where $\omega_c$ is the Chandler angular frequency. This causes a mutual torque $L_w$ and consequent energy dissipation $-dE_w/dt$ given by

$$-\frac{dE_w}{dt} = L_w\,\Delta\omega_w = K_w(\Delta\omega_w)^2 = K_w\omega_c^2\alpha^2 \qquad (3.55)$$

so that the time constant for decay of wobble *amplitude* is obtained by relating Equation 3.55 to Equations 3.43 or 3.48:

$$\tau_w = \frac{2E_w}{(-dE_w/dt)} = \frac{AH}{K_w}\left(\frac{\omega}{\omega_c}\right)^2 \qquad (3.56)$$

For $K_w \approx K_R = 2.6 \times 10^{28}$ J sec, $\tau_w \approx 6 \times 10^4$ years, which is several hundred times the longest of the spectrally estimated decay times. As Rochester and Smylie (1965) noted, this conclusion is incompatible with an appreciable contribution to damping by core-mantle coupling. The suspicion that a vital factor has been omitted from the consideration of coupling has been kept alive by Runcorn (1970), who argued that brief ("impulsive") electromagnetic torques resulting from eddies that reached the core-mantle boundary (as sunspots breaking out on the surface of the Sun) were capable of exciting the wobble. This is not absolutely forbidden by the foregoing argument, assuming an independent sink of wobble energy (such as the oceans), but looks sufficiently improbable that it should be considered seriously only if all other mechanisms fail.

The forced annual wobble, which occurs with an amplitude comparable to the 14-month Chandler wobble, is almost certainly excited by seasonal movements of the atmosphere. Munk and Hassan (1961) concluded that the spectral power in the continuum of atmospheric movements was too weak at the 14-month period to maintain the Chandler wobble, but this conclusion has been questioned by Wilson and Haubrich (1974). Since the wobble decays in about $n$ wobble periods ($n = 23$ according to Currie, 1974, or 190 from Graber's 1976 data), the magnitude of a repeated random shift in atmospheric mass that is required to maintain a wobble of angular amplitude $\bar{\alpha}$ is about $n^{-1/2}$ times the movement required to displace the principal inertia axis by this angle. If this is accomplished by movement of a mass $\Delta m$ through a distance equal to the radius of the Earth then we can represent its effect as a moment of inertia increment

$$\Delta I = a^2\,\Delta m = (C - A)\bar{\alpha}/n^{1/2} \qquad (3.57)$$

For $\alpha = 0.14$ arc sec, we obtain $\Delta m = 9 \times 10^{14}$ kg or $3 \times 10^{14}$ kg for the two alternative values of $n$ above, or approximately $10^{-4}$ of the total atmospheric mass. It is also just possible to consider a mass of snow of this magnitude deposited over a continental area in one season. Alternatively a global wind pattern of the kind considered by Lambeck and

Cazenave (1973, 1974), but in a north-south sense, that is, normal to the Earth's rotation, would suffice if it could be represented by a mean angular velocity $\Delta\omega$ for the whole atmosphere given by

$$C_A\,\Delta\omega = (C - A)\bar{\omega}\alpha/\sqrt{n} \tag{3.58}$$

where $C_A = 1.4 \times 10^{32}$ kg m$^2$ is the atmospheric moment of inertia and $\omega$ is the Earth's angular velocity. This gives $\Delta\omega/\omega = 3 \times 10^{-4}$ or $1 \times 10^{-4}$ corresponding to a maximum wind speed of 0.9 or 0.3 m sec$^{-1}$, significantly less than the zonal wind required to explain the lod variations. It therefore appears that the atmosphere cannot be discounted.

The possibility that the Chandler wobble is excited by earthquakes has also been discounted (Munk and MacDonald, 1960b, pp. 163–164) and subsequently revived (Mansinha and Smylie, 1967, 1968; Smylie and Mansinha, 1971). Both convincing observations and realistic calculations indicating adequacy of the earthquake excitation have been difficult to produce. Mansinha and Smylie (1973) insist that contrary arguments are in error but it must be rated at least as likely that the wobble is excited by large-scale mantle movements, of which earthquakes are a particular feature, and that the instantaneous displacement fields themselves are not entirely relevant (Myerson, 1970). The earthquake excitation is clearly easier to accept if $Q_W \approx 600$, instead of previously considered smaller values. The essential point is that earthquakes are random impulses which cause mass displacements and produce small changes in the moments and products of inertia of the Earth, thereby shifting its axis of symmetry with respect to the axis of rotation, essentially in the manner postulated for atmospheric excitation except that the displaced masses may be larger and the displacements much smaller. Very large earthquakes are disproportionately effective, so that we obtain an indication of the magnitude of the required movements by supposing that a single very large earthquake occurring once per year shifts the axis by an angle $\Delta\alpha = 0.010$ arc sec ($5 \times 10^{-8}$ rad), that is, about $1/\sqrt{190}$ of the ambient wobble amplitude, this being sufficient to maintain the wobble against decay with a 225-year time constant. With favorable orientation of the fault this requires a moment of inertia increment

$$\Delta I = (C - A)\,\Delta\alpha = 1.3 \times 10^{28}\ \text{kg m}^2 \tag{3.59}$$

where $C$, $A$ are the principal moments of inertia of the Earth.

To assess the plausibility of earthquake excitation of this magnitude, we may appeal to the dislocation theory of the earthquake mechanism (Section 5.2) to obtain an order of magnitude estimate of the changes in moments or products of inertia that would result from a large earthquake. Since the moment of inertia of the Earth about any selected

($z$) axis is

$$I = \int (x^2 + y^2)\, dm \tag{3.60}$$

integrated over all mass elements $dm$ of the Earth, the increment resulting from displacements $\Delta x \ll x$ in the $x$ direction is

$$\Delta I = 2 \int x\, \Delta x\, dm \tag{3.61}$$

$x$ being measured from the $z$ axis through the center of mass, which is unaffected by the displacements, that is,

$$\int \Delta x\, dm = 0 \tag{3.62}$$

As Press (1965) emphasised, the displacements associated with earthquakes are not confined to the immediate focal volumes, but extend outward indefinitely with diminishing amplitude; in calculating changes in moment of inertia the displacements at remote distances from an earthquake become important. The general representation in a semi-infinite medium is lengthy and, in a realistic earth model, very involved indeed, but for an order-of-magnitude estimate we may consider a simple relative movement of blocks having dimensions indicated by observed earthquake displacements. Taking the probably rather generous dimensions of a dislocation model of the 1964 Alaskan earthquake (Fig. 5.11) to be representative of the occasional very large shocks, we have blocks $800\ \text{km} \times 200\ \text{km} \times 200\ \text{km}$ and density $3000\ \text{kg m}^{-3}$ (total mass $10^{20}$ kg), whose centers of mass are separated by a distance $x = 200$ km, oppositely displaced by $\Delta x = 22$ m, so that Equation 3.61 gives

$$\Delta I = 8.5 \times 10^{26}\ \text{kg m}^2 \tag{3.63}$$

which is well short of the requirement (Eq. 3.59).

The factor by which teleseismic displacements increase this estimate is indicated by that fact that with increasing radial distance $r$ from a fault the mass considered increases as $r^3$ and the separation $x$ of oppositely displaced masses increases as $r$ while the displacement $\Delta x$ (in a homogeneous, infinite medium) varies only as $r^{-2}$*, so that a simple minded approach suggests that the moment change increases as the square of the extent of the dislocated medium. Since this is approximately 10 times the block dimensions considered above it gives a dramatic upward revision of $\Delta I$ at least to the order of magnitude required by Equation 3.59. Realistic calculations require account to be

*This is indicated by a simple dimensional consideration. If we put $\Delta x \propto Sab/r^n$ for a fault of area $a \times b$, across which there is a slip $S$, then $n$ must be 2.

taken of boundary conditions at the free surface and the core interface as well as variable elasticity and density, and also gravity, and the present argument suffices only to indicate that if earthquake displacements are inadequate to excite the wobble then the inadequacy is likely to be no more than a small factor. It is probable that more subtle, irregular, but large-scale mantle movements associated with earthquakes, but not specifically coinciding with them, make a major contribution. Stuart and Johnston (1974) have reported measurements with an array of tiltmeters near the San Andreas fault in California, which indicate that for shocks in the magnitude range 3 to 4.4, associated preshock and postshock movements accumulate to seismic moments about 10 times those of the earthquakes themselves. If this observation is general and applies also to the very large shocks and if the relevant movements occur rapidly enough, that is, within half a wobble period or seven months, then the seismic excitation is certainly adequate to explain the wobble; a reexamination of geodetic survey data for the area of the 1906 San Francisco earthquake led Thatcher (1974) to suggest that this may be so.

It may be noted that the earthquake excitation mechanism does not explain the $Q$ of the wobble and no excitation theory can be very satisfying unless we also have an explanation for the damping. Section 10.4 presents an argument that suggests that the Chandler wobble is damped by the accompanying motion of the ocean (pole tide) but that mantle damping is also important if the $Q$ of the wobble is of order 600.

# 4

# Gravity and Tides

"It is not unlikely that the first remark of many who see my title will be that so small a subject as the Tides cannot demand a whole volume; . . ."

G. H. DARWIN, 1898, *THE TIDES*, p. vi

## 4.1 GRAVITY AS GRADIENT OF THE GEOPOTENTIAL

Gravitational acceleration on the geoid can be related directly to the flattening, so that one can be determined from the other without appealing to independent evidence of the moments of inertia of the Earth. The first-order theory suffices for many purposes and is given here; a derivation of the second-order terms may be found in Jeffreys (1970). Satellites have provided more precise values of the low-order harmonics in gravitational potential than can possibly be obtained from surface gravity data, so that gravity surveys can now be referred to an ellipsoid derived from the satellite work and internationally adopted as standard (International Union of Geodesy and Geophysics, 1967).

Gravity $g$ is obtained by differentiating the total geopotential, as given by Equations 3.1 and 3.6:

$$U = -\frac{GM}{r} + \frac{GMa^2}{2r^3} J_2(3 \sin^2 \phi - 1) - \tfrac{1}{2}\omega^2 r^2 \cos^2 \phi \tag{4.1}$$

Since

$$g = -\text{grad } U$$

we have

$$g = -\left[\left(\frac{\partial U}{\partial r}\right)^2 + \left(\frac{1}{r}\frac{\partial U}{\partial \phi}\right)^2\right]^{1/2} \tag{4.2}$$

But the normal to the geoid departs from the radial direction only by a small angle $(\phi_g - \phi)$, which is of order $f$, as in Figure 3.1, so that to the first order in small quantities the second term in Equation 4.2 is negligible, and therefore

$$-g = \frac{\partial U}{\partial r} = \frac{GM}{r^2} - \frac{3}{2}\frac{GMa^2}{r^4}J_2(3\sin^2\phi - 1) - \omega^2 r(1 - \sin^2\phi) \tag{4.3}$$

Now we may substitute the value of $r$ on the geoid, at arbitrary latitude $\phi$:

$$r = a(1 - f\sin^2\phi) \tag{4.4}$$

where $a$ is the equatorial radius and $f$ is the flattening given by Equation 3.25:

$$f = \frac{3}{2}J_2 + \tfrac{1}{2}m \tag{4.5}$$

We substitute for $r$ from (4.4) in (4.3) and use the binomial expansion

$$(1 - f\sin^2\phi)^{-n} = (1 + nf\sin^2\phi\ldots) \tag{4.6}$$

This allows products of small quantities to be neglected, and since the second and third terms in Equation 4.3 are themselves of order $f$ times the first term, the expansion is applied only to the first term:

$$-g = \frac{GM}{a^2}(1 + 2f\sin^2\phi) - \frac{3}{2}\cdot\frac{GM}{a^2}J_2(3\sin^2\phi - 1) - \omega^2 a(1 - \sin^2\phi) \tag{4.7}$$

The equatorial gravity is therefore

$$-g_e = \frac{GM}{a^2}(1 + \tfrac{3}{2}J_2 - m) \tag{4.8}$$

since, by (3.23),

$$m \approx \frac{\omega^2 a^3}{GM}$$

Then, again to the first order in $f$, the gravity at latitude $\phi$ is given in terms of the equatorial value $g_e$, from Equations 4.7 and 4.8:

$$g = g_e[1 - (\tfrac{9}{2}J_2 - 2f - m)\sin^2\phi] \tag{4.9}$$

which, by Equation 4.5, takes two more useful forms:

$$g = g_e[1 + (2m - \tfrac{3}{2}J_2)\sin^2\phi] \tag{4.10}$$

or

$$g = g_e[1 + (\tfrac{5}{2}m - f)\sin^2\phi] \tag{4.11}$$

The basic result, Equation 4.11, is known as Clairaut's theorem. Historically its value has been to provide an estimate of ellipticity independently of astrogeodetic surveys. By retaining higher-order terms, the following result is obtained in terms of geographic latitude, $\phi_g$:*

$$g = g_e\left[1 + (\tfrac{5}{2}m - f - \tfrac{17}{14}mf)\sin^2\phi_g + \left(\frac{f^2}{8} - \tfrac{5}{8}mf\right)\sin^2 2\phi_g + \cdots\right] \quad (4.12)$$

With recent values of the constants, this gives the geodetic standard to which gravity surveys should be referred (International Union of Geodesy & Geophysics 1967):

$$g = 9.780318\,(1 + 0.0053024\sin^2\phi_g - 0.0000059\sin^2 2\phi_g)\,\mathrm{msec}^{-2} \quad (4.13)$$

An earlier international gravity formula, adopted in 1930, may continue in use for some time, where gravity surveys have been based on it, making a change inconvenient:

$$g = 9.780490(1 + 0.0052884\sin^2\phi_g - 0.0000059\sin^2 2\phi_g)\,\mathrm{m\,sec}^{-2} \quad (4.14)$$

These equations refer to gravity on an ideal geoidal (sea-level) surface. Gravity survey data are referred to the standard latitude variation, Equation 4.13 or 4.14, with corrections for the elevation of the Earth's surface, where measurements are actually made. Such corrections imply some knowledge of the crustal structure and therefore, to some extent, beg the question that a gravity survey attempts to answer. However, surveys show local departures from the standard of reference amounting to 30 times the third term of the reference formula and thus give clear evidence of density variations in the crust. The large-scale features of the gravity field, which reflect deep-seated variations, are best indicated by analyses of satellite orbits.

## 4.2 THE SATELLITE GEOID

In Section 3.2 the torques exerted by the Sun and Moon on the Earth's equatorial bulge were shown to cause precession of the Earth. Equal and opposite torques are, of course, exerted by the Earth on the Sun and Moon; in the case of the Moon the influence on its orbit is appreciable. A precisely similar torque is exerted by the equatorial bulge on artificial satellites whose masses are too small to influence appreciably the motion of the Earth, but whose orbits provide the most

*The second-order theory is given by Jeffreys (1970, Chapter 4), but note that he uses a slightly different definition of $m$ from that in Equation 3.23, which results in an additional term in Equation 4.12.

precise evidence of the large-scale departures of the Earth from spherical symmetry.

First consider an axially symmetric Earth whose external gravitational potential is of simple ellipsoidal form, being represented by a second-order zonal harmonic as in Equation 3.6:

$$V = -\frac{GM}{r} + \frac{GMa^2J_2}{2r^3}(3\sin^2\phi - 1) \tag{4.15}$$

To first order, a satellite's motion is controlled by the central force, given by the first term in Equation 4.15, and its orbit is therefore an ellipse with the center of the Earth at one focus. It is convenient to deal with the special case of a circular orbit of radius $r$. The effect of the second term can then be treated as a perturbation by calculating the torque $L$ exerted as in Section 3.2. Writing the satellite mass as $m$,

$$L = -m\frac{\partial V}{\partial\phi} = -\frac{3GMa^2J_2m}{r^3}\sin\phi\cos\phi \tag{4.16}$$

We may consider this torque to cause a precession of the satellite orbit, an effect known as regression of the nodes, the succession of points at which the orbit crosses the Earth's equatorial plane (as seen in stationary coordinates, that is, not in the Earth's rotating coordinates). The torque acts on the angular momentum $p$ of the satellite,

$$p = mr^2\omega_s \tag{4.17}$$

where $\omega_s$ is its orbital angular velocity given by Kepler's third law (as in Equation 4.41):

$$\omega_s^2r^3 = GM \tag{4.18}$$

As in the analysis of Section 3.2, the instantaneous angular rate of the precessional motion of the satellite is

$$\omega_p = \frac{L}{p\sin\phi} = -\frac{3GMa^2J_2}{r^5\omega_s}\cos\phi \tag{4.19}$$

and, if the inclination of the plane of the orbit to the equatorial plane is $i$, the mean precessional rate is

$$\bar{\omega}_p = -\frac{3}{2}\frac{GMa^2J_2}{r^5\omega_s}\cos i \tag{4.20}$$

Thus the angular change $\Delta\Omega$, per orbital revolution, in the position of a node is given by

$$\frac{\Delta\Omega}{2\pi} = \frac{\bar{\omega}_p}{\omega_s} = -\frac{3}{2}\frac{a^2}{r^2}J_2\cos i \tag{4.21}$$

where the simplification is effected by Equation 4.18. The factor $J_2$ is thus determined, in principle, by very simple observations on regression of nodes. For a low altitude satellite at $i = 45°$, $\Delta\Omega \approx 0.4°$.

In practice the situation is considerably more complicated.* Orbits are elliptical and the ellipticity enters the equation for nodal regression. However, there is also a steady motion of the perigee and apogee of the orbit which can be useful as a check. Of greater interest is the fact that the geoid has small departures from the simple ellipsoidal form. These are smaller by factors of 1000 or more than the ellipticity but are nevertheless observable by means of satellite orbits. Still regarding the Earth as an axially symmetric body, the gravitational potential can be represented as an infinite series of zonal harmonics (in terms of Legendre polynomials, $P_l$, see Appendix C):

$$V = -\frac{GM}{r}\left[1 - \sum_l \left(\frac{a}{r}\right)^l J_l P_l(\sin\phi)\right] \tag{4.22}$$

The harmonics contribute to nodal regression by amounts depending on the inclination $i$ of a satellite orbit, and the lower-order coefficients $J_l$ can be estimated from the rates of regression for satellites with different orbital inclinations. Rotation of the Earth ensures that, averaged over a sufficient time, it appears to an external observer or satellite to have rotational symmetry and, at least in principle, it is a reasonably straightforward matter to deduce the coefficients $J_l$ from the long-term nodal drifts of satellites with a variety of orbital inclinations. More than 20 zonal coefficients are known with reasonable reliability.

Of greater interest is the more general study of geoid undulations that vary with longitude as well as latitude. The departures from axial symmetry are less easily (and therefore less accurately or reliably) determined because they cause shorter period perturbations in satellite orbits, for example an oscillation in the rate of nodal regression. The total gravitational potential is represented by an infinite series of spherical harmonics:

$$V = -\frac{GM}{r}\left\{1 + \sum_{l=2}^{\infty}\left(\frac{a}{r}\right)^l \sum_{m=0}^{l} P_l^m(\sin\phi)[C_l^m \cos m\lambda + S_l^m \sin m\lambda]\right\} \tag{4.23}$$

where $P_l^m$ $(\sin\phi)$ is the associated Legendre polynomial (see Appendix C). Note that the reversal of signs between Equations 4.22 and 4.23 means that $C_l^0 = -J_l$. The effects on satellite orbits of geoid undulations and the estimation of geoid coefficients from satellite observations are

*Apart from a more complicated gravitational field and orbit ellipticity, atmospheric drag and lunar and solar accelerations must be allowed for.

discussed by Kaula (1966). Recent evaluations that made use also of surface gravity data are by Gaposchkin and Lambeck (1971), who tabulated coefficients to degree $l$ and order $m$ (16, 16) and Rapp (1973), who gave a coefficient set complete to (20, 20). Although improvements in the numerical values may be expected from more sophisticated satellites, this is about as fine a detail as can reasonably be represented in spherical harmonic form. The lower-degree coefficients from the set by Rapp (1973) are listed in Table 4.1.

The coefficients in Table 4.1 are completely dominated by the term representing the ellipticity, $\bar{C}_2^0$. The value in the table refers to fully normalized harmonic functions, which are defined as having an r.m.s. value of unity over the surface of a sphere (see Appendix C). In terms of the conventional (unnormalized) Legendre polynomial $P_2$ (which has an r.m.s. value $1/\sqrt{5}$) this coefficient is $J_2 = 1082.64 \times 10^{-6}$. Almost all of this is accounted for by the equilibrium ellipticity due to rotation. Subtracting the equilibrium value of $\bar{C}_2^0$ from the value in the value in the table we are left with $-4.7 \times 10^{-6}$, which is more nearly in line with the magnitudes of the other coefficients, all of which indicate departures of the mass distribution within the Earth from hydrostatic equilibrium.

The form of the geoid, referred to the equilibrium ellipsoid and plotted from the Gaposchkin and Lambeck (1971) coefficients, is shown in Figure 4.1. This shows that the equator is, on average, a geoidal "high," indicative of the slight ellipticity excess, although the most striking "low" in the figure appears almost on the equator, south of India. But the extreme departures from equilibrium are only about $\pm 100$ $m$, which, as considered in the following section, is very much less than would appear if the surface features of the Earth were protrusions from an otherwise uniform Earth. The large scale features of the geoid show little correlation with the surface elevation of the crust (continents versus oceans), but Kaula (1969b), who made a study of the relationship between the geoid "highs" and "lows" and the active geological (tectonic) zones, concluded that there is a correspondence to tectonic features, in other words that the geoid is, at least in part, an expression of the deep-seated mantle movements (convection) that are responsible for geological activity (as discussed in Chapter 10).

An indication that the mass "anomalies" responsible for the geoid features are in the upper mantle, at depths of a few hundred kilometers, is given by an analysis of the stresses required to support them. Higbie and Stacey (1970) supposed that the geoid was due to undulations of a single nearly spherical (or ellipsoidal) internal surface with a density contrast, $\Delta\rho$. Each harmonic term $U_l^m$ in the gravitational potential at the Earth's radius $a$ was attributed to the corresponding harmonic $h_l^m$ in the

**Table 4.1** Harmonic Coefficients of the Earth's Gravitational Potential[a]

| $l$ \ $m$ | 0 | 1 | 2 | 3 | 4 | 5 | 6 | 7 | 8 |
|---|---|---|---|---|---|---|---|---|---|
| 2 | − 484.172 | — | 2.426 | | | | | | |
| | | | − 1.386 | | | | | | |
| 3 | 0.958 | 2.017 | 0.919 | 0.719 | | | | | |
| | | 0.251 | − 0.617 | 1.420 | | | | | |
| 4 | 0.547 | − 0.532 | 0.354 | 0.974 | − 0.167 | | | | |
| | | − 0.444 | 0.662 | − 0.220 | 0.312 | | | | |
| 5 | 0.068 | − 0.069 | 0.657 | − 0.472 | − 0.315 | 0.149 | | | |
| | | − 0.082 | − 0.317 | − 0.231 | 0.028 | − 0.679 | | | |
| 6 | − 0.161 | − 0.089 | 0.068 | 0.017 | − 0.101 | − 0.293 | 0.038 | | |
| | | − 0.020 | − 0.368 | − 0.024 | − 0.453 | − 0.508 | − 0.230 | | |
| 7 | 0.092 | 0.252 | 0.339 | 0.259 | − 0.270 | − 0.007 | − 0.329 | 0.065 | |
| | | 0.131 | 0.085 | − 0.216 | − 0.086 | 0.053 | 0.150 | 0.036 | |
| 8 | 0.062 | 0.024 | 0.049 | − 0.024 | − 0.240 | − 0.093 | − 0.037 | 0.051 | − 0.091 |
| | | 0.092 | 0.066 | − 0.074 | 0.068 | 0.084 | 0.301 | 0.073 | 0.097 |

[a]Coefficients to degree and order (8, 8) from the more extensive set of Rapp (1973). These refer to the fully normalized harmonic functions $p_l^m$ defined in Appendix C. For each value of $(l, m)$ the coefficients given are $\bar{C}_l^m$ followed by $\bar{S}_l^m$ in units of $10^{-6}$. Uncertainties are of order 0.02.

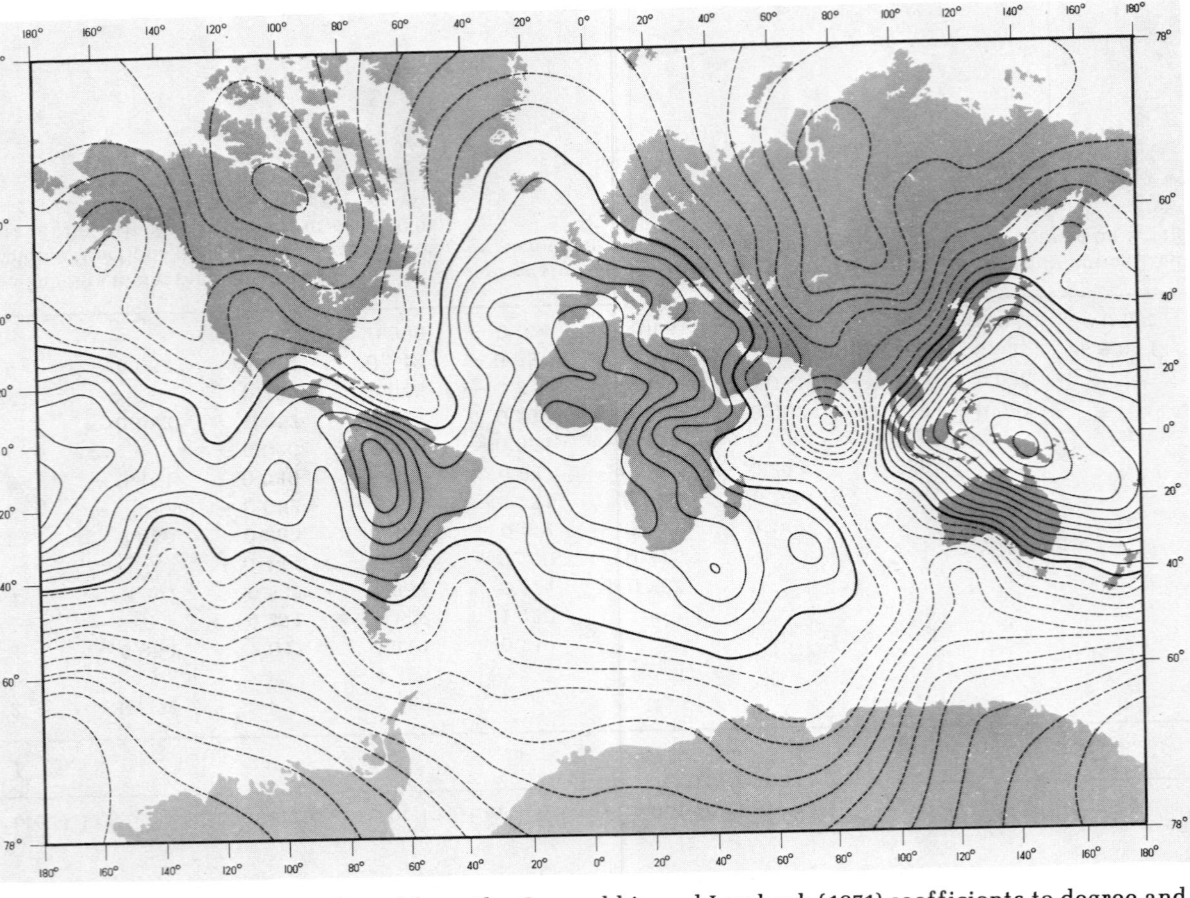


**Figure 4.1.** Geoid contours plotted from the Gaposchkin and Lambeck (1971) coefficients to degree and order (16, 16). The contour interval is 10 m and negative contours are represented by broken lines. The reference ellipsoid, from which this figure shows the departures, has the equilibrium flattening, 1/299.76.

elevation of the boundary at mean radius $r$:

$$\Delta\rho\, h_l^m = \tfrac{1}{3}\bar{\rho} a (2l+1) \left(\frac{a}{r}\right)^{l+2} U_l^m \tag{4.24}$$

where $\bar{\rho}$ is the mean Earth density. The magnitude of the stress required to support each harmonic is approximately

$$\sigma_l^m \simeq \tfrac{1}{2} g_r\, \Delta\rho\, h_l^m = \tfrac{1}{6} g_r \bar{\rho} a (2l+1) \left(\frac{a}{r}\right)^{l+2} U_l^m \tag{4.25}$$

where $g_r$ is the gravity at radius $r$. However, we are not concerned with the stresses associated with individual harmonics of the geoid but with the stresses associated with particular "wavelengths" of the geoid pattern. To approximate this situation, Higbie and Stacey (1970) used values of $U_l$ given by

$$U_l = \left\{ \sum_{m=0}^{l} [(\bar{C}_l^m)^2 + (\bar{S}_l^m)^2] \right\}^{1/2} \tag{4.26}$$

which is the r.m.s. value of that part of the potential that is due to all harmonics of degree $l$, and they obtained corresponding stresses, $\sigma_l$, for each of a series of assumed values of $(a/r)$. Figure 4.2 is a plot of $\sigma_l$ as a function of $l$ for each of a series of depths of the boundary, $(a-r)$. The purpose was to seek a depth for which $\sigma_l$ was independent of $l$; with this particular set of coefficients, which were taken from Rapp (1968), the preferred depth was 655 km, but alternative data sets gave depths in the range 250 to 900 km. An extension of this analysis by McQueen and Stacey (1976), using the improved and extended data set by Rapp (1973), shows that a single boundary is inadequate and that negatively correlated undulations of two boundaries are required to fit the data. The negative correlation simply means that highs of one boundary are generally superimposed on the lows of the other. It is natural to attribute the boundaries to the phase transition zones of the upper mantle, which occur at depths of 420 km and 670 km in the Earth model in Appendix G, and this is consistent with the data. The product of density contrast and undulation amplitude is of order $10^6$ kg m$^{-2}$ for each layer, that is, if a density contrast is 300 kg m$^{-3}$, the undulation amplitude is 3 km (for each harmonic degree) and the corresponding mantle stresses are about 50 bars ($5 \times 10^6$ Pa). As noted in Section 10.3, this is consistent with the energetics of mantle convection.

The supposition that the undulations are stress limited, and therefore that the stresses corresponding to each harmonic degree are similar, is formally equivalent to the assumption that the undulations of each of the boundaries have white spatial spectra over the wavelength range of the available geoid coefficients (down to 900 km at $l = 20$). Such spectra

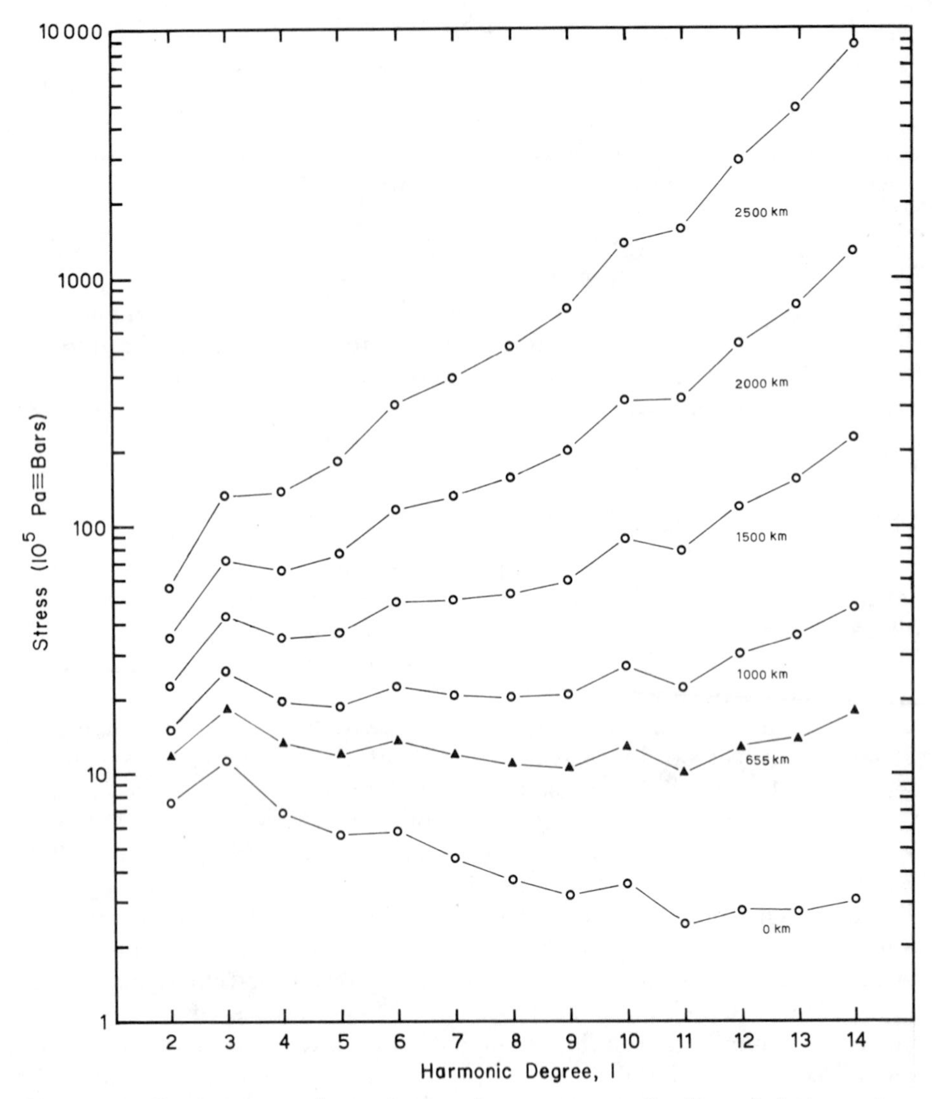


**Figure 4.2.** Rms stresses due to harmonic components in the undulations of an internal boundary. At each assumed depth the form of the boundary is that which would give the observed geoid (with the measured ellipticity subtracted), according to harmonic coefficients by Rapp (1968). Figure after Higbie and Stacey (1970).

could be a consequence of random, sharp protrusions of the upper mantle phase boundaries, as probably occur where the Benioff subduction zones of the mantle convection pattern intersect them.

Second- and third-degree harmonics, included in Figure 4.2, were omitted from the later analysis; the value in Figure 4.2 for $l = 2$ neglects the nonequilibrium ellipticity, which may be due partly to delayed rebound from polar glaciation (McConnell, 1968; O'Connell, 1971) and relatively greater south polar flattening has been proposed as a cause of the strong third-degree zonal harmonic (Khan and O'Keefe, 1974).

The form of a phase boundary at 350 km depth, which would give the gravitational potential coefficients of Gaposhkin and Lambeck (1971), is shown in Figure 4.3. In comparison with Figure 4.1 this gives a stronger relative emphasis to the higher harmonics, making sharper features apparent and the correlation with geologically active zones clearer, especially the coincidence of highs with "subduction zones" or lines of crustal convergence. This is considered further in Chapter 10.

## 4.3 CRUSTAL STRUCTURE AND THE PRINCIPLE OF ISOSTASY

The distinctness of continents and oceans is made apparent by plotting the areas of the Earth's solid surface at different levels above and below sea level. Such a plot is known as a hypsometric curve and is given as a histogram in Figure 4.4. This shows a bimodal distribution with a peak near to sea level, corresponding to continents, and another at about 5 km depth, corresponding to the ocean basins. Mountain ranges and deep ocean trenches occupy very small fractions of the surface area, but the most significant feature of the hypsometric curve is the smallness of the area between elevations of $-1$ km and $-3$ km. In other words, the continents have sharp submarine margins, separating them from the ocean basins. This is also apparent in the smooth curve in Figure 4.4, which is the integral of the data in the histogram and is thus an averaged characteristic of the surface elevation.

Sea level is necessarily the reference level for Figure 4.4, but it is not the demarcation between continental and oceanic structures. If sea level had happened to be, say, 1.5 km lower, it would have marked the true continental margins reasonably well and we might have supposed that the sea level was in some way responsible for continental structure. However, particularly in view of the leveling effect of erosion processes, the general forms and distribution of continents cannot be results simply of forces at the Earth's surface but must have a deep-seated origin. This is considered further in Chapter 10. By including submerged areas of continents, which we may define roughly as the areas where the sea is

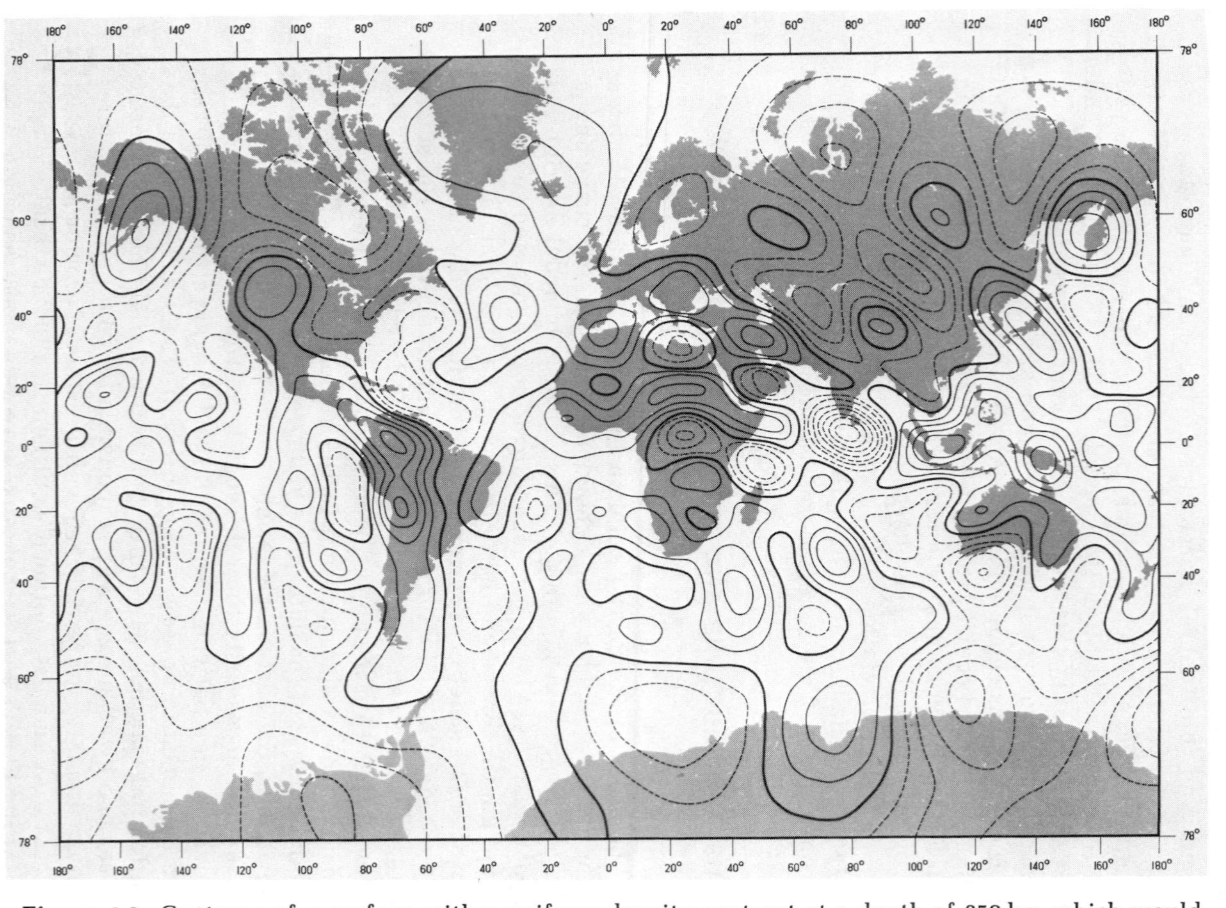


**Figure 4.3.** Contours of a surface with a uniform density contrast at a depth of 350 km which would produce the geoid represented in Figure 4.1. The contour interval would be 1000 m for a density contrast of 500 kg m$^{-3}$ (which is a plausible value for a phase boundary). This is similar to a plot by Higbie and Stacey (1971) but with a smaller contour interval.

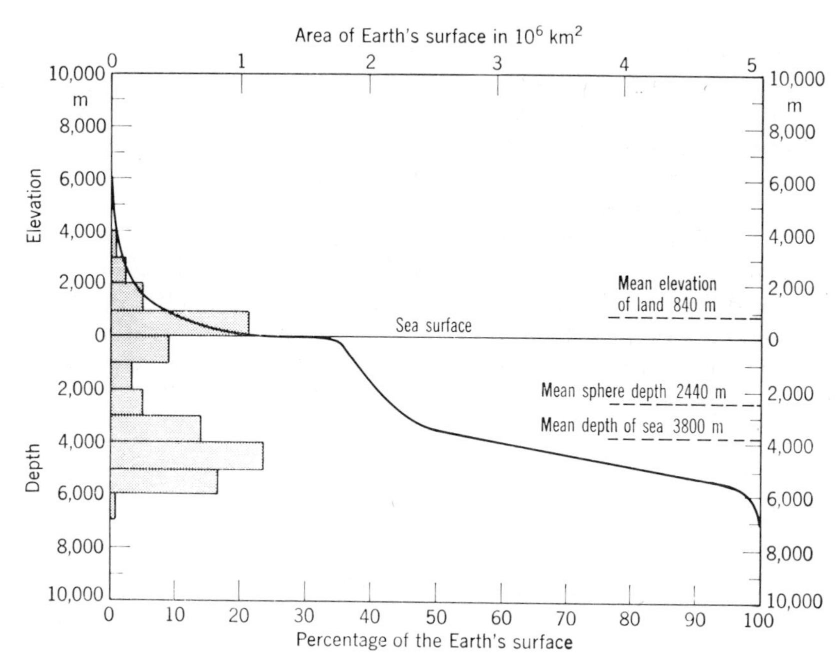


**Figure 4.4.** Histogram of areas of the Earth's solid surface in intervals of 1 km in elevation with the integral curve showing the area above any depth. After Sverdrup, Johnson, and Fleming (1942) who used the compilations of E. Kossinna. Updating with more recent ocean data (Menard and Smith, 1966) does not significantly alter the figure.

less than 1 km deep, the total area of the continents is found to be nearly 40% of the Earth's surface, whereas only 29% is actually above sea level.

Since, as we see from Figure 4.4, the continental masses are distinct, standing 5 km above the ocean floor, and their individual areas are large, the fact that their existence is not apparent in the low-order harmonics of the geoid requires mass compensation at depth to quite a high degree of adjustment. The principle of mass compensation has been known for over a century, since a geodetic survey across North India showed that the Himalayas caused much less deflection of the vertical than if they were simply a prominence on a uniform Earth. This discovery was the birth of the theory of isostasy, according to which the total mass of rock (and sea where it occurs) in any vertical column of unit cross section is constant. Such columns may be considered based at a particular level of "compensation." below which the Earth can be assumed uniform. Thus variations in elevation of the crust are supported hydrostatically. The balance is nowhere perfect, but on a continental scale it is very nearly so.

The distribution of mass at depth cannot be determined from surface gravity data without additional evidence and, even with the approximate location by seismic methods of boundaries between layers of different densities, there is generally a range of admissible alternatives. Heiskanen and Meinesz (1958) and Garland (1965) have reviewed the subject and presented computed density profiles for different regions, in which the seismic evidence is taken into account. These incorporate in different proportions the two rival hypotheses of isostasy that grew directly from the original survey work in India. In 1854 J. H. Pratt suggested that the higher parts of the crust were elevated by virtue of their lower densities, as in Figure 4.5*a*, and in 1855 G. B. Airy proposed

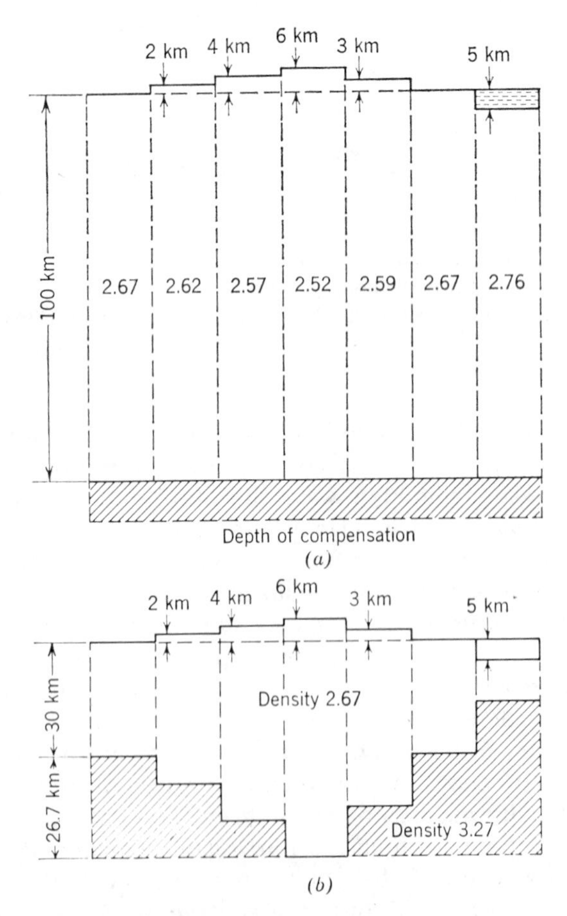


**Figure 4.5.** Isostatic compensation according to (*a*) J. H. Pratt and (*b*) G. B. Airy, with numerical values of density given by W. A. Heiskanen. Figures based on Heiskanen and Meinesz (1958).

the scheme represented in Figure 4.5*b*. Airy visualized the crustal masses as logs (all of the same density) floating in water. A log appearing higher out of the water than its neighbors must extend correspondingly deeper. This accords with the seismic evidence that the continental crust (35 to 40 km average) is thicker than the oceanic crust (about 5 km) by an amount greatly exceeding the difference in surface elevations. In the considered opinion of W. A. Heiskanen (see Heiskanen and Meinesz, 1958), 63%, on the average, of the isostatic balance of the crust is achieved by the Airy principle of depth compensation and 37% by density differences, as envisaged by Pratt. But this is almost certainly an oversimplification as seismic studies (e.g., Dorman et al., 1960; Sipkin and Jordan, 1975) indicate lateral heterogeneities, in particular differences between continental and oceanic structures extending to depths of several hundred kilometers; also Dorman and Lewis (1972) interpreted North American gravity data as an association of crustal loads with density decrease at about 100 km depth, which is partially counteracted by a density increase at greater depth, of order 450 km.

The observation that the form of the geoid is not correlated with continents and oceans, and that its departures from an ellipsoid of revolution are slight compared with what we would expect if the continents were simply superimposed on an ellipsoidally layered Earth, allows us to estimate the degree of exactness of the isostatic balance of the continents. It is convenient to discuss their effect on the geoid in terms of the geometrically simple pair of continents represented in Figure 4.6. First consider the effect on the satellite geoid of a pair of circular continents, radius $r$, height $h$, and density $\rho_c$, superimposed on an otherwise spherically symmetrical earth, as in Figure 4.6*a*. The resulting flattening of the geoid is calculable from the difference between the moments of inertia of the Earth about axes 1 and 2, by Equation 3.21, neglecting the second (rotational) term, which is not relevant here. The model is otherwise spherically symmetrical; thus we need consider only the moments of inertia of the continents, which are, for $r \ll a$,

$$\begin{aligned} I_1 &\approx 2(\pi r^2 h \rho_c)\frac{r^2}{2} \\ I_2 &\approx 2\int_a^{a+h} (\pi r^2 \rho_c) x^2\, dx = 2\pi r^2 \rho_c a^2 h \ (h < a) \end{aligned} \tag{4.27}$$

so that, approximately,

$$I_1 - I_2 = -2\pi r^2 \rho_c a^2 h \tag{4.28}$$

The geoid flattening $f_g$ is, therefore,

$$f_g = \frac{3}{2}\frac{I_1 - I_2}{Ma^2} = -\frac{9}{4}\frac{r^2 h}{a^3}\frac{\rho_c}{\bar{\rho}} \tag{4.29}$$

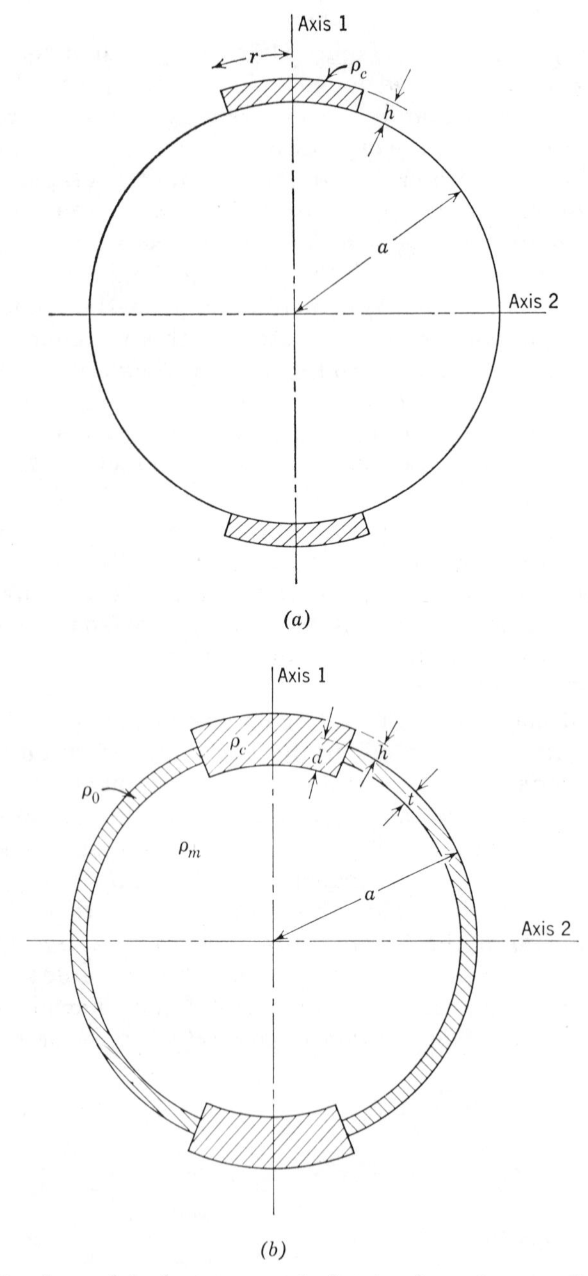


**Figure 4.6.** Simple model of a symmetrical pair of continents, illustrating their effect on the geoid. (*a*) Continents superimposed on an otherwise spherically symmetric Earth. (*b*) Continents of density $\rho_c$ overlaying a mantle of density $\rho_m$ and isostatically balanced with an oceanic crust of density $\rho_0$.

where $\bar{\rho}$ is the mean density of the Earth. The negative sign implies that the ellipsoid is prolate and elongated along axis 1. The elevation of the geoid in the direction of axis 1, relative to axis 2, is therefore

$$h_g = f_g a = \frac{9}{4}\frac{r^2}{a^2}\frac{\rho_c}{\bar{\rho}} h \tag{4.30}$$

With appropriate values, $r = 2000\ \text{km} = a/3$, $\rho_c/\bar{\rho} = \frac{1}{2}$ and $h = 5\ \text{km}$, we have

$$h_g = 0.625\ \text{km}$$

that is, the geoid is elevated about $\frac{1}{8}$km per 1 km thickness of continental material superimposed on an otherwise uniform Earth. The observed effect of continents on the geoid is evidently less than 50 m and perhaps much less, so that the masses of the continents must be compensated at depth, and, in spite of the fact that they stand 5 km above the ocean floor, they appear gravitationally as much less than 400 m of "surplus" crust.

The isostatic balance can be examined more closely in terms of the model represented in Figure 4.6*b*, in which the continents have roots and there is also an oceanic crust of thickness $t$ and density $\rho_0$. The continents reach to a height $h$ above the top of the oceanic crust but also extend a depth $d$ below it, and we can establish a relationship between $d, h, t, \rho_c, \rho_0$ and $\rho_m$, the mantle density, from the requirement that the continents give zero ellipticity to the geoid. This requirement is satisfied by making the moment of inertia of the continents equal to that of the oceanic crust and mantle, which would replace them to produce a spherically symmetrical Earth:

$$2\int_{a-d}^{a+h} (\pi r^2 \rho_c) x^2\, dx = 2\int_{a-d}^{a-t} (\pi r^2 \rho_m) x^2\, dx + 2\int_{a-t}^{a} (\pi r^2 \rho_0) x^2\, dx \tag{4.31}$$

from which, with $d, t, h \ll a$,

$$\rho_c(h + d) = \rho_m(d - t) + \rho_0 t \tag{4.32}$$

Equation 4.32 expresses the principle that the total masses in all vertical columns are the same, which must therefore be valid to better than about 8% of 5 km or 400 m of crust, as noted previously. This is the general case of isostatic balance, without preference for the Pratt or Airy principles. The balance would be achieved by Pratt's method if $d = t$, so that

$$\rho_c(h + t) = \rho_0 t \tag{4.33}$$

or by Airy's method if $\rho_c = \rho_0$, in which case

$$\rho_c(h + d - t) = \rho_m(d - t) \tag{4.34}$$

The fact that the continental scale departures from isostatic balance are very small must be interpreted as a consequence of the limited strength of the mantle or, since the geoid features are presumed to be dynamically maintained, we may refer to the mantle as having a moderate (i.e., not very high) viscosity, of order $10^{21}$ decapoise. But much steeper gravity anomalies can occur on a smaller scale. This can be understood in terms of a *lithosphere*, 70 to 150 km thick, of relatively high strength, overlying the *asthenosphere* or weak layer, which may extend to 300 or 400 km. By virtue of its strength the lithosphere can support small scale mass "anomalies" but a mass excess over a large area would depress it into the asthenosphere, where flow reestablishes isostatic balance. We can assign a strength sufficient to support the large-scale features of the geoid to the *mesosphere*, a zone of increasing strength below the asthenosphere, but arguments based on static stresses cannot be taken too far because of the evidence that the Earth's extensive gravity anomalies are dynamically maintained by mantle convection (Chapter 10). The most striking gravity anomalies are found over island arcs, which mark the subduction of the lithosphere along lines of convective downturn.

The restoration of isostatic equilibrium in an area where crustal movements have occurred is a matter of considerable interest in connection with the rheological properties of the upper mantle (Chapter 10). The steady rise of a large area around the Gulf of Bothnia (Fig. 10.15) is interpreted as a process of isostatic rebound from the depression of the area caused by heavy glaciation during the last ice age. A similar rise is observed in the St. Lawrence valley-Great Lakes area of Canada. In Section 4.2 it is pointed out that, on the same basis, the former greater general glaciation of both polar regions must be responsible for part of the excess ellipticity of the Earth and Khan and O'Keefe (1974) suggest that the pear shape of the geoid is due to a proportionately greater unrecovered depression of Antarctica. It is supposed that the movements associated with isostatic adjustment extend to a depth of 100 to 300 km, at which the mantle behaves as a fluid with a viscosity of order $10^{20}$ decapoise (Section 10.2); if Newtonian viscosity is an appropriate concept for the mantle, this is not an unreasonable figure.

## 4.4 TIDES

The rotation of the Earth in the gravity fields of the Moon and Sun imposes periodicities in the gravitational potential at any point on the surface. The most obvious effect is the marine tide but there are also deformations of the solid Earth (the earth tides). The study of tides owes much to the pioneering work of G. H. Darwin, whose 1898 monograph is still fascinating reading and has recently been reprinted (Darwin, 1962). Melchior (1966) has reviewed the observation and interpretation of tidal strains in the solid Earth.

Consider the potential at an arbitrary point on the Earth's surface, as in Figure 4.7, due to the combination of the Moon's gravity and rotation with orbital angular velocity $\omega_L$ about the axis through the common centre of mass (distant $r$ from $P$):

$$W = -\frac{Gm}{R'} - \tfrac{1}{2}\omega_L{}^2 r^2 \tag{4.35}$$

where $m$ is the mass of the Moon. The distance $R'$ can be represented in terms of the Earth-Moon distance $R$, the Earth radius $a$, and the angle $\psi$ between the radius to $P$ and the Earth-Moon axis:

$$(R')^2 = R^2 + a^2 - 2aR\cos\psi \tag{4.36}$$

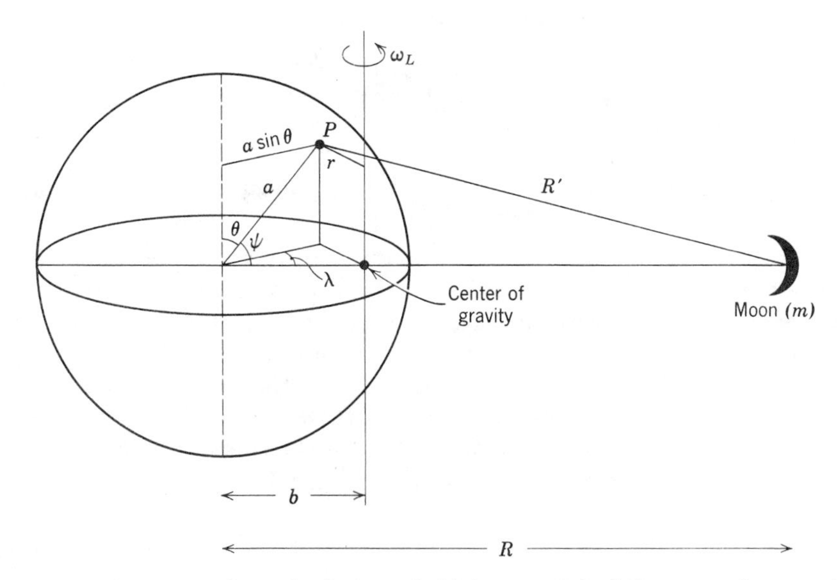


**Figure 4.7.** Geometry for calculation of tidal potential of the moon (mass $m$) at distance $R$ from the centre of the Earth and $R'$ from the arbitrary point $P$ on the surface of the Earth. The intersection of the lunar orbital plane with the Earth is shown as an ellipse and the angular coordinates of $P$ are $\theta$, referred to the normal to this plane and $\lambda$ measured within the plane from the Earth-Moon axis. The center of gravity of the system, about which both the Earth and Moon orbit, is at a distance $b$ from the center of the Earth, slightly less than the Earth radius, $a$.

so that to the second order in the small quantity $a/R$,

$$(R')^{-1} = R^{-1}\left(1 - \tfrac{1}{2}\frac{a^2}{R^2} + \frac{a}{R}\cos\psi + \tfrac{3}{2}\frac{a^2}{R^2}\cos^2\psi + \ldots\right) \tag{4.37}$$

We also have the trigonometric relationships

$$\cos\psi = \sin\theta\cos\lambda \tag{4.38}$$

and

$$\begin{aligned} r^2 &= b^2 + (a\sin\theta)^2 - 2b(a\sin\theta)\cos\lambda \\ &= b^2 + a^2\sin^2\theta - 2ba\cos\psi \end{aligned} \tag{4.39}$$

where

$$b = \frac{m}{M+m}R \tag{4.40}$$

The relationship between orbital angular velocity, $\omega_L$, about the common center of mass and $R$ is obtained by equating the centripetal force on either body to their mutual gravitational attraction (assuming a circular orbit):

$$\omega_L{}^2 R^3 = G(M+m) \tag{4.41}$$

which is the special case of Kepler's third law (see Appendix A). Substituting (4.37), (4.39), (4.40), and (4.41) in (4.35), collecting terms and rearranging, we obtain

$$W = -\frac{Gm}{R}\left(1 + \tfrac{1}{2}\frac{m}{M+m}\right) - \frac{Gma^2}{R^3}\left(\tfrac{3}{2}\cos^2\psi - \tfrac{1}{2}\right) - \tfrac{1}{2}\omega_L{}^2 a^2\sin^2\theta \tag{4.42}$$

The three terms in Equation 4.42 are readily identified. The first is simply a constant, being the value of the gravitational potential due to the Moon at the center of the Earth, with a small correction arising from the mutual rotation. The second is a second-order zonal harmonic (see Appendix C) and represents a deformation of the equipotential surface to a prolate ellipsoid aligned with the Earth-Moon axis. Rotation of the Earth within this potential is responsible for the tides. It is convenient to separate it and refer to it as the tidal potential (Fig. 4.8):

$$W_2 = -\frac{Gma^2}{R^3}\left(\tfrac{3}{2}\cos^2\psi - \tfrac{1}{2}\right) \tag{4.43}$$

The third term of Equation 4.42 is the rotational potential of the point $P$ about an axis *through the center of the Earth* normal to the orbital plane. It is responsible for an oblate ellipsoidal deformation of the equipotential surface. This is larger by a factor nearly 30 than the ellipticity due to $W_2$, but does not have a tidal effect because it is associated with axial

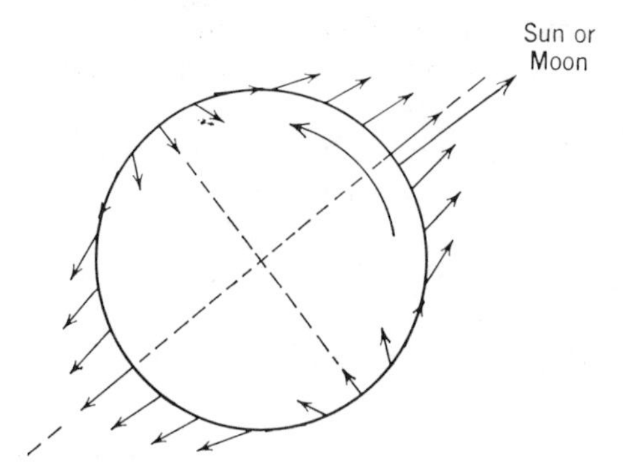


**Figure 4.8.** Tidal forces due to the Moon (or Sun). Apart from a phase lag due to dissipative processes (Section 4.5), the pattern is symmetrical about the Earth-Moon (or Sun) line.

rotation and merely becomes part of the equatorial bulge of rotation, angular momenta being vectorially added.

The ellipticity of the tidal bulge gives a predominantly semidiurnal tide (with two peaks per day), but the rotation of the Earth about an axis inclined to the orbital plane imparts a diurnal asymmetry to the tides (the "tidal inequality"). *A* and *B* in Figure 4.9 represent successive positions of a point on the surface at an interval of about 12 hours. At *A* the point is very close to the maximum "high" tide, but at *B* there is hardly any tide. At this stage in the lunar month the point sees a predominantly diurnal tide, although it becomes semidiurnal when the Moon is above the equator. The ellipticity of the lunar orbit imposes further periodicities and superposition of solar tides results in a quite

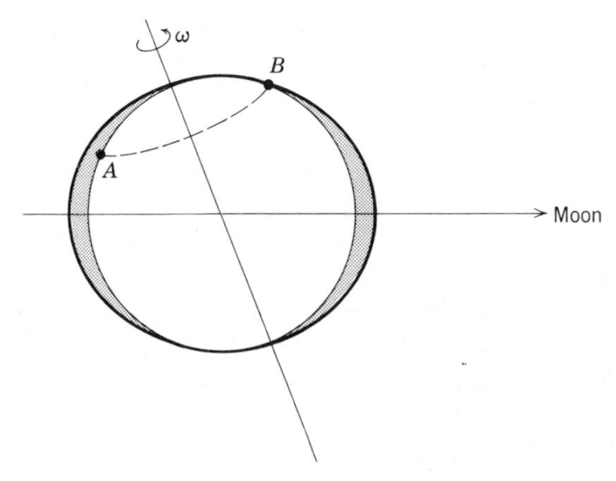


**Figure 4.9.** Rotation of the Earth about an axis inclined to the lunar orbital plane introduces an asymmetry to the tides—the tidal inequality—which is apparent as a diurnal tidal component.

complex waveform. But from a fundamental geophysical viewpoint the important feature is the prolate ellipsoidal bulge represented by Equation 4.43.

If the Earth were rigid then the tidal variation in gravity would be given by the radial variation in $W_2$

$$\Delta g = -\frac{\partial W_2}{\partial a} = \frac{Gma}{R^3}(3\cos^2\psi - 1) \tag{4.44}$$

and the circumferential component of the gravity disturbance

$$\Delta g_\psi = -\frac{1}{a}\frac{\partial W_2}{\partial \psi} = -\frac{3}{2}\frac{Gma}{R^3}\sin 2\psi \tag{4.45}$$

Related to gravity $g$ on the undisturbed Earth (mass $M$)

$$g = -\frac{GM}{a^2} \tag{4.46}$$

the fractional change in intensity of the gravity field is

$$\frac{\Delta g}{g} = -\frac{m}{M}\left(\frac{a}{R}\right)^3 (3\cos^2\psi - 1) \tag{4.47}$$

(which is close to the limit of sensitivity of instruments used for gravity surveys) and the deflection of the vertical is $\alpha$, given by

$$\tan\alpha = \frac{\Delta g_\psi}{g} = \frac{3}{2}\frac{m}{M}\left(\frac{a}{R}\right)^3 \sin 2\psi \tag{4.48}$$

For the lunar tides $(m/M)(a/R)^3 = 5.6 \times 10^{-8}$ and for the solar tides the corresponding quantity is 0.45 times this value. Alternatively we may represent the variation in height of the equipotential surface (still for a hypothetical rigid Earth) by

$$\Delta a = \frac{W_2}{g} = \frac{m}{M}\left(\frac{a}{R}\right)^3 a\left(\tfrac{3}{2}\cos^2\psi - \tfrac{1}{2}\right) \tag{4.49}$$

the peak-to-trough amplitude of which is 0.535 meter for the lunar tide.

The fact that both the solid Earth and the oceans are deformed by the tidal forces modifies the observed tidal potential. The deformation, which assumes the form of $W_2$, is conveniently represented in terms of the values of two dimensionless parameters $h$ and $k$, which were introduced by A. E. H. Love and are referred to as Love numbers, and a third $l$ due to T. Shida, defined as follows:

$h$ is the ratio of the height of the body tide to the height of the equilibrium (static) marine tide, that is, the height of the sea if it had time to come to equilibrium with the tidal potential.

$k$ is the ratio of the additional potential produced by the redistribution of mass to the deforming potential.
$l$ is the ratio of horizontal displacement of the crust to that of the equilibrium fluid tide.

For the hypothetical rigid Earth all three numbers are zero and for a fluid Earth in tidal equilibrium, $h_f = 1$, $l_f = 1$ by definition and $k_f$ is a function of the density profile, similar to the equilibrium flattening $f_H$ (Eq. 3.29)

$$k_f = 3J_2/m = \left(2\frac{f_H}{m} - 1\right) \tag{4.50}$$

For a hypothetical fluid Earth of uniform density we would have $k_f = \frac{3}{2}$ and for the actual density profile $k_f = 0.937$.

Observations of various tidal effects give various combinations of the Love numbers, but there are quite significant discrepancies between alternative estimates. Some of these are real in the sense that the Earth, being inhomogeneous and including in particular a fluid core and fluid oceans in which the form of the motion may be rate dependent, cannot be represented by a single set of constants for different kinds of deformation. So we distinguish by the symbol $k_2$ the Love number describing deformation in response to the tidal potential, $W_2$ (Eq. 4.43). The most direct (although technically difficult) determination of $k_2$ is from the tidal deformation of the Earth's potential field as seen by satellites. Kolenkiewicz et al. (1973) reported a satellite value, $k_2 = 0.245 \pm 0.005$, which is at the lower end of the range of estimates from other observations, almost certainly because it is a whole Earth value, including the oceans, whereas other estimates refer to solid Earth deformation (Lambeck et al., 1974). It is interesting to note that this value of $k_2$ indicates an inverted tide in the oceans. This is in accord with the dynamical theory of the marine tide, which is a shallow water wave of natural velocity $\sqrt{gh}$ (Eq. 5.47), less than the rotational speed $(\omega - \omega_L)a$, of the driving force (i.e., the Moon's gravity). If $\sqrt{gh}$ were much greater than $(\omega - \omega_L)a$, the marine tide would be an equilibrium tide following the tidal potential; if $\sqrt{gh} = (\omega - \omega_L)a$ then a world-encircling tide would be resonant and on the other side of resonance, with $\sqrt{gh} < (\omega - \omega_L)a$, the tide is inverted, that is, low tides occur where equilibrium theory gives highs and vice versa (see Lamb, 1963, pp. 268–9; Officer, 1974, pp. 176–8). In fact, marine tides are very variable, owing to the complexity of sea floor geometry.

In principle quite simple observations suffice to determine $h$ and $k$ but the variability of marine tides introduces considerable uncertainty, which can be overcome only by using very long period tides (monthly, annual, etc.) for which the oceans may be close to hydrostatic equilibrium. Then the tidal potential as seen by the ocean is $W_2$ plus the potential due to the response of the Earth, $k_2W_2$. But the marine tide is measured with respect to the solid Earth, which is itself deformed by $hW_2$, so that the observed marine tidal amplitude is

$$\Delta z = (1 + k - h)\frac{W_2}{g} = \frac{\gamma W_2}{g} \tag{4.51}$$

Since $W_2$ is known, $\gamma$ is determined from the tidal measurements. Similarly there are three contributions to the observed variations in the strength of the gravity field: the disturbing potential, $W_2$, the tidal displacement of the observing site, and the potential of the deformed mass of the Earth. This gives another combination of Love numbers:

$$\Delta g = -(1 + h - \tfrac{3}{2}k)\frac{\partial W_2}{\partial r} = -\delta\frac{\partial W_2}{\partial r} \tag{4.52}$$

The deformation of the Earth in response to the Chandler wobble is also conveniently represented in terms of the Love number $k$. The ratio of the observed wobble period $T_0$ to the period $T_R$ expected for a rigid Earth with the same dynamical ellipticity is given by Munk and MacDonald (1960b; see also Kaula, 1968, pp. 188–190):

$$\frac{T_0}{T_R} = \frac{k_f}{k_f - k} \tag{4.53}$$

Since $k_f$, $T_0$, and $T_R$ are all known, the elastic lengthening of the wobble period gives an estimate of $k$:

$$k = k_f\left(1 - \frac{T_R}{T_0}\right) = 0.277 \tag{4.54}$$

Although the Love numbers cannot be represented by single, unambiguous values, representative values are

$$h = 0.59$$
$$k = 0.27$$
$$l = 0.04$$

Takeuchi (1950) calculated Love numbers from Earth model data (Chapter 6) and concluded that the rigidity of the core must be less than $10^9$ Pa ($10^4$ bars).* Fluidity of the core is indicated by the fact that shear waves are not propagated through it, but it is of interest to confirm that there is a real absence of rigidity and not merely strong shear wave attenuation.

*The mean effective rigidity for the whole Earth is $1.45 \times 10^{11}$ Pa or, for the solid part of the Earth alone, $1.73 \times 10^{11}$ Pa.

## 4.5 TIDAL FRICTION AND THE HISTORY OF THE LUNAR ORBIT

If the tides were perfectly frictionless, then, even allowing for inertia in the marine tides, the average tidal bulge of the Earth would be perfectly aligned with the Earth-Moon axis (or Earth-Sun axis in the case of the solar tide). But tidal dissipation causes a lag, with the result that the tidal bulge is slightly misaligned, as in Figure 4.10. The bulge gives asymmetry to the geopotential, thereby exerting an orbital accelerating torque on the Moon (or Sun) and slowing the Earth's rotation. Integrated over geological time this is a first-order effect; it is vital to our understanding of the history of the Earth and Moon, as has been recognized particularly clearly since the inclusion of tidal friction in orbital calculations by H. Gerstenkorn (see review by Alfvén, 1965) suggested that the Moon was captured.

As noted in Section 4.4, the most direct estimate of the deformation of the geopotential field by tides is from the perturbation of satellite orbits. Newton (1968) reported the first measurements of this kind and more recent data are by Kolenkiewicz et al. (1974) who concluded that the tidal ellipticity was represented by a second-degree Love number $k_2 = 0.245 \pm 0.005$ and a phase lag $\phi_2 = (3.2 \pm 0.5)$ degrees. This value of $k_2$ is significantly lower than some others, but for the present purpose uncertainty in $k_2$ is less important than the value of $\phi_2$. It is of interest that the satellite estimate of $\phi_2$ coincides with that obtained by Smith

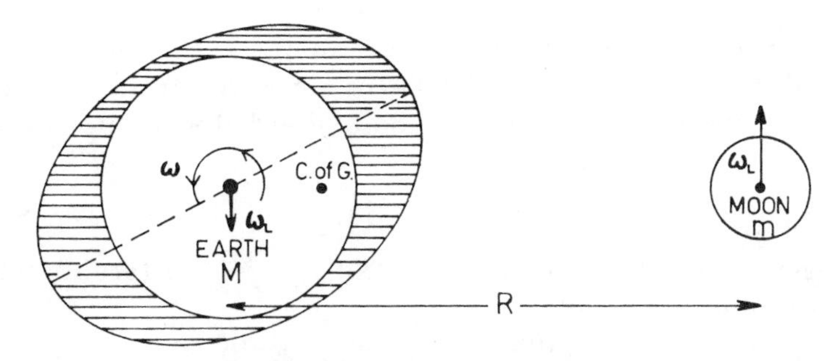


**Figure 4.10.** Origin of the tidal torque. The gravitational gradient of the Moon raises a tidal bulge in the Earth, and in particular in the sea, but dissipative processes cause a lag in the tidal response so that on average the high tides* occur at points on the Earth that were the sublunar (or antilunar) points a few minutes before. The gravitational potential of the tidal ellipticity thus exerts an orbital accelerating force on the Moon and slows the Earth's rotation. The Moon orbits with angular velocity $\omega_L$ about the common centre of gravity but maintains a constant face to the Earth, while the Earth has an axial rotation $\omega > \omega_L$. [*Note. The marine tide is, in fact, inverted, as discussed in Section 4.4.]

and Jungels (1970) from area strain deduced from linear earth strain-meter records (3° ± 1°). We can estimate the lunar tidal torque directly from these values.

The tide-raising potential of the Moon, mass $m$ at distance $R$, on the surface of the Earth, radius $a$, is $W_2$, given by Equation 4.43. Due to the deformation an additional potential $k_2 W_2$ results, so that at the distance of the Moon the form of the potential due to the tidal deformation of the Earth is

$$V_T = k_2 W_2 \left(\frac{a}{R}\right)^3 = -\frac{Gma^5}{R^6} k_2(\tfrac{3}{2}\cos^2\psi - \tfrac{1}{2}) \tag{4.55}$$

We can now identify $\psi$ with $\phi_2$ the angle between the Earth-Moon line and the axis of the tidal bulge to obtain the tidal torque exerted on the Moon ($m$):

$$L = m\left(\frac{\partial V_T}{\partial \psi}\right)_{\psi=\phi_2} = \frac{3}{2}\frac{Gm^2a^5k_2}{R^6}\sin 2\phi_2 \tag{4.56}$$

With the numerical values of $k_2$ and $\phi_2$ by Kolenkiewicz et al. (1974) we obtain $L = (4.8 \pm 0.8) \times 10^{16}$ newton-meters (N-m). This torque causes an orbital acceleration of the Earth and Moon about their common center of mass; an equal and opposite torque exerted by the Moon on the tidal bulge slows the Earth's rotation.

$L$ must be equated to the rate of change of the orbital angular momentum $a_L$, which is

$$a_L = \frac{Mm}{M+m} R^2 \omega_L \tag{4.57}$$

the angular velocity $\omega_L$ and distance $R$ of the Moon being related by Kepler's third law (given by Eq. 4.41 for a circular orbit), which allows the rates of change of $\omega_L$ and $R$ to be expressed separately since

$$a_L = \frac{G^{1/2}Mm}{(M+m)^{1/2}} R^{1/2} = \frac{G^{2/3}Mm}{(M+m)^{1/3}} \omega_L^{-1/3} \tag{4.58}$$

The present value of $a_L$ is $2.85 \times 10^{34}$ kg m$^2$ sec$^{-1}$. Differentiating with respect to time

$$L = \frac{da_L}{dt} = -\frac{a_L}{3\omega_L}\frac{d\omega_L}{dt} = \frac{a_L}{2R}\frac{dR}{dt} \tag{4.59}$$

which, with the above value of $L$, gives

$$\frac{d\omega_L}{dt} = -1.3_4 \times 10^{-23} \text{ rad sec}^{-2}$$

$$= -(28 \pm 5) \text{ arc sec century}^{-2} \tag{4.60A}$$

$$\frac{dR}{dt} = 1.3 \times 10^{-9} \text{ m sec}^{-1} = 4 \text{ cm year}^{-1} \tag{4.61A}$$

The corresponding retardation of the axial rotation of the Earth, assuming conservation of total angular momentum in the Earth-Moon system, is

$$\frac{d\omega}{dt} = -\frac{L}{C} = -6.0 \times 10^{-22}\,rad\;sec^{-2} \qquad (4.62A)$$

$C = 8.0378 \times 10^{37}$ kg m$^2$ being the axial moment of inertia.

Recent astronomical determinations of $d\omega_L/dt$ give values appreciably larger than Equation 4.60A, but these observations are of two kinds which give disparate results themselves. We must note that fluctuations in the Earth's rotation rate due to internal redistribution of angular momentum (Section 3.4) prevent any use of observations based upon the length of the day. Newton (1969) used ancient eclipse data, thereby in effect appealing to a long record of the comparison of orbital angular velocities of the Earth-Moon and Earth-Sun and obtained values of $(d\omega_L/dt)$ for different epochs, $(-41''.6 \pm 4''.3)$ century$^{-2}$ and $(-42''.3 \pm 6''.1)$ century$^{-2}$. Oesterwinter and Cohen (1972) used half a century of instrumented observations of the Moon and planets to obtain $(-38'' \pm 8'')$ century$^{-2}$. Van Flandern (1975) considered the best estimate from all observations to be $(-38'' \pm 4'')$ century$^{-2}$. The reality of the difference between this estimate and that in Equation 4.60A must be doubted. Another disparity arises from comparison with independent measurements of the Moon's motion in terms of time as recorded by atomic clocks, from which Van Flandern (1975) deduced the value $(-65'' \pm 18'')$ century$^{-2}$. He ascribed the difference to a time-dependence of the gravitational constant, $G$. Both the data interpretation and the implications will obviously attract close scrutiny. In the following geophysically oriented discussion the conventional astronomical estimate will be assumed valid, that is,

$$\frac{d\omega_L}{dt} = -38''\,\text{century}^{-2} = 1.85 \times 10^{-23}\,\text{rad sec}^{-2} \qquad (4.60B)$$

with corresponding revisions of Equations 4.61A and 4.62A:

$$\frac{dR}{dt} = 1.8 \times 10^{-9}\,\text{m sec}^{-1} = 5.6\,\text{cm year}^{-1} \qquad (4.61B)$$

$$\frac{d\omega}{dt} = -8.2 \times 10^{-22}\,\text{rad sec}^{-2} \qquad (4.62B)$$

A smaller contribution to the slowing of the Earth's rotation arises from the solar tide, which has an amplitude (proportional to $m/R^3$ for the tide-raising body) 0.45 times the amplitude of the lunar tide. If the tidal dissipation is a linear process then it is proportional to the square of tidal amplitude and the rotational energy lost to the solar tide is about

0.2 times the lunar tidal dissipation. Most of the dissipation is due to marine tides, in which turbulence is probably important, so that the linearity assumption may not be a good one; the dissipation then depends more strongly on tidal amplitude, and the lunar and solar tidal dissipations interact. But we have no direct evidence for this effect and extrapolations of tidal friction to the remote past have assumed linearity. With this assumption, the estimates of lunar tidal friction from either the satellite-determined tidal bulge or from the eclipse data are independent of solar tidal effects, but if there is a significant nonlinear interaction it could account for a difference between the two estimates, as in Equations 4.60A and 4.60B.

It is of particular interest to calculate the rotational energy dissipation. The total energy is a sum of three terms due to axial rotation of the Earth, rotation of Earth and Moon about the common center of gravity, and mutual potential energy:

$$E = \tfrac{1}{2}C\omega^2 + \tfrac{1}{2}R^2\omega_L^2\left(\frac{mM}{M+m}\right) - \frac{GMm}{R} \tag{4.63}$$

which, by Equation 4.41, reduces to

$$E = \tfrac{1}{2}C\omega^2 - \tfrac{1}{2}\frac{GMm}{R} \tag{4.64}$$

Thus

$$\frac{dE}{dt} = C\omega\frac{d\omega}{dt} + \tfrac{1}{2}\frac{GMm}{R^2}\frac{dR}{dt} = -L(\omega - \omega_L) \tag{4.65}$$

Using the tidal estimate of $L$, the dissipation amounts to $3.4 \times 10^{12}$ W; assuming validity of the eclipse data (Eq. 4.60B) the total dissipation is $4.6 \times 10^{12}$ W. The appearance is of a shortfall in the tidal bulge estimate by $1.2 \times 10^{12}$ W (but with a large uncertainty). Precessional dissipation is unlikely to amount to more than about $10^{-5}$ of this.

From data on tidal currents Miller (1966) estimated the total marine dissipation to be only $1.5 \times 10^{12}$ W but Lambeck (1975) finds that satellite orbit data indicate marine tidal dissipation very close to the total tidal friction required by the astronomical data. The estimate for the solid part of the Earth is much smaller (Section 10.4) and the atmospheric tide results in an insignificant torque (Siebert, 1961; Chapman and Lindzen, 1970). The role of the core appears to be negligible (Lambeck, 1975), although Houben et al. (1975) have questioned this.

Extrapolating to the remote past the present rate of lunar recession (about 5.6 cm/year by Equation 4.61B) and the corresponding changes in $\omega$ and $\omega_L$, we see that over geological time the orbit has undergone dramatic evolution. Attempts to treat the problem rigorously are faced not only with mathematical difficulties but with uncertainties in the assumptions about how tidal friction behaved in the past. But a quite simple approach suffices to give the important conclusions.

In one respect, treatment of the long time scale becomes simpler than present-day observations. The fluctuations in rate of rotation caused by core-mantle interactions (Section 3.4) become trivial because a departure $\Delta\omega$ from equilibrium rotation $\omega$ by a few parts in $10^8$ represents a time of only about 100 years in the steady slowing (Eq. 4.62B). We can therefore relate $\omega_L$ and the "observed" $\omega$ by the conservation of angular momentum*:

$$a = C\omega + \frac{Mm}{M+m} R^2\omega_L = \text{constant} = (1+K)C\omega_0 \tag{4.66}$$

where $K = 4.912$ is the present ratio of orbital angular momentum $a_L$ (Eq. 4.57) to the angular momentum of axial rotation, $\omega_0$ being the present rotational angular velocity. The axial angular momentum of the Moon is neglected, as is the friction of the solar tide which slowly transfers angular momentum to the Earth-Moon orbit about the Sun. Solar tides account for 20% of the present frictional loss but proportionately much less in the past when the Moon was closer. Substituting for $R$ by Equation 4.41 in Equation 4.46 we relate $\omega$ and $\omega_L$:

$$\omega_L[(1+K)C\omega_0 - C\omega]^3 = \frac{G^2M^3m^3}{M+m} \tag{4.67}$$

Consider now the condition $\omega_L = \omega$, that is, the Moon revolving about the Earth synchronously with the rotation of the Earth but with total angular momentum equal to the present value. Putting $\omega_L = \omega$ in Equation 4.67 and dividing by $C^3\omega_0^4$ to make the equation conveniently dimensionless, we obtain

$$\frac{\omega}{\omega_0}\left(1+K-\frac{\omega}{\omega_0}\right)^3 = \frac{G^2M^3m^3}{C^3\omega_0^4(M+m)} \tag{4.68}$$

which is, with insertion of numerical values,

$$\frac{\omega}{\omega_0}\left(5.912-\frac{\omega}{\omega_0}\right)^3 = 4.240 \tag{4.69}$$

The two real solutions of (4.69) are $\omega/\omega_0 = 4.96$ and 0.0207, corresponding to rotational periods of 4.84 hours and 48.2 days (i.e., present-time days) and lunar orbital distances of $1.4_6 \times 10^4$ km (2.3 Earth radii) and $5.6_4 \times$

*The conservation of angular momentum in the Earth-Moon system is a much better approximation than has sometimes been supposed. Runcorn (1964) and Weinstein and Keeney (1973) added a term to Equation 4.66 to account for the loss of angular momentum by solar tides, but they supposed the ratio of solar tidal torque to lunar tidal torque to be constant instead of varying (as $R^{-6}$) and therefore greatly overestimated the solar effect.

$10^5$ km (88 Earth radii), compared with the present orbit at 60.3 Earth radii. These two orbital states are asymptotic limits for the system since there is no tidal friction when the rotations are synchronous. The Moon is moving away from the close synchronous orbit toward the distant one.

We may note that a lunar orbit at 2.3 Earth radii is physically impossible because at this distance the Moon would be torn apart gravitationally. Inside a critical distance, the Roche limit, the self-gravitation of a satellite is inadequate to hold it together against the deforming (tidal) forces of its larger parent. For the Moon the Roche limit is at about 3 Earth radii (Appendix B). Thus the Moon could never have been very close to the inner synchronous orbit. This difficulty also besets more rigorous calculations, which include both inclination and ellipticity of the lunar orbit. Several authors have reported such calculations, but with conflicting conclusions, generally favoring capture of the Moon, either into an initially retrograde orbit, as in the original discussion by H. Gerstenkorn, or directly into a prograde orbit. But these calculations have never convinced proponents of the alternative theories of lunar formation in orbit or even fission of the Earth. A collection of reviews of these conflicting ideas has conveniently been published together (Cameron, 1970; O'Keefe, 1970; Singer, 1970b).

Of critical importance to this discussion is the estimation of the rate of evolution of the lunar orbit over geological time. Since the lunar tidal amplitude is proportional to $R^{-3}$ and dissipation varies as the square of amplitude if the process is linear, or more strongly if it is nonlinear, we obtain a lower bound to the rate of orbital evolution in the past by supposing it to be proportional to $R^{-6}$. The dissipation becomes startlingly more rapid as we go back in time and $R$ decreases. Whether the Moon was captured or formed in orbit the event would have been geologically violent if an approach within 20 Earth radii or so occurred. As a simplifying approximation (but not a bad one) we can ignore the variations of $\omega$, $\omega_L$ with time to write the variation of lunar orbital radius with time

$$\frac{dR}{dt} = R_0'\left(\frac{R_0}{R}\right)^6 \tag{4.70}$$

where $R_0$, $R_0'$ are the present values of $R$, $dR/dt$. Integrating from $R = R_1$ at $t = 0$ to the present time ($t = T$) we obtain the result

$$T = \frac{1}{7}\frac{R_0}{R_0'}\left[1 - \left(\frac{R_1}{R_0}\right)^7\right] \tag{4.71}$$

Thus it hardly matters what value of $R_1$, (i.e., initial value $R$) we assume if $R_1 \ll R_0$; the time of orbital evolution is

$$T = \frac{1}{7}\frac{R_0}{R_0'} \approx 10^9 \text{ years} \tag{4.72}$$

More rigorous calculations, using the recent estimates of the present rate (Eqs. 4.60B, 4.61B, 4.62B) come very close to the same result. But the conclusion that the Moon was close to the Roche limit about $10^9$ years ago is securely discounted by geological evidence. Its implausibility in terms of terrestrial evidence (see, for example, Panella, 1972) is strengthened by the observation that the Moon has been geologically dormant for $3 \times 10^9$ years, although the massive tide raised in the Moon by the Earth would have been even more dramatic than the terrestrial tide.

We must conclude that tidal friction was very much less in the remote past than we would deduce on the basis of present-day observations. Counts of daily growth "rings" of ancient corals confirm that, over the past 500 million years, the rate of orbital evolution has varied significantly (Panella, 1972). Sensitivity of corals to tides imparts monthly cycles as well as the daily and annual growth increments, so that the number of days per month and per year can be estimated for past periods. The trend is illustrated by considering a particular example. Accepting, as a well-determined value (Panella, 1972), the number of days per year in the Mississipian period (340 my ago) as 397, and a constant value for the length of the year, the Earth's rotation rate at that time is found to be greater by $\Delta\omega = 6.4 \times 10^{-6}$ rad sec$^{-1}$. Extrapolating Equation 4.62 gives $\Delta\omega = 9.6 \times 10^{-6}$ rad sec$^{-1}$, indicating that, even over the past 340 my, intrinsic friction of the tides has increased. The friction of pre-Cambrian tides must have been very much less effective still.

Since the present-day tidal dissipation is dominated by marine tides, it is this feature that must have changed. The greater part of the present dissipation is attributed to the shallow seas, marginal to the continents. But both the continental crust and the ocean water are products of the progressive differentiation of the mantle over geological time (Section 2.3), so that, relative to the continental blocks, sea level has gradually risen, invading the continents. If sea level were, say, 1 km lower, then, as can be seen from Figure 4.4, the area of shallow sea would be greatly reduced—the relatively steep continental slopes would be the true continental margins. Thus tidal friction is clearly consistent with the geochemical evidence for progressive evolution of the continents, atmosphere, and oceans.

Since orbital calculations alone do not suffice to indicate the primeval state of the Earth-Moon system, other evidence must be invoked. A useful, if rather qualitative approach is to appeal to observations on planetary rotation, to guess the plausible range of the rotation rate with which the Earth was formed. Rotations of asteroids and of the major planets suggest that a six- to eight-hour rotation period is normal (Hartman and Larson, 1967). Combining Equations 4.66 and 4.41 to relate

$R$ to $\omega$, we obtain

$$R = \frac{M+m}{GM^2m^2} C^2[(1+K)\omega_0 - \omega]^2 \tag{4.73}$$

This relationship is plotted as Figure 4.11. Allowance for loss of angular momentum from the Earth-Moon system by solar tides, which give an additional torque $0.2(R/R_0)^6$ times the lunar tidal torque, yields the solid lines in the figure; clearly the difference is not significant. The lunar capture alternative is readily allowed for by a simple adjustment of the same data, since if the Earth captured the Moon as a result of close approach on a parabolic orbit with a closest distance $R_{min}$, it captured angular momentum equal to $\sqrt{2}$ times the orbital angular momentum of a circular orbit of radius $R_{min}$. Then $R_{min}$ is related to the rotational angular velocity of the Earth by

$$R_{min} = \frac{M+m}{2GM^2m^2} C^2[(1+K)\omega_0 - \omega]^2 \tag{4.74}$$

Values of $R_{min}$ are also plotted in Figure 4.11. This figure makes capture in an initially retrograde orbit implausible in terms of the required initial rotation of the Earth. For the favored alternative, accretion of the Moon at a distance of 10 to 23 Earth radii corresponds to an initial terrestrial rotation period of six to eight hours.

The obliquity of the ecliptic, or inclination of the Earth's rotational axis with respect to the orbital plane, may appear as a reason for contemplating capture of the Moon. But a survey of the planets shows that such inclinations are normal. Evidently the final stage in the accretion of the planets was the infall of substantial "planetismals," some of which may have been captured as transient satellites on retrograde orbits, transferring their angular momenta as they plunge in. Singer (1970a) offered this as the explanation for the very slow rotation of Venus. It is presumed that such a late infall tipped the Earth's rotational axis. On the other hand, friction of the solar tide accounts for the slow rotation of Mercury and similarly friction of the tide raised in the Moon by the Earth has completely stopped its rotation relative to the Earth.

Detailed theories of the formation of the Moon as an adjunct to the Earth are constrained by the significant differences in composition of the two bodies (Ringwood, 1970a; Mason, 1971). Most obviously the Moon is deficient in iron, lacking a core. It is easier to envisage its formation from an extended early terrestrial atmosphere or "sediment ring," as strongly advocated by Ringwood (1970a) than either capture of a small planet that was formed independently but at the same distance from the Sun or fission of the Earth, provided that when it formed the Moon was

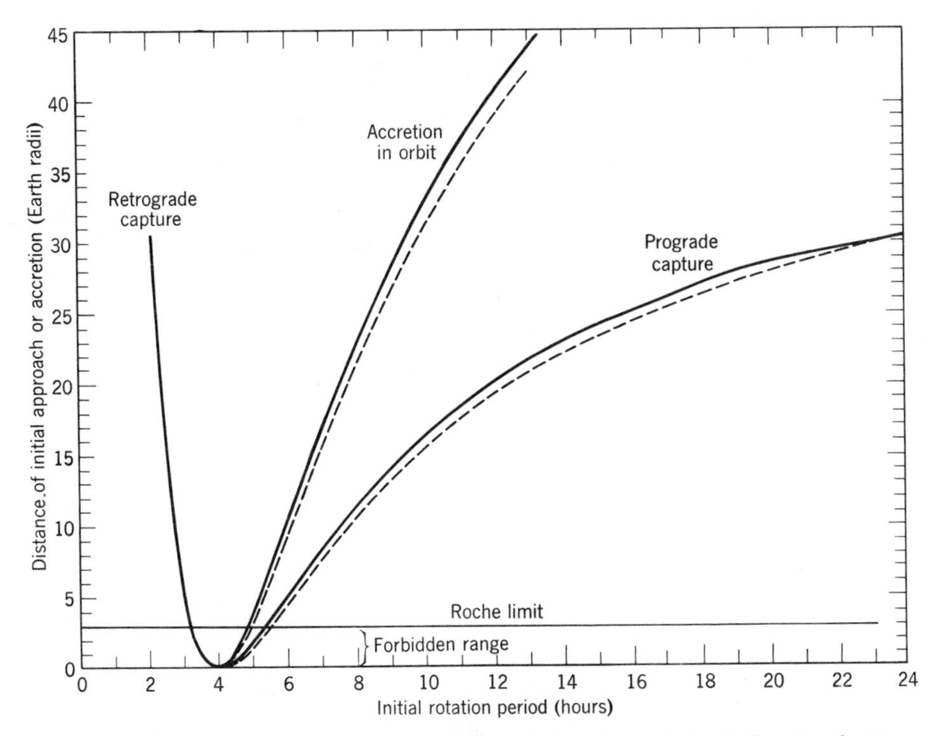


**Figure 4.11.** Dependence on the Earth's rotational period of the Earth-Moon distance for a circular orbit ("accretion in orbit" curve) and the distance of initial approach for capture from a parabolic orbit. Broken lines represent Equation 4.73 or 4.74. Solid lines represent results corrected for solar tidal friction.

in the Earth's equatorial plane. Rather than appeal to planetary interactions to bring the Moon toward the ecliptic plane, it may be supposed that the Moon accreted in an equatorial orbit virtually simultaneously with the formation of the Earth and that late capture of another body subsequently shifted the Earth's rotational axis.*

In the very remote future, the Earth and Moon will approach a state of coherent rotation at the 48-day period, presenting constant faces to one another. However, the solar tide will continue a slow transfer of the rotational angular momentum of the Earth-Moon pair to orbital motion about the Sun so that the Moon will again approach the Earth, eventually coalescing with it.

*Note the comparison with the satellites of Uranus, which orbit in the equatorial plane of the planet, although this is inclined at 98° to the ecliptic plane (i.e., the rotation is just retrograde).

# 5

# Seismicity and the Earthquake Mechanism

"The problem of the origin of the forces which drive the faults is left as an exercise for the reader"

BENIOFF, 1962, p. 131

## 5.1 SEISMICITY OF THE EARTH

By seismicity* we mean the geography of earthquakes, particularly their magnitudes (or energies) and their distribution over the surface of the Earth and within the Earth. The general geographical distribution of earthquakes was established in early compilations of F. de Montessus de Ballore; the word seismicity is associated particularly with the classic work of Gutenberg and Richter (1954).

The worldwide distribution of earthquakes is apparent in Figure 5.1. The most intense activity is around the circum-Pacific belt; according to the tabulations of Gutenberg and Richter (1954), 75.4% of the energy release by shallow earthquakes during the period 1904 to 1952 occurred there. A further 22.9% was released around the trans-Asiatic or Alpide belt, which extends from Indonesia through the Himalayas to the Mediterranean, leaving less than 2% for the rest of the world. Thus seismic energy release is strongly concentrated in what are now recognised as zones of crustal convergence (Section 10.1). The energy release in the intermediate- (70 to 300 km) and deep-focus (>300 km) earth-

*Seismicity ≡ seismic activity.

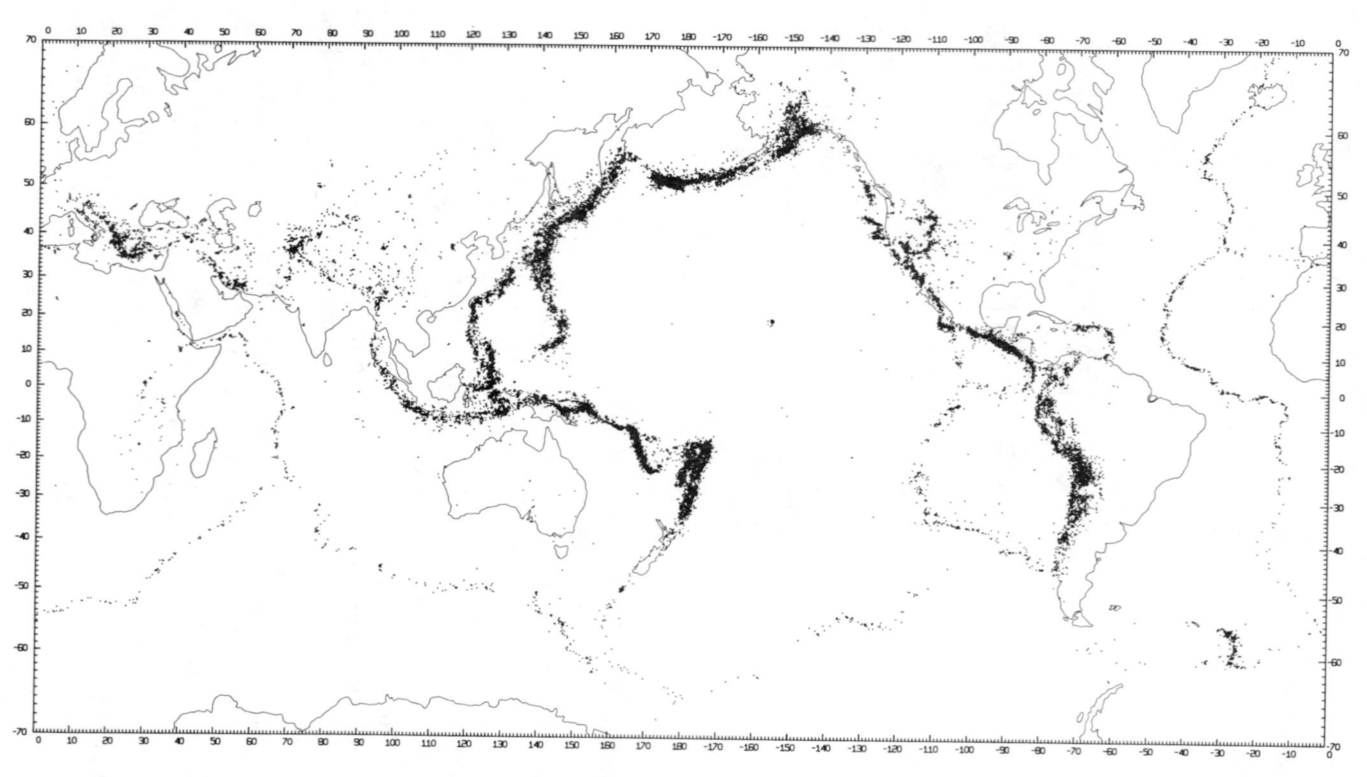

**Figure 5.1.** Epicenters of 29,000 earthquakes 1961–1967, depths 0–700 km, plotted by Barazangi and Dorman (1969) and reproduced, by permission, from their manuscript.

quakes is even more strongly concentrated in the circum-Pacific belt. The pattern of energy release is dominated by a few large earthquakes but the belts of relatively minor seismicity are also important indicators of geological activity. Clearly apparent on Figure 5.1 are the ocean ridges, best known of which is the Mid-Atlantic ridge, which are centers of a pattern of ocean-floor spreading.

The most intense activity occurs along arc-shaped geographic features, which, in the most striking cases, show a common pattern of deep ocean trench, volcanic island arc, and an associated steep gravity anomaly, with earthquake foci distributed about a plane dipping at an angle of about 45°; in cases where the nearest continent is close, the plane dips under the continent. Figure 5.2 is a plot of earthquake foci under the Tonga arc. As numerous authors have noted, the system of ocean ridges and ocean trench-island arc structures is very suggestive of

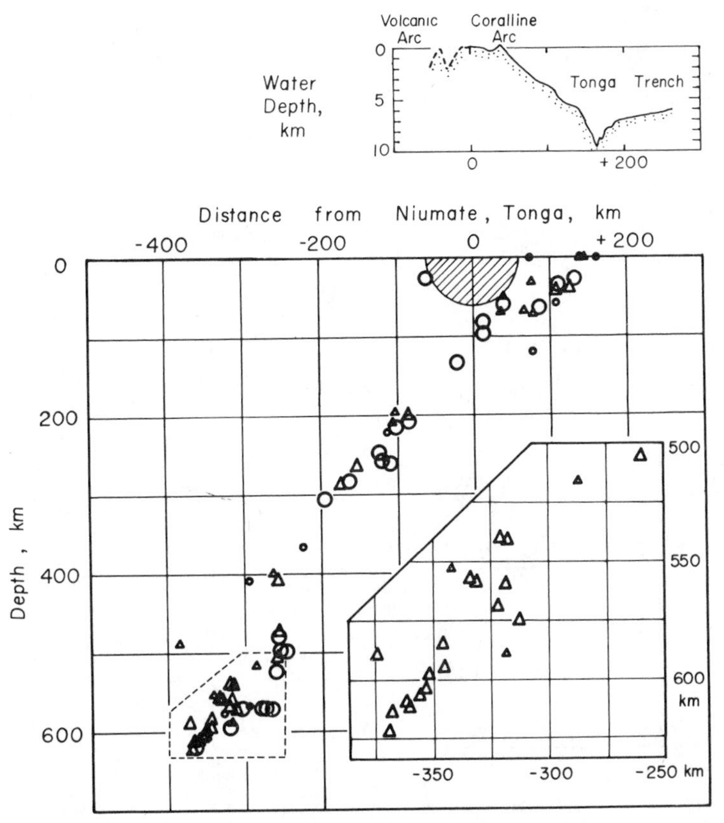


**Figure 5.2.** Dipping plane of earthquake foci under the Tonga Island arc. Reproduced, by permission, from Sykes et al. (1969).

a mechanism of ocean floor spreading and reabsorption by an underlying pattern of mantle convection. These observations have now crystallised into the hypothesis of *plate tectonics*, which is the subject of Section 10.1.

Deep-focus earthquakes are fewer in total number and more limited in geographical distribution than are shallow earthquakes. They are almost confined to the island arc areas of the Pacific region. Relative numbers of deep and shallow earthquakes for the most active island arcs are indicated in Figure 5.3. The deepest-known shock was at 720 km and it is probable that the deeper parts of the Earth are completely aseismic. However, earthquakes at all depths down to 700 km are evidently parts of an overall pattern of movement in the mantle. The distribution is not random—deep and shallow shocks are closely associated geographically

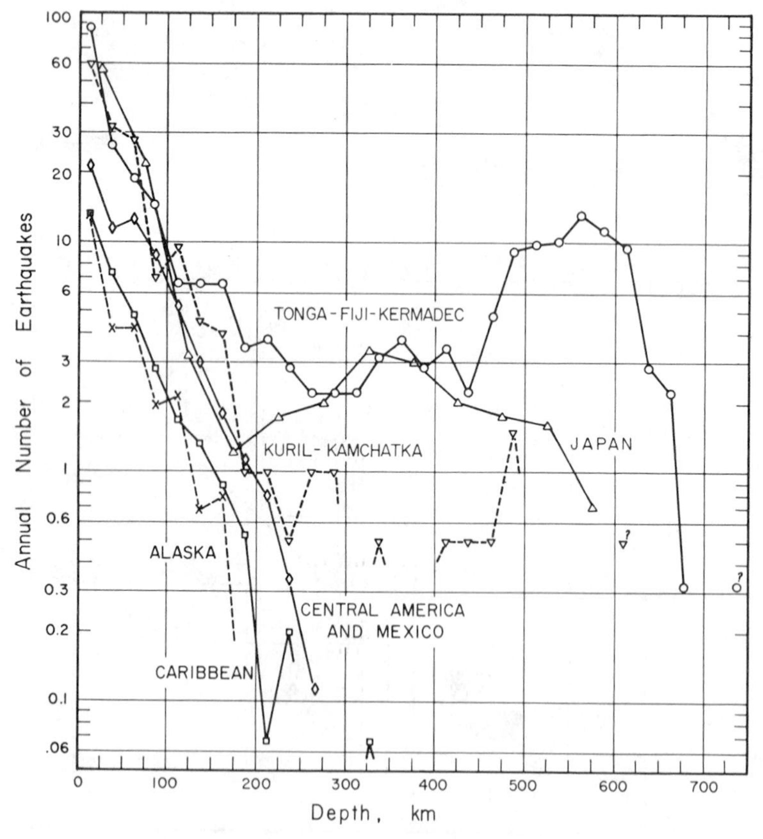


**Figure 5.3.** Variations in earthquake frequency with depth for six regions. Figure reproduced, by permission, from Isacks et al. (1968).

and it can hardly be doubted that the same deep-seated process is responsible.

The *magnitude* of an earthquake is a quantitative measure of its size, determined from the amplitudes of the elastic waves that it generates. The scale of magnitudes now in universal use was first developed by C. F. Richter for local earthquakes in California and subsequently improved and generalized to earthquakes at any distance. It is known as the Richter magnitude scale; the historical development of its present form has been well summarized in Richter's (1958) own book and the whole problem of magnitudes has been reviewed by Båth (1966). As Richter points out, the success of the magnitude classification is due to the logarithmic scale, which gives a fine subdivision while allowing an enormous range of sizes between the largest and the smallest measurable earthquakes to be represented.

Following Båth (1966), magnitude $M$ may be defined by the equation

$$M = \log_{10}\left(\frac{a}{T}\right) + f(\Delta, h) + C \tag{5.1}$$

where $a$ is the amplitude of the ground motion (in microns) for a particular type of wave [strictly surface waves; if body waves (Section 6.1) are used, a different magnitude estimate $m$ is obtained], $T$ is the dominant wave period (in seconds), $\Delta$ is the distance, measured as the angle subtended at the center of the Earth, between the earthquake and the seismometer, and $h$ is the depth of the focus or origin of the earthquake. $f(\Delta, h)$ is a term found from a study of many recordings, which accounts for the diminution of wave amplitude with distance, due principally to geometrical spreading of the wave, as studied by Carpenter (1966), but also partly to anelastic attenuation (as discussed in Chapter 10), and $C$ is a station correction to adjust the observations for local peculiarities of seismometer siting. The function $f(\Delta, h)$ has the effect of reducing all observations to a standard epicentral distance, originally taken to be 100 km, at which the wave amplitudes are directly comparable; the selection of a very small standard amplitude to correspond to magnitude zero then fixes the scale. The largest earthquake recorded in modern times was in Assam in 1952 and had a Richter magnitude of 8.7. The Alaska shock of March 1964 had a magnitude of 8.4. Modern, highly sensitive seismometers can record local shocks down to magnitude $-2$ or $-3$. With reasonable control, that is, a good spread of reporting stations, magnitudes are determined to 0.1, so that the classification is indeed a sensitive one, although uncertainties may amount to $\pm 0.5$. The usual magnitude scale is defined in terms of surface wave amplitudes; body waves are now in more general use, being less dependent on earthquake depth (surface waves are not generated effectively by deep focus events), and a slightly different scale leads to a body wave magnitude $m$, but $M$ and $m$ are directly related, so that this is a problem of detail and not of principle.

Fundamental interest in the magnitude scale arises primarily from the direct relationship between magnitude and the total elastic wave energy of an earthquake. Here the significance of the factor $a/T$ in Equation 5.1 becomes apparent, since this ratio is a measure of the actual ground strain in a seismic wave. We can thus represent the relationship between total energy and magnitude in terms of an obvious empirical equation

$$\log_{10} E = A + BM \tag{5.2}$$

Noting that wave energy per unit volume of rock is proportional to the square of strain, if earthquakes of different magnitudes produced wave trains of similar forms we would have $B = 2$. However, both the spectra and lengths of wave trains are functions of earthquake magnitude and the appropriate values of $A$ and $B$ must be determined by comparing magnitudes with integrated wave energies for earthquakes with a range of magnitudes. Values of the constants have varied considerably since the first estimates but recently have converged to better agreement; with the values preferred by Båth (1966), Equation 5.2 is, for $E$ in Joules ($10^7$ ergs)

$$\log_{10} E = 1.44\, M + 5.24 \tag{5.3}$$

or, with the relationship between body-wave and surface-wave magnitudes (Richter, 1958):

$$M = 1.59m - 3.97 \tag{5.4}$$

we have

$$\log_{10} E = 2.3m - 0.5 \tag{5.5}$$

Equations 5.4 and 5.5 must be construed as approximate relationships expressing the average behavior for shallow shocks. Evernden et al. (1971) plotted $M$ versus $m$ on a log-log scale, on which the best-fitting linear relationship had a gradient slightly greater than unity with equal values at $M = m = 4$, which is in rather poor agreement with Equation 5.4. They also showed, as has been recognized for some years, that explosions give values of $M$ consistently lower by about unity than earthquakes of similar body wave magnitudes $m$.

These relationships are good approximations for the range $4 < M < 7$, but probably become poorer for very small or very large shocks. They certainly suffice as a quantitative basis for comparison of earthquakes and for assessing their significance in the global energy balance. Thus substitution of $M = 8.7$ (for the largest well-recorded earthquake) in

Equation 5.3 gives the prodigious energy of $6 \times 10^{17}$ J, which is about 0.06% of the annual energy dissipation by flow of heat from the entire Earth. We can also estimate the average annual energy release by all earthquakes from the data of Gutenberg and Richter (1954) on numbers of earthquakes as a function of magnitude. They represented the number $\Delta N$ of shocks in magnitude intervals of 0.1 for $4 < M < 8$ by the relationship

$$\log_{10}(\Delta N) = -0.48 + 0.90(8 - M) \tag{5.6}$$

This is more conveniently represented as

$$\log_{10}\left(\frac{1}{10}\frac{dN}{dM}\right) = 6.72 - 0.90\,M \tag{5.7}$$

which is equivalent to

$$\frac{dN}{dM} = 5.25 \times 10^{7} \exp(-2.07\,M) \tag{5.8}$$

where $dN$ is the number of shocks per year in the magnitude range $dM$.* Equation 5.3 is rewritten in a similar form

$$E = 1.74 \times 10^{5} \exp(3.32\,M)\,\mathrm{J} \tag{5.9}$$

which combines with Equation 5.8 to give the total energy in Joules per year, $d\epsilon$, for the $dN$ earthquakes in the magnitude range $dM$:

$$d\epsilon = E\,dN = 9.13 \times 10^{12} \exp(1.25\,M)\,dM \tag{5.10}$$

Equation 5.10 integrates to give the energy in any magnitude interval $M_1$ to $M_2$:

$$\epsilon_{12} = 7.35 \times 10^{12}[\exp(1.25\,M_2) - \exp(1.25\,M_1)] \tag{5.11}$$

It is apparent from Equation 5.11 that the smaller magnitudes contribute negligibly to the total energy release in spite of their much larger numbers. The numerical value is determined by the upper limit, which must be assumed for the distribution in Equation 5.6. If we take the upper magnitude limit to be $M_2 = 8.7$, then the total energy is $4 \times 10^{17}$ J; this may be a slight overestimate since the statistical result (5.6) cannot strictly be applied to the very largest shocks, but it is apparent that there is as much energy in the very occasional magnitude 8.7 earthquake as the annual average for all earthquakes.

The strength of ground movement during an earthquake is measured

*Plots by Chinnery and North (1975) indicate that a linear relationship between log $N$ and $M$ is valid only up to $M \approx 7.5$, but that log $N$ is linear in seismic moment, an alternative measure of earthquake "size" defined by Equation 5.22, up to the largest well-recorded earthquakes.

by strong motion instruments or accelerometers where these are available, but the use of reports on the effects on humans and buildings gives much more complete coverage of populated areas. To allow comparisons to be made on a semiquantitative basis, several scales of local earthquake intensity have been devised. The one most generally used was due originally to G. Mercalli and has subsequently been modified by a number of authors; the 1956 version is given by Richter (1958). It classifies effects on a scale of 12 distinguishable intensities, so that contours of equal intensity, or isoseismals, may be drawn on a map of the epicentral region of an earthquake to indicate its extent and the area of greatest intensity. A further elaboration by S. V. Medvedev, W. Sponheuer, and V. Karnik has been presented as the M.S.K. intensity scale. Since the object of the intensity scale is to make a quantitative assessment in terms of qualitative observations, the scale intensity, $I$, was related approximately to ground acceleration, $a$, by Gutenberg and Richter (1956):

$$\log_{10} a = \frac{I}{3} - \frac{5}{2} \tag{5.12}$$

$I$ being an integer in the range 1 to 12 expressing the local severity of structural damage and $a$ being in m sec$^{-2}$.

The magnitude and intensity scales are independent and assess different aspects of an earthquake; in particular, the Mercalli intensity represents a rather qualitative assessment and each earthquake has a range of intensities, depending on where it is observed. Nevertheless, there is a correlation between magnitude and the maximum intensity $I_{max}$ for shallow shocks, which Karnik (1961) represented by the empirical relationship:

$$M = 0.67 I_{max} + 1.7 \log_{10} h - 1.4 \tag{5.13}$$

where $h$ is the depth of focus in kilometers.

## 5.2 THE SEISMIC FOCAL MECHANISM

Elucidation of the immediate causes and mechanisms of earthquakes is one of the most significant problems in geophysics. The physical processes of elastic strain accumulation or concentration and the triggering mechanism are basic to the still-unsolved problem of earthquake prediction (Section 5.3); the term *earthquake mechanism* conventionally refers simply to fault orientation, the displacement and stress release patterns and the dynamic process of seismic wave generation.

The focus of an earthquake is the point within the Earth from which the elastic waves radiate. The point source is, of course, an idealized concept and in considering the earthquake mechanism we are dealing with a focal volume, but still refer to the focus as the point from which the initial wave radiates. The term epicenter refers to the point on the Earth's surface immediately above the focus. A large, shallow earthquake is normally accompanied by substantial deformation of the ground over hundreds of kilometers and this indicates the volume of rock from which stress is released. Comparison of extensive geodetic observations in California before and after the San Francisco earthquake of 1906 led Reid (1911) to present the elastic rebound theory of earthquakes, the principle of which had been discussed previously by a number of seismologists but without general agreement. Following Reid's documentation of the Californian evidence it became the central idea in virtually all theories of the earthquake mechanism.

The sequence of events envisaged in the elastic rebound theory is shown in Figure 5.4, the progressive buildup of strain energy over several years being represented by the development from (*a*) to (*c*). In Reid's terms, the break occurs when the strain becomes "greater than the strength of the rock can withstand" and elastic strain is replaced by the discontinuous fault displacement in (*d*). It now appears that the triggering mechanism is a subtle and perhaps complicated process that is very inadequately understood and that simple fracture and sliding friction do not satisfactorily describe the fault movement itself. Nevertheless, apart from the possible release of energy along the fault plane by crushing, or perhaps even local melting, of rock, the stored elastic energy in the surrounding rock is converted to elastic wave energy which is radiated outward to all parts of the Earth.

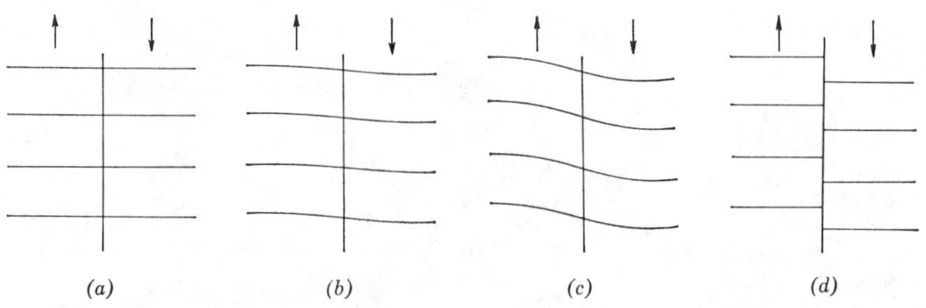


**Figure 5.4.** Sequence of events in the elastic rebound theory of an earthquake. Due to regional shearing movement in the sense shown, elastic strain is slowly built up from the unstrained state (*a*) to state (*c*), at which it is suddenly released across the fault, producing displacements as in (*d*), releasing the stored elastic energy.

A striking feature of the 1906 San Francisco earthquake (and of some others, but by no means all) is that it broke the surface, so that the fault displacement was obvious over a considerable distance (the estimated total length of the break is about 300 km). Relative horizontal movements of the kind occurring in California and illustrated in Figure 5.5, occur on *transcurrent* faults. Both this illustration and the main San Andreas fault on which San Francisco lies are right lateral, or dextral, faults; that is, if one stands on one side of the fault and looks across it, the opposite side is seen to move to the right. Fault movements in the reverse sense are termed left lateral or sinistral. At least as important but generally less obvious visually are relative vertical or dip slip movements, as illustrated in Figure 5.6. Progressive movements in the same area are found to be in the same sense, in accord with Reid's (1911) conclusion that earthquakes are transient expressions of larger scale crustal movements, but the apparent dominance of observed horizontal movements helped to obscure for many years the underlying mechanism, which is now attributed to convection in the (solid) mantle of the Earth (Section 10.1), the essential feature of which is relative vertical motion.

**Figure 5.5.** Transcurrent displacement of rows of orange trees that occurred during an earthquake in Imperial Valley, California, in September 1950. Photograph by David Scherman, *Life Magazine* © Time Inc.

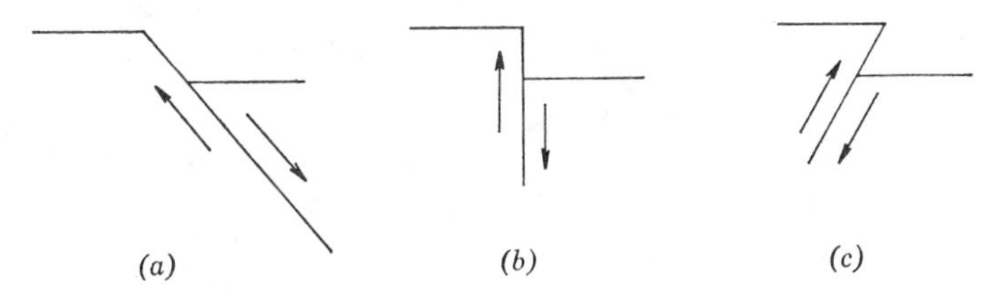


**Figure 5.6.** (*a*) Normal, (*b*) vertical, and (*c*) reverse (or overthrust) faults of the dip-slip type. In general, fault movements may have both dip-slip and transcurrent components and, if neither is dominant, the faulting is termed "oblique."

The San Francisco earthquake is also the starting point of the dislocation theory of earthquakes, a more detailed mathematical development of the elastic rebound theory, in which the stresses and strains associated with fault displacement are explicitly considered. If a known displacement occurs over a known area of fault face, then the change in stress, strain, or displacement at any point in the medium is calculable, but even in such apparently simple geometrical cases the mathematics is tedious and resort to numerical methods is necessary. One convenient starting point is to integrate the effects of an array of point sources (nuclei of strain) of types tabulated by Mindlin and Cheng (1950). Applications to earthquakes by Chinnery (1961), Maruyama (1964), and Press (1965) have considered geometries appropriate for real earthquakes. However, a simple model suffices to demonstrate the principle. In several cases of shallow, transcurrent faulting, including San Francisco, 1906, displacements near to a fault are adequately explained by assuming the fault face to be infinite in one dimension and this greatly simplifies the problem by reducing it to a two-dimensional one. It is worth considering the reasonableness of this approximation in the case of the San Francisco earthquake, which was accompanied by movement over a 300-km length of fault, although significant displacements were confined to a zone only about 10 km wide on either side. Clearly this situation is explicable only if the depth of fault movement was very slight compared with the length and the assumption of an infinite fault is therefore a very good approximation, except near to the ends.

Displacements during shallow, transcurrent faulting are those characteristic of screw dislocations, familiar in solid-state physics (Cottrell, 1953; Friedel, 1964; see also Kittel, 1971) and represented in Figure 10.11*b*. In a single, simple dislocation the displacement is uniform across the fault face, as in Figure 5.7*a*. This model of a fault is mathematically convenient as a starting point in dislocation theory, but is not a physically plausible model because the displacement discon-

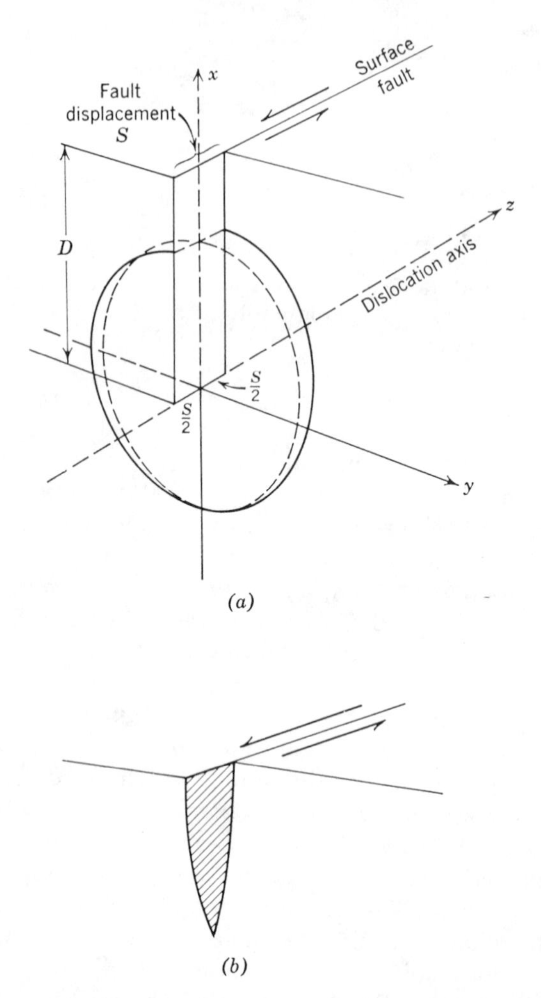


**Figure 5.7.** Section across a mathematical model of a transcurrent fault. In (*a*) the simple case of uniform relative displacement $S$ between opposite faces of the fault is represented. The dashed circle is displaced to the solid, twisted circle. However, the discontinuity in slip at the bottom of the fault is physically unrealistic and demands that the fault movement vary with depth rather as in (*b*).

tinuity at the bottom of the fault (the dislocation axis) is a singularity, implying infinite stresses. The model must be improved by grading the fault displacement to zero at the edges of the fault plane, as in Figure 5.7*b*. This is referred to as a compound dislocation because it may be treated as a sum of many small dislocations with progressively displaced axes.

In a simple screw dislocation in an infinite medium, the shear is uniform about a circle centered at the axis of the dislocation, as in Figure 5.7*a*. If the circle has a radius $r = (x^2 + y^2)^{1/2}$, then the shear strain is $(S/2\pi r)$ for a fault slip $S$. We are here interested in the resolved component of the shear strain across any plane, $y$ = constant, which is $(S/2\pi r).\, x/r$. The effect of a free surface at $x = D$ is to modify the stress field so that there are no stresses across the surface, and this is represented by the addition of the stress field of a hypothetical image dislocation situated at a distance $D$ above the surface (at $x = 2D$). Then the shear strain across any plane ($y$ = constant) parallel to the fault plane ($y = 0$) is, with the following substitution of $x = D$ to obtain the surface strain,

$$\epsilon_D = \frac{S}{2\pi}\left[\frac{x}{x^2+y^2} + \frac{2D-x}{(2D-x)^2+y^2}\right]_{x=D} = \frac{SD}{\pi(D^2+y^2)} \tag{5.14}$$

In the first expression the first term gives strain due to the dislocation itself and the second term accounts for the free surface in terms of the image dislocation. Displacements of surface points are obtained by integrating with respect to $y$, noting the singularity of the model at $y = 0$:

$$\text{Displacement} = \int_{\infty}^{y} \epsilon\, dy = \frac{S}{2}\left(1 - \frac{2}{\pi}\tan^{-1}\frac{y}{D}\right) \tag{5.15}$$

This inverse tangent form for displacement appeared first in a 1956 thesis by M. G. Rochester.

In Figure 5.8, Equation 5.15 is plotted on a scale that gives a reasonable fit to the geodetically observed displacements accompanying the San Francisco earthquake of 1906. A particular significance of the fit is that it provides an estimate of the depth of fault movement from the pattern of surface displacement. More elaborate calculations, using models of the type represented in Figure 5.7*b* (e.g., Shamsi and Stacey, 1969), do not affect the displacement curve sufficiently to demonstrate an improvement in matching the observations. In fact, the simple dislocation model gives satisfactory displacement fields at points well removed from the singularities at the boundaries of a fault plane.

The Alaskan earthquake of 1964 is believed to have been caused by a reverse (thrust) fault movement of the type represented in Figure 5.6*c*, but at a very shallow dip angle (9°) and probably not breaking the surface, although if it did so the break would have been out at sea. Fault plane solutions based on observations of seismic waves (discussed below) are ambiguous and, for the Alaskan shock, the preferred shallow dipping plane cannot be distinguished from one dipping steeply, perpendicular to it. The compelling evidence is obtained from surface displacements (Plafker, 1965; Small and Parkin, 1967). Dislocation models of both

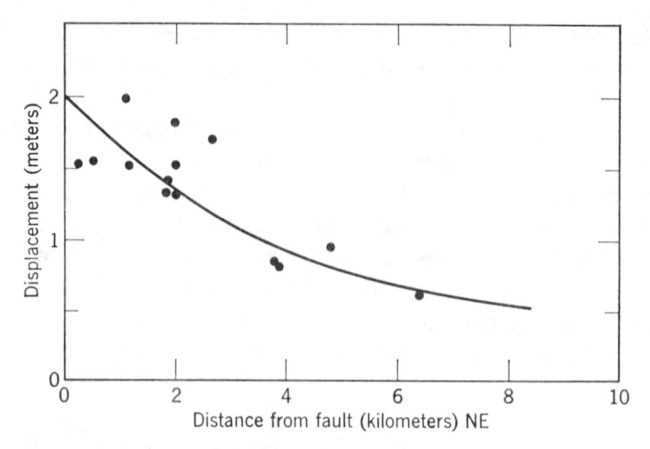


**Figure 5.8.** Displacement as a function of distance from a transcurrent fault of depth 3.5 km and slip 4 m, according to Equation 5.15, with data from the comparison of geodetic surveys before and after the 1906 Californian earthquake (NE side of fault only). The displacements were measured with respect to distant points that were unaffected by the earthquake and the curve represents the deformation of a straight line drawn normal to the fault immediately before the earthquake. Since the earthquake must be presumed to have released, not produced, strain energy, the measured strains are the inverse of the strains released by the shock.

types, that is, with steeply dipping fault planes (Press and Jackson, 1965) and nearly horizontal fault planes (Savage and Hastie, 1966; Stauder and Bollinger, 1966), have been fitted to the vertical displacement data in Figure 5.9, but only the latter are compatible with the horizontal movements in Figure 5.10. Surface movements of a dislocation model of this earthquake are shown in Figure 5.11.

Reliable surface indications of the direction of fault movement in a particular earthquake are not generally available, but the direction can often be deduced from studies of the seismic waves which arrive at observatories distributed all around the focus. This (fault plane solution) method is based on a development by P. Byerly and has been reviewed by Stauder (1962). The essential idea is represented in Figure 5.12, in which the directions and relative magnitudes of *first motions* of the seismograph traces are represented vectorially as a function of azimuth. In this simple picture the fault movement is assumed to be synchronous over the fault face, which is equivalent to assuming a point source, and the first wave to arrive is either a compression or a rarefaction according to the orientation of the observing station with respect to the fault movement; the division of first motions into quadrants gives the *quad-*

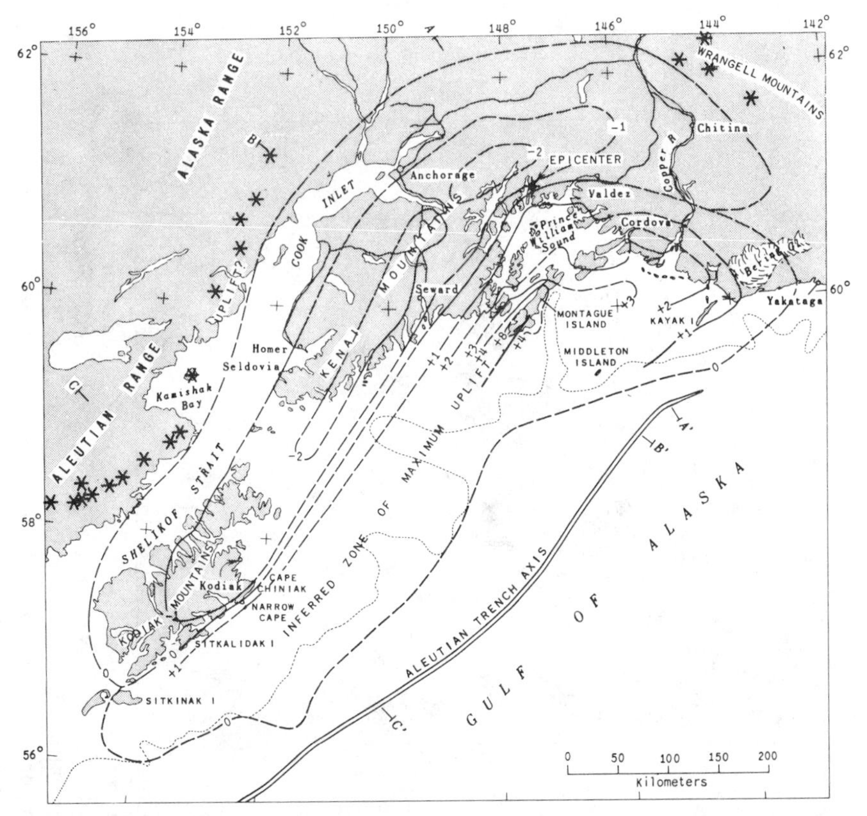


**Figure 5.9.** Contours of changes in surface elevation that occurred as a result of the Alaskan earthquake of March 1964. Reproduced, by permission from Plafker (1966). The contours are marked in meters and shown by broken lines where they are inferred. The dotted line, approximating the edge of the continental shelf, represents 200 m water depth and volcanoes are indicated by asterisks. (Copyright 1966 by the American Association for the Advancement of Science.) Somewhat larger vertical displacements of the sea floor were reported by Malloy (1964).

*rantal pattern*, which is always sought. The quadrants are separated by two nodal planes, the fault plane and an *auxiliary plane* normal to it. The method has an essential ambiguity in that it does not distinguish between the fault plane and the auxiliary plane, but frequently secondary effects, such as aximuthal variations in surface waves, or other arguments (as in the Alaskan case considered above) make only one of the alternatives plausible. In general, of course, the fault plane and the direction of shear may have any orientations and the fault plane problem

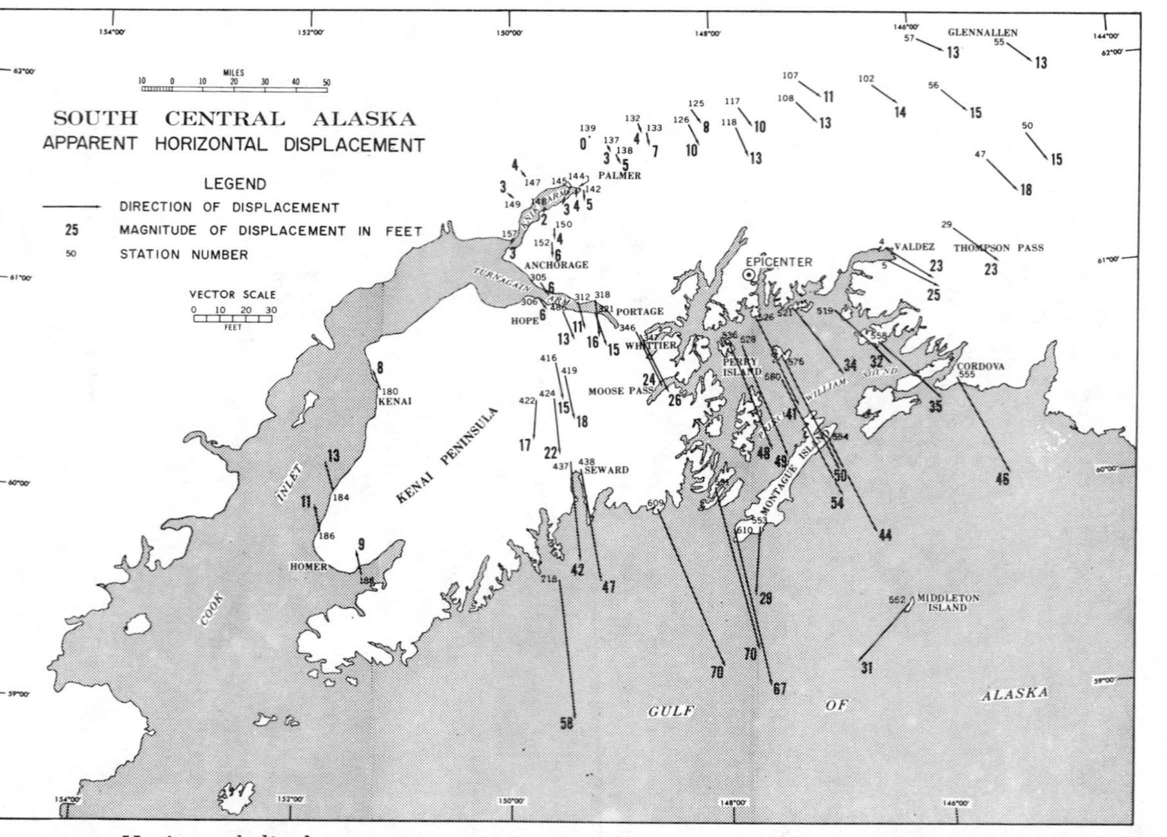


**Figure 5.10.** Horizontal displacements accompanying the great Alaskan earthquake of March 1964. Figure reproduced from Small and Parkin (1967), by permission of the Coast and Geodetic Survey, Environmental Science Services Administration, U.S. Department of Commerce. These data are of interest not only because they show the importance of horizontal motion in this earthquake, but because they are the largest reported geodetic displacements associated with any earthquake. Horizontal movements are difficult to measure accurately but the broad features of the gross movements shown here are not in doubt. It is concluded that the relative movement between fault faces amounted to about 40 m

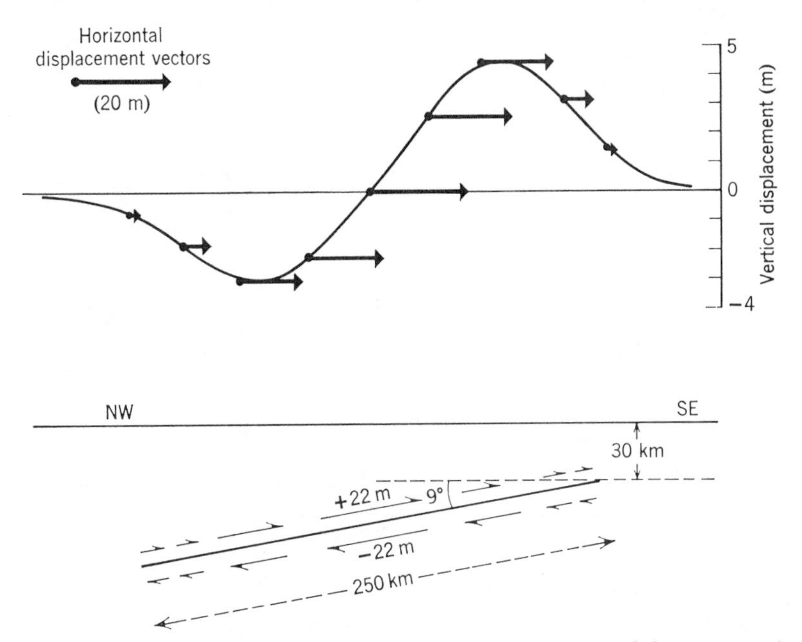


**Figure 5.11.** Surface displacements due to a dislocation model representing the Alaskan earthquake and fitted to the data in Figures 5.9 and 5.10. The fault plane is here assumed to be so long in the direction normal to the figure that two-dimensional equations of edge dislocations (see Figure 10.10*a*) apply. In general a dislocation has both edge and screw components and can only be approximated by one of these when the fault plane is so long that effects at its ends are neglected.

is a three-dimensional one, with observations made on the spherical surface of the Earth. The detailed deductions are complicated by the refraction of seismic waves (see Sections 6.1 and 6.2) and consequent refraction of the nodal planes, but this is a complication of detail and raises no difficulty of principle.

The point source concept implicit in the simple first motion observations used in fault plane solutions is valid if long period waves are used, such that their wavelengths are substantially longer than the largest dimension of a fault and thus record the major movements, unaffected by local diffraction or structural complexities of seismic sources. Records from the long-period instruments of the worldwide network stations installed by the U.S. Coast and Geodetic Survey are particularly well suited to this work and give an impressive consistency to fault plane interpretations (Sykes, 1967).

The concept of fracture followed by implied dry frictional sliding of

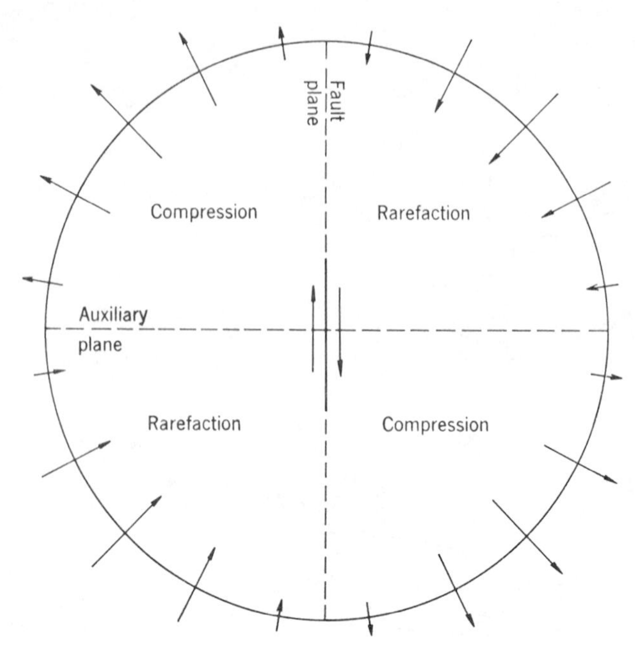


**Figure 5.12.** Quadrantal pattern of first motions in seismic waves radiated from an idealized (point source) fault movement.

rock faces is implausible, even for shallow shocks. The coefficient of friction between any rock faces is of order unity and the observed shear stress release in shallow shocks, such as San Francisco, 1906 (about 100 bars or $10^7$ Pa) corresponds to possible frictional sliding not deeper than about 1 km. However this situation is significantly changed by the introduction of fluid pore pressure, as was recognized by Hubbert and Rubey (1959) and dramatically demonstrated at Denver, Colorado, by a sequence of earthquakes triggered by waste fluids that were pumped down a deep well (Healy et al., 1968). The role of pore water in controlling seismic stress release is now a subject of intensive study in connection with earthquake prediction (Section 5.3). The general processes are described by the term *dilatancy* (Frank, 1965; Nur, 1972). The essential feature of the dilatancy hypothesis is that many (perhaps all) shallow earthquakes are preceded by progressive opening of cracks and pores in critically strained rock, as a consequence of which the rock *dilates*. But so long as the pores are dry, frictional contact between grains or blocks suffices to prevent the occurrence of an earthquake; the dilation is accompanied by *dilatancy hardening*. The earthquake only occurs after the pores have been filled with water, which reduces the friction across a fault and allows it to move. Although some observations appear to contradict this hypothesis it does explain well many others and provides a valuable focus for current ideas on earthquake prediction.

Major earthquakes are followed by sequences of aftershocks which usually die away in a characteristic manner. Nur and Booker (1972) showed that this behavior is also explicable in terms of the redistribution of pore fluid resulting from a main shock. They argued that although a main shock releases stress overall, compressions and dilations of the surrounding region leave some volumes dilatancy hardened and holding substantial residual stress. As these are weakened by diffusion of water, the aftershocks occur. The initial variation of aftershock frequency with time as $t^{1/2}$ appears to be in accord with a diffusion mechanism.*

The occurrence of deep focus earthquakes remains a problem to which dilatancy effects appear to be irrelevant. Various hypotheses of unstable (self-accelerating) *creep* (i.e., progressive inelastic deformation) have been offered to explain them. One possibility is thermally unstable creep, which occurs at a rate that is very strongly temperature dependent, especially near to the melting point of a material, so that a zone of initial relative weakness is selectively further weakened by the heat release of deformation, which may under appropriate conditions lead to catastrophic failure (Stacey 1963; Griggs and Baker, 1968). This constitutes a plausible (but insecure) explanation of the phenomenon of "stick-slip," a well-observed laboratory effect in which sliding rock faces under confining pressure exhibit earthquakelike stress drops (Byerlee and Brace, 1968; Byerlee, 1970). It is also possible that the concept of a single fault is too simple for deep focus earthquakes. Static balance of the stresses before and after an earthquake demands that the stress pattern be represented not by a single couple, which has a moment, but by a double couple, that is two couples with equal and opposite moments (Fig. 5.13). In this connection it is of some interest that deep focus earthquakes commonly occur in pairs with similar foci and a time separation of a day or less, presumably on fault planes that are quite differently oriented.

Fault movement during an earthquake is not synchronous over the active fault face but is initiated at a particular point and spreads at a speed that is generally slightly less than the shear wave velocity in the medium (Press et al., 1961) and may not be constant, so that the seismic

*The time $t$ taken for a diffusing fluid or physical property (such as heat) to reach a specified concentration at distance $x$ from its source is proportional to $x^2$; conversely the distance reached by a specified concentration varies as $\sqrt{t}$.

**Figure 5.13.** (*a*) Single couple, with moment, and (*b*) double couple without moment. Since the single couple has a moment it implies rotational acceleration; seismic stresses must be represented by double couples (or more complicated stress patterns).

radiation may be both complicated and orientation dependent. At least some very large earthquakes may be better regarded as rapid sequences of smaller, related but distinct events (Sweetser and Cohen, 1974). Brune (1970) argued that the azimuthally averaged spectrum must be the same as that of an equivalent earthquake in which the fault displacement was synchronous. This allows us to use observed (preferably averaged) spectra to estimate fault dimensions, since in a simple-minded way we expect such an earthquake to generate a shear wave pulse having a characteristic half-wavelength, $\lambda/2$, comparable to the linear dimension of the fault (diameter $d$ if it is assumed to be circular). The corresponding characteristic frequency of a shear wave (velocity $V_S$) having a wavelength of order $2d$ is thus

$$f_c \approx 0.5\ V_S/d \tag{5.16}$$

Brune's (1970) analysis gave a far-field amplitude spectrum of the form

$$A \propto \frac{1}{f^2 + f_c^2} \tag{5.17}$$

where

$$f_c = 0.7\ V_S/d \tag{5.18}$$

This spectrum is "white," that is, has amplitude independent of frequency, for low frequencies, $f \ll f_c$, but falls off as $1/f^2$ for high frequencies, $f \gg f_c$. $f_c$ is referred to as the *corner frequency*. A "white" low frequency spectrum is a characteristic of a single pulse and reflects the fundamental fact that the Fourier transform or spectrum of a delta function is white. Similarly the spectrum of any single pulse has power per unit frequency interval independent of frequency for frequencies that are low compared with the reciprocal of the pulse duration. Randall (1973) notes that Brune's spectral theory is more general than the model on which it was originally based.

Seismic spectra must also depend on rupture velocity (Archambeau, 1968), the probable variability of which ensures that any theory is no more than an approximation to observations. Linde and Sacks (1972) reported measurements of both compressional and shear wave spectra from the same earthquakes that gave "dominant" frequencies higher for $P$ waves by the factor expected from their greater speed, as is seen by substituting $V_P$ for $V_S$ in Equation 5.16 or 5.18. This is in accord with the source dimension control on the radiated spectrum, but the appearance of spectral peaks, instead of corner frequencies below which the spectra were white, is incompatible with single displacement pulses. If they are real and general phenomena, they may indicate that the time signatures of fault movements (or at least the ones which Linde and Sacks observed, for South American deep focus events) are not well approximated by the single "square-wave" pulse assumed by Brune (1970) but are actually oscillatory and "overshoot" the unstressed state, perhaps more than once. This interpretation compels the inference that stress release is complete. It also suggests that the simple-minded interpretation represented by Equation 5.16 may be nearer to correct than Equation 5.18.

Having an estimate of earthquake energy $E$ (from magnitude by Equations 5.3 or 5.5 or, better, by an integration of total wave energy) and focal volume $V \approx (\pi/6)\, d^3$ from the spectrum by Equation 5.16, we can estimate the effective stress release $\sigma$, since

$$E \approx \tfrac{1}{2}\frac{\sigma^2}{\mu} V \approx \frac{\pi}{12}\frac{\sigma^2}{\mu} d^3 \tag{5.19}$$

where $\mu$ is the rigidity of the medium. With Equation 5.16 this gives

$$Ef_c^3 \approx \frac{\pi}{96}\frac{V_S^3}{\mu}\sigma^2 \tag{5.20}$$

Data by Linde and Sacks (1972) indicate that earthquake energy increases rather more strongly with "dominant" wave period ($f^{-1}$) than Equation 5.20 would suggest, assuming constant $\sigma$ (which appears to imply that a greater proportion of the strain energy is dissipated as heat across the fault faces of smaller earthquakes, as considered further below), but that for magnitudes in the range 5 to 6, Equation 5.20 is a good approximation. With the numerical value from their $S$-wave data

$$Ef_c^3 = 6.6 \times 10^8 \,\mathrm{J\,sec^{-3}} \tag{5.21}$$

For events at 550 to 600 km depth, to which these data refer, $\mu = 1.4 \times 10^{11}$ Pa, $V_S = 6.6 \times 10^3$ m sec$^{-1}$, giving $\sigma = 10^7$ Pa (100 bar) which coincides with independent estimates of stress release in shallow shocks (e.g., Brune, 1970).

The concept of *seismic moment* is important in connection with the long period spectra of earthquakes (Aki, 1966, 1972) and with the accumulated seismic displacement in an extended fault zone (Brune,

1968). The general expression for seismic moment due to slip $S$ over a fault area $A$ in a medium of rigidity $\mu$ (with both $S$ and $\mu$ varying over the fault) is

$$m_0 = \int_A \mu S \, dA \tag{5.22}$$

Assuming constant $\mu$, the average slip $\langle S \rangle$ is defined to give

$$m_0 = \mu A \langle S \rangle \tag{5.23}$$

This is equivalent to the value of the moment of force of each couple of the double-couple stress field, since if we consider a fault of dimension $l \times b$ $(l > b)$ so that $A = lb$, then the elastic strain release, $\epsilon$, is $S/b$ and the stress release is

$$\sigma = \mu \langle S \rangle / b = m_0 / Ab \tag{5.24}$$

or

$$m_0 = A\sigma b \tag{5.25}$$

The shearing force across the fault plane is $A\sigma$ (stress $\times$ area) and the lever arm of this force is $b$, that is, the moment of the stress field may be represented by forces at $b/2$ on either side of the fault.

A comparison of seismic moment $m_0$ and strain energy $E_S$ gives the stress release $\sigma$ since energy is force $\times$ distance moved, that is,

$$E_S = (\sigma A)\langle S \rangle \tag{5.26}$$

which combines with Equation 5.23 to give

$$\sigma = \mu E_S / m_0 \tag{5.27}$$

The significance of these relationships is that seismic moments can be estimated from the long period spectra of earthquakes, that is, the teleseismic displacement amplitudes at periods long compared with the propagation time of a wave (or of fault rupture) across the dimension of the fault plane. This spectral amplitude is directly proportional to seismic moment (independently of frequency). Thus earthquake energies, as deduced from magnitudes by Equations 5.3 or 5.5, and moments from long period spectra give values of stress release independently of corner frequency estimates (Eq. 5.20) or geodetically observed displacement fields. Values obtained are generally about $10^7$ Pa (100 bar) (Aki, 1966; Brune, 1970) at least for shocks in the magnitude range $5 \leqslant M \leqslant 6$. However the method is only very approximate, as is apparent from Aki's (1972) consideration of the moment-magnitude relationship. For magnitude $M < 6$ Aki's data give the result

$$\log_{10} m_0 \approx M + 11.6 \tag{5.28}$$

($m_0$ in newton-meters) but for larger magnitudes $m_0$ increases more rapidly with $M$. A plot indicating a slightly different numerical relationship is given by Chinnery and North (1975). Combined with Equation 5.3, (5.28) gives

$$E \propto m_0^{1.44} \quad (5.29)$$

which is consistent with (5.27) only if stress release increases rather implausibly with magnitude. It is more reasonable to infer that Equation 5.3 gives the radiated wave energy $E_R$ but Equation 5.27, refers to the total energy $E_S$ including that dissipated across the fault face. The locally dissipated energy is proportionately more important for small shocks, becoming insignificant for $M > 6$. Thus for a rough but representative calculation of stress release we should take $M = 6$, which by Eq. 5.28, gives $m_0 = 4 \times 10^{17}$ N-m and, by Equation 5.3, $E = 7.5 \times 10^{13}$ J, so that with $\mu \approx 5 \times 10^{10}$ Pa (for very shallow shocks) we obtain $\sigma \approx 10^7$ Pa (100 bars).

These are crude, empirical approximations, useful for order-of-magnitude calculations but lacking the precision necessary for a closer examination of details of fault behavior, such as the actual energy dissipation in a fault. An upper bound to the dissipation along the San Andreas fault is imposed by the absence of a significant local heat flow anomaly (Brune et al., 1969), but this suffices only to demonstrate that frictional losses do not greatly exceed that represented by a fault moving at 5 cm/year under a stress of about 100 bars; it does not preclude the appearance in the fault itself of a large fraction of the energy calculated above. If the radiated energy, $E_R$, is given in terms of magnitude by Equation 5.3 and, from Equation 5.27, the total energy, that is the strain energy released $E_S$, is

$$E_S = m_0 \epsilon \quad (5.29)$$

where $\epsilon = 2 \times 10^{-4}$ is the elastic strain for values of $\sigma$ and $\mu$ used above, then Equations 5.28 and 5.29 relate $E_S$ to magnitude:

$$\log_{10} E_S = M + 7.9 \quad (5.30)$$

and hence to radiated energy:

$$\log_{10} E_R = 1.44 \log_{10} E_S - 6.14 \quad (5.31)$$

This is a simple empirical relationship, obtained from the seismic moment estimates of Aki (1972) and is not based on any theoretical model of fault properties. It applies only up to $M \approx 6$ ($E_S$, $E_R \approx 8 \times 10^{13}$ J), above which $E_R$ and $E_S$ are not distinguishable. The relationship between them is shown in Figure 5.14.

The precise timing of an earthquake is a consequence of still

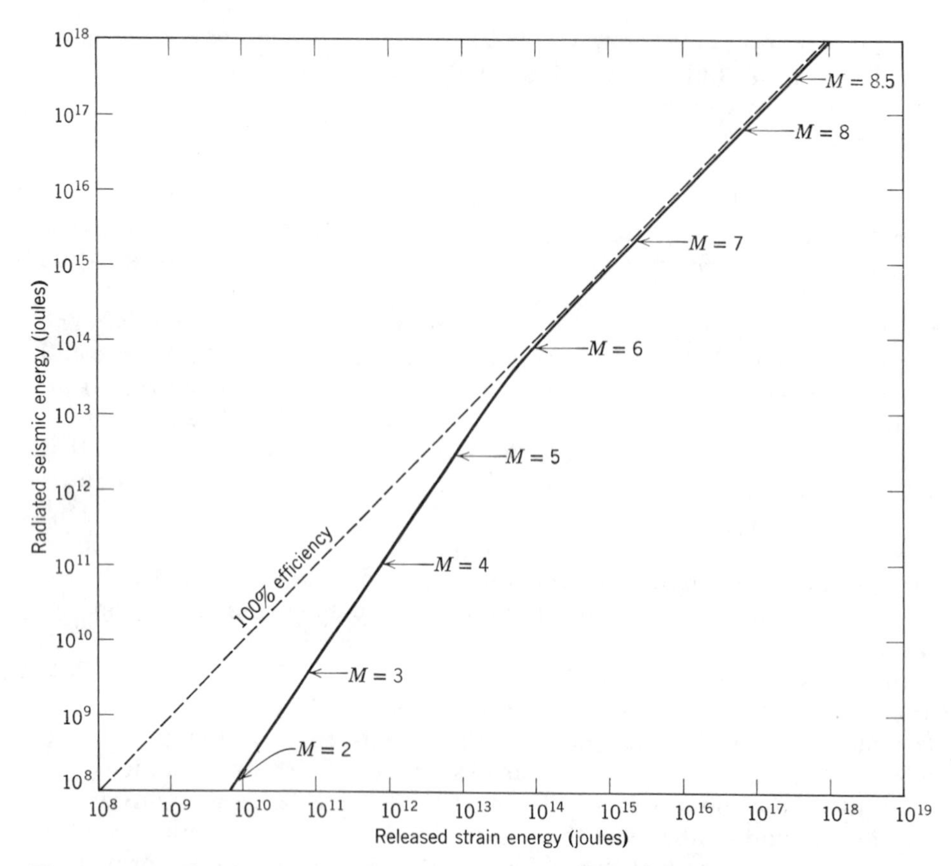


**Figure 5.14.** Variation of radiated seismic energy with total strain energy release obtained from Aki's (1972) magnitude-seismic moment relationship. The difference represents energy lost in the fault zone.

ill-understood mechanical and (at least for many shallow shocks) hydraulic processes following a self-determined course that does not appear to be influenced readily by external effects. The possibility of controlling earthquakes by introducing or withdrawing fluids to or from fault zones has been canvassed but its general effectiveness is problematical at best. Statistical studies (e.g., Knopoff, 1964b; Simpson, 1967) have failed to find evidence of triggering by tidal stresses in spite of the high rate of change of tidal stress relative to the *average* rate of stress buildup between earthquakes. Writing the value $\sigma_T$ of the shear component of tidal stress as an oscillatory function of time $t$

$$\sigma_T = \sigma_{T_0} \sin 2\pi \frac{t}{\tau} \tag{5.32}$$

where the amplitude $\sigma_{T_0} = 2.5 \times 10^3$ Pa corresponds to a tidal strain of $5 \times 10^{-8}$ in material with crustal rigidity $5 \times 10^{10}$ Pa and $\tau = 4.5 \times 10^4$ sec is the semidiurnal lunar period, the maximum rate of change is

$$\left(\frac{d\sigma_T}{dt}\right)_{\max} = \frac{2\pi}{\tau}\sigma_{T_0} = 0.3 \text{ Pa sec}^{-1} \tag{5.33}$$

This is substantially greater than the average rate of stress buildup between earthquakes, for which we may take $10^{-2}$ Pa sec$^{-1}$ ($10^7$ Pa in 30 years) as representative. The lack of correlation of earthquake occurrences with tidal phase shows that this apparently significant tidal stress modulation is not relevant to the triggering mechanism. However, precisely the opposite conclusion is drawn from the data from lunar seismometers (Latham et al., 1971; Lammlein et al., 1974). Moonquakes occur preferentially at the perigee of the lunar orbit and there is no serious alternative to a tidal explanation. The tidal deformation of the Moon by the Earth is, of course, much larger than the lunar tidal strain of the Earth (by a factor of about 130, the mass ratio squared times radius ratio cubed—see, for example, Equation 4.49). But the variation of this tidal deformation and the motion of the body of the Moon with respect to it arise only from the radial motion of the Moon on its slightly elliptical orbit and its libration or oscillatory rotation relative to the Earth-Moon axis by virtue of the variability of its orbital angular velocity with respect to the fixed axial angular velocity. Thus tidal stresses in the Moon are quite small by comparison with the seismic stresses in the Earth. We may suppose that the tidal stresses are merely triggering moonquakes and are not the principal cause.

## 5.3 THE EARTHQUAKE PREDICTION PROBLEM

Our present inability to give specific predictions of earthquakes is a measure of the basic lack of understanding of them. But after a long, discouraging history the prospect of reliable and eventually routine predictions now looks brighter. An important turning point was a Japanese report prepared in 1961 and published in an English edition the following year (Tsuboi et al., 1962). At that time even the intrinsic predictability of earthquakes was doubted and geophysicists were reluctant to risk their reputations in a subject tainted with a crackpot fringe of science. The report insisted that inability to predict earthquakes was due entirely to scientific and technological inadequacy and not to an intrinsic unpredictability and it proposed wide-ranging instrumental developments. More recently attention has focussed on precursory anomalies in the travel times of waves from microforeshocks, first observed in the U.S.S.R. (Semenov, 1969; Nersesov et al., 1969).

We can hardly doubt that the energy released during an earthquake is stored immediately before the shock as elastic strain energy in the focal region. Most approaches to prediction have been based on observations of earth strain with the expectation, for which there was some evidence, that linear strain or tilt underwent characteristic changes before an earthquake. The magnitudes of the anticipated changes, stresses of order $10^7$ Pa (100 bar) or elastic strains of $2 \times 10^{-4}$, present no instrumental difficulty of observation. But in practice considerable difficulty arises in distinguishing seismically significant effects from irrelevant ones and in obtaining adequate instrumental coverage.

Let us imagine that we are required to design an earthquake prediction system for a particular area, based on the deployment of an array of instruments of a certain type, with telementry links to a nerve center, and that the system must be adequate to predict with certainty any shallow earthquake in the area having a Richter magnitude of 4 or greater. This corresponds to a radiated wave energy of $10^{11}$ J, or perhaps a total strain energy release of about $8 \times 10^{11}$ J if Figure 5.14 is approximately correct. By Equation 5.19 the stressed volume is

$$V \approx 2 E_S \mu / \sigma^2 = 8 \times 10^8 \text{ m}^3 \approx (1 \text{ km})^3 \tag{5.34}$$

so that, while recognizing that the strains extend beyond the nominal focal volume itself, but allowing for the necessity for instrumental redundancy to be sure that an observed effect is real, the instrumental spacing must be about 1 km. Even so, a magnitude 4 shock at a depth of a few kilometers would be missed.

A further difficulty with strain measurements is that strain is an ambiguous quantity. It may be elastic, and therefore seismically dangerous, or it may be inelastic and harmless. Aseismic movement or creep is well known and has been monitored on sections of the San Andreas fault, California, for a number of years (Steinbrugge et al., 1960; Tocher, 1960; Radbruch et al., 1966). Ideally, what is needed to demonstrate that an observed strain is elastic is a measurement of the corresponding stress, which is much more difficult. Some indirect methods of inferring stress changes are under examination, but only one is really independent of dilatancy effects (which have become a major study of their own) and this is the geomagnetic method, based on the piezomagnetic effect or stress dependence of magnetization of rock. The strongly magnetic minerals, such as magnetite, are magnetostrictive, that is, they change dimensions by a small factor when magnetized (a few parts in $10^5$ for magnetization to saturation). The thermodynamic converse of this is the piezomagnetic effect, by which magnetic susceptibilities and remanent magnetizations of the important magnetic minerals both change by 1 to 3

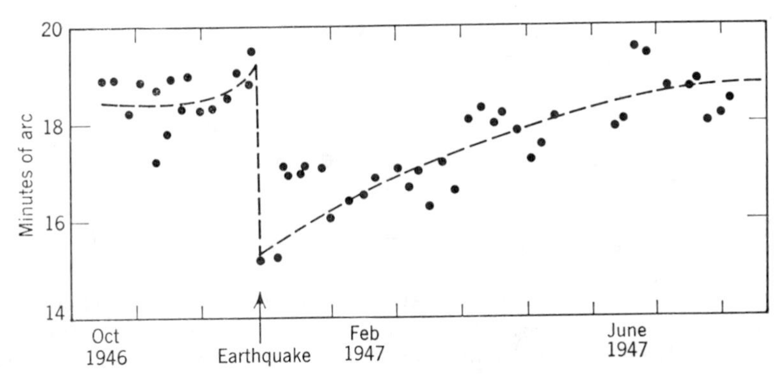


**Figure 5.15.** Differences between values of magnetic declination at two Japanese observatories 80 km and 460 km from an earthquake. Redrawn from Kato and Utashiro (1949). Subsequent work has caused the reality of this apparent piezomagnetic effect to be doubted (it could also arise from slight movement of the landmarks used for magnetic declination sightings), but the effect is now being sought with improved instruments in closer networks.

parts in $10^4$ per bar of directed stress or 1 to 3 parts in $10^9$ per Pascal (Stacey and Johnston, 1972) and therefore by 1 to 3% under representative seismic stresses. This is sufficient to produce a measurable time-dependent magnetic anomaly (Stacey, 1964) but convincing observations of the seismomagnetic effect have been hard to obtain. Nagata (1969) has reviewed the evidence. Data from an early report, which has encouraged the expectation of finding significant effects, is reproduced in Figure 5.15. More recently convincing magnetic changes were found to precede and accompany eruptions of New Zealand volcanoes (Johnston and Stacey, 1969a,b) and the piezomagnetic effect of ground loading by a man-made lake has also been observed (Davis and Stacey, 1972).

In spite of the intrinsic difficulties with strain observations and allowing for a greater readiness to publish encouraging data than discouraging or inconclusive results, there are sufficient reports of anomalous preearthquake strains and tilts to demonstrate the relevance of measurements of this kind (Figs. 5.16, 5.17, 5.18). Geodetic measurements of movements occurring over baselines up to 20 km have also given some encouraging results (Hofmann, 1968) but none of these experiments has come near to a system for reliable prediction.

A systematic study of a wide range of geophysical effects led Russian geophysicists to find characteristic precursive variations of arrival times of waves from microforeshocks originating in the focal regions of subsequent larger earthquakes (Nersesov et al., 1969; Semenov, 1969). This has been followed by both verifications (Aggarwal

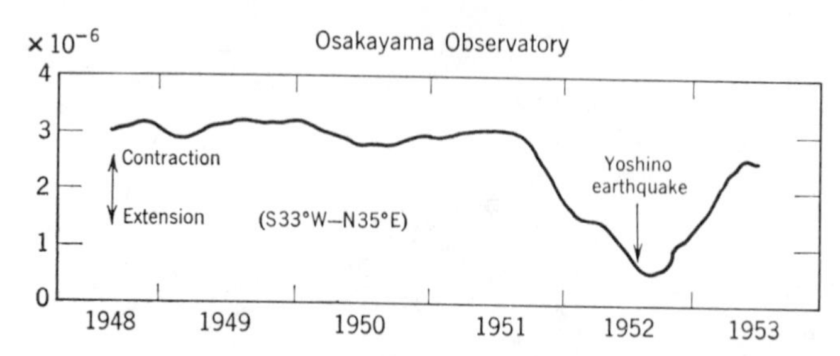


**Figure 5.16.** Variation in ground strain at an observatory 90 km from the epicenter of the Yoshino earthquake in 1952. From Sassa and Nishimura (1956). This report triggered much subsequent work, but consistent data, usable for prediction, have not been obtained and it is now clear that records of individual instruments are not usable.

et al., 1973; Whitcomb et al., 1973; Scholz et al., 1973) and contradictions (McEvilly and Johnson, 1974; Cramer and Kovach, 1974; McGarr, 1974). The nature of the original evidence is indicated by Figure 5.19, in which are shown departures from the normal or average values for particular areas of the ratios of *S*- and *P*-wave travel times to an array of close seismic stations. These observations were explained by Nur (1972) in

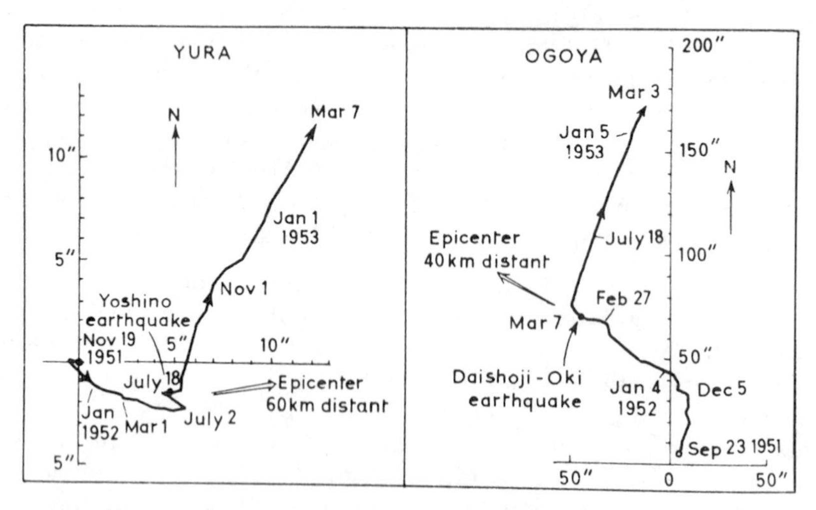


**Figure 5.17.** Vector plots of progressive changes in ground tilt at Japanese observatories. The distance of each experimental point from the origin of the figure represents the magnitude of the tilt accumulated since the starting date and the direction is the direction of downward tilt. A tangent to such a vector curve represents the instantaneous direction of tilting movement. From Sassa and Nishimura (1956).

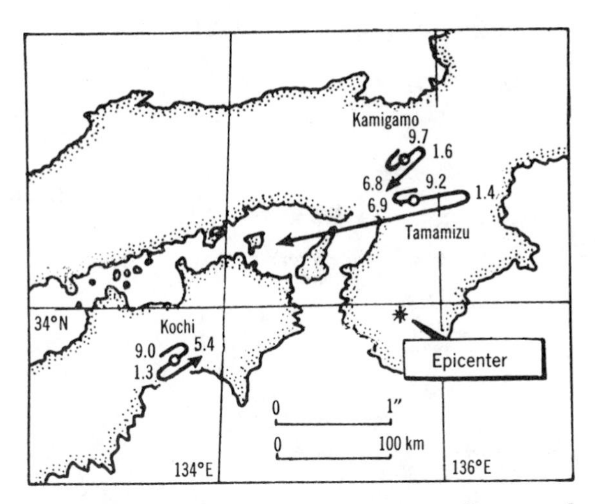


**Figure 5.18.** Rapid tilting giving an $S$-shaped vector diagram during a 10-hour interval before the Nanki earthquake in 1950, as measured at observatories up to 200 km away. Numbers on the curves represent hours before the shock. These data encourage the hope that tilt records may give a "crash" warning hours before a major shock. From Sassa and Nishimura (1956).

terms of the dilatancy effect, as discussed briefly in Section 5.2. The precursory decrease in $t_s/t_p$ was attributed primarily to an increase in $t_p$, supposedly arising from reduced $P$-wave velocity as pores and cracks opened in the focal volume of the impending shock [although the magnitude of the effect makes this interpretation appear doubtful (Hadley, 1975)]. The $P$-wave velocity, and hence $t_s/t_p$, return to normal when the pores have filled with water, which is the necessary preliminary to the main shock. The duration of the anomaly, $t$, is determined by the competition between extension of the cracking and diffusion of ground water into the cracks, being related to the dilatant volume, $V$, by the diffusion equation, which gives

$$t \propto d^2 \propto V^{2/3} \tag{5.35}$$

where $d$ is the linear extent of the volume. Then, assuming that the elastic strain energy per unit volume is independent of volume or magnitude, the total strain energy $E_S$ is directly related to volume, that is,

$$t \propto E_S^{2/3} \tag{5.36}$$

Data given by four reports of precursory travel-time anomalies allow direct comparisons of the durations $t$ with magnitudes $M$ or wave energies $E_R$. In Figure 5.20 these data are converted to released strain

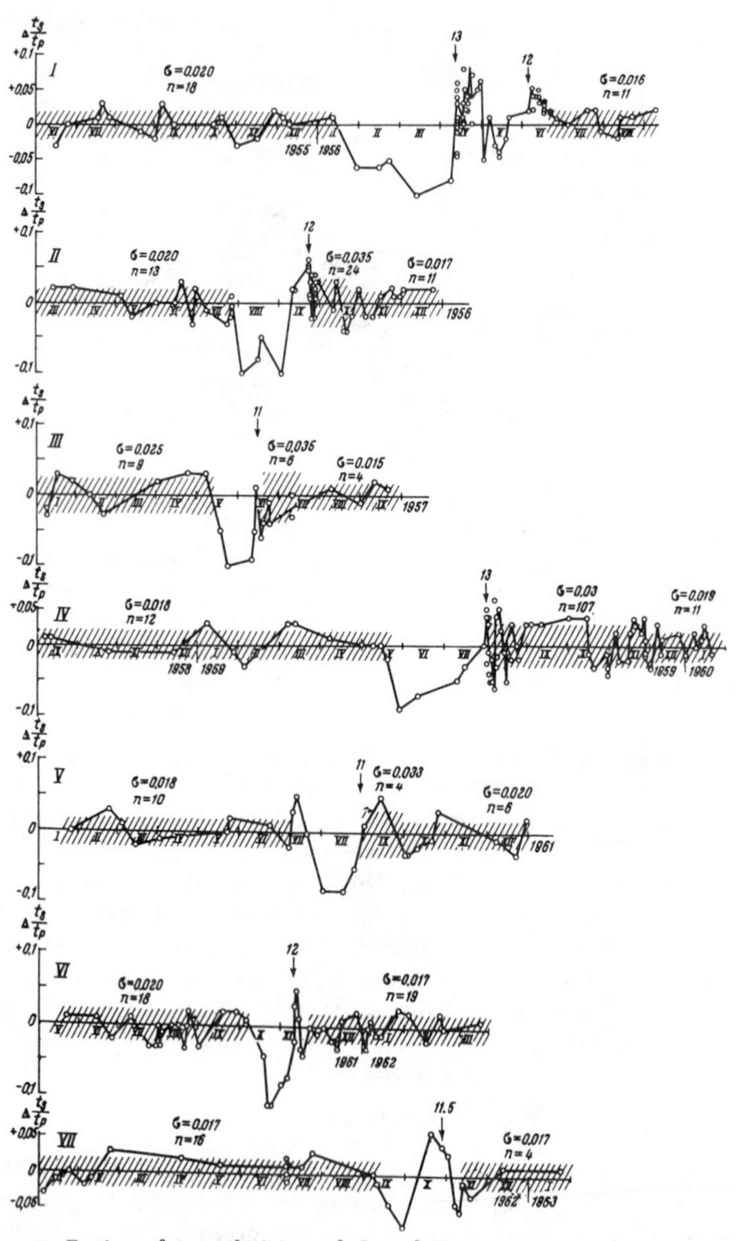


**Figure 5.19.** Ratios of travel times of $S$ and $P$ waves, $t_s/t_p$, from small shocks originating in the focal volumes of larger shocks indicated by arrows (with numbers indicating energies in joules). This shows that $t_s/t_p$ becomes anomalously low for a time, increasing with the magnitude of the impending shock. The major earthquake occurs only after $t_s/t_p$ has returned to normal. Hatching indicates the "normal" range of variation with standard deviations, $\sigma$, for the $n$-indicated events. From Semenov (1969).

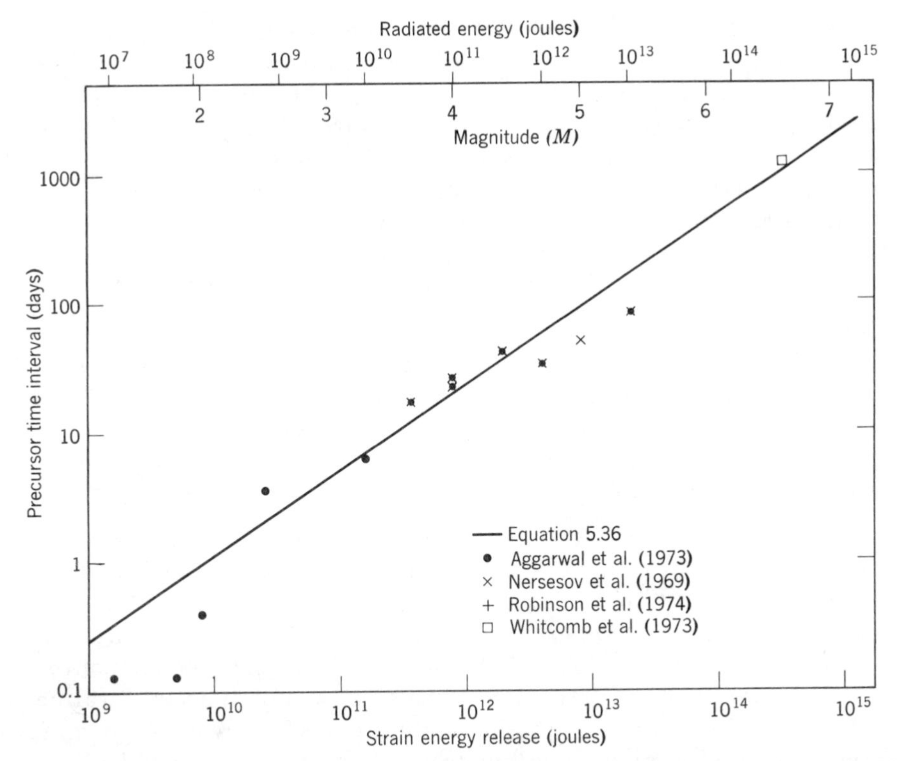


**Figure 5.20.** Durations of anomalous travel times (as represented by Fig. 5.19) before earthquakes with a wide range of magnitudes according four independent sets of data. Original representations in terms of magnitude or radiated wave energy are here converted to total strain energy release assuming validity of Figure 5.14, for comparison with deductions from the dilatancy hypothesis, as represented by Equation 5.36, shown as the solid line.

energies $E_S$, assuming validity of Figure 5.14, the corresponding values of $E_R$ and $M$ being shown at the top of the figure. The apparent excellent accord with Equation 5.36 led to the widespread hope that specific prediction of earthquakes in time, place, and magnitude is at hand.

However, indications that the usual interpretation of the travel-time anomalies has, in some respects, been too simple have come from several sources. Extensive and careful measurements of travel times from quarry blasts (McEvilly and Johnson, 1974; Boore et al., 1975) showed that no such travel-time anomalies appear for these controlled wave sources, even though the waves had propagated through the focal regions of impending earthquakes. Similarly, teleseismic waves received in the epicentral region of the Bear Valley earthquake of 1972 were found to be unaffected (Cramer and Kovach, 1974) although travel times of local sources showed the $t_s/t_p$ anomaly (Robinson et al., 1974). It is also noted that travel-time anomalies did not precede a magnitude $3\frac{3}{4}$ shock in a deep mine in South Africa (McGarr, 1974), although this shock may have been quite unrepresentative of natural shallow earthquakes for two related reasons: there was a virtual absence of groundwater, and for this reason the "breaking" stress may have been higher, and second, the focal volume would have been correspondingly smaller.

The original conclusion of Nersesov et al. (1969) has a direct relevance to the problem:

*"The observed sharp differences in the values of the ratio $t_s/t_p$ ... are not explained by physico-mechanical properties of the material in the path of the seismic ray between the focus and the recording station ... (but) can be assigned to the earthquake focus, in which there is an absence of axial symmetry in the excitation of radiation."*

Just such an asymmetry appears in a theory of seismic sources in strained media by Minster and Archambeau (1974), who found a frequency-dependent group delay for certain waves. At first, this appears also to explain the apparent anisotropy of $S$-wave velocity before earthquakes (Gupta, 1973). But the magnitude of the $t_s/t_p$ effect ($\sim 5\%$), and the fact that the variation in this ratio (but not its duration) is independent of the size of the impending shock, suggests the need for a completely different explanation. Intrinsic, stress-induced velocity variations do occur and may be measured in the ground with sophisticated equipment (e.g., Gladwin and Stacey, 1974a), but appear inadequate to explain many earthquake $t_s/t_p$ observations (Hadley, 1975).

Acceptance of the concept of dilatancy does not depend on the precise interpretation of travel time observations. The idea that fluid diffusion controls the time scale of precursive effects, as in Figure 5.20, agrees well with many observations, although it is probably applicable only to depths less than about 10 km.

## 5.4 MICROSEISMS

In addition to the discrete seismic events associated with earthquakes, seismic records show a more or less continuous background of movement, having a dominant period usually of 5 to 10 sec and variable amplitude of ground motion with a maximum exceeding 10 $\mu$m (Brune and Oliver, 1959). There is a variety of sources for these microseisms, some of which are demonstrably local in origin (wind action on trees, etc.), but the most important are due to the action of storm waves at sea (Deacon, 1947; Darbyshire, 1962). There are two microseismic effects of ocean waves: the beating of waves on shorelines, which can be appreciably coherent by virtue of the refraction of waves which tend to turn parallel to coastlines, and the interference of sea waves over deep water. The two effects may be distinguished by the frequencies of the microseisms that they generate (Haubrich et al., 1963; see also Hasselmann, 1963, for a theoretical discussion). The effect of waves on a coastline is to generate microseisms with the same period as the waves and is readily understood, but the interference effect gives microseisms of half the ocean-wave period. Interfering water waves cause a pressure fluctuation at the bottom, even in deep water; this is not apparent in the first-order theory of wave motion but was found as a second-order effect by M. Miche, whose calculations were developed into a complete theory of microseism generation by Longuet-Higgins (1950). The simple treatment by Longuet-Higgins and Ursell (1948) is followed here.

Consider the case of a sinusoidal standing wave (produced by two waves traveling in opposite directions) in which the vertical displacement $\eta$ of the water surface is given by

$$\eta = a \cos kx \cos \omega t^{*} \tag{5.37}$$

Profiles of the water surface at $t = (2\pi/\omega)n$, $(2\pi/\omega)(n + \frac{1}{4})$, $(2\pi/\omega)(n + \frac{1}{2})$, where $n$ is any integer, are shown in Figure 5.21. The total potential energy of the column of water bounded by the planes $x = 0$, $2\pi/k$, normal to the plane of the diagram, oscillates between the undisturbed value in position $B$ to a higher value in positions $A$ and $C$, the difference appearing as kinetic energy of the water in position $B$. But although total energy is conserved, a force must be exerted on the body of water to change its potential energy and this is apparent as an oscillatory pressure, which is calculable from the oscillation in potential energy. This is simply the work done in raising water within the volume marked by horizontal shading to the areas marked by vertical shading. Consider a "thickness" $b$ perpendicular to the plane of Figure 5.21. If the height of the water in a segment $dx$ of the wave at $x$ is $\eta$, the average height of the water standing above the reference plane is $\eta/2$ and its mass is $\rho\eta b\,dx$ (which is negative where $\eta$ is negative). Thus the difference in potential

*Wavelength $(2\pi/k)$ and frequency $(\omega/2\pi)$ are, of course, related but water waves are dispersive and the relationship is not obvious. For the general case, in water of depth $h$,

$$\omega^2 = gk \tanh(kh)$$

where $g$ is gravity. This is of interest also in Section 5.5.

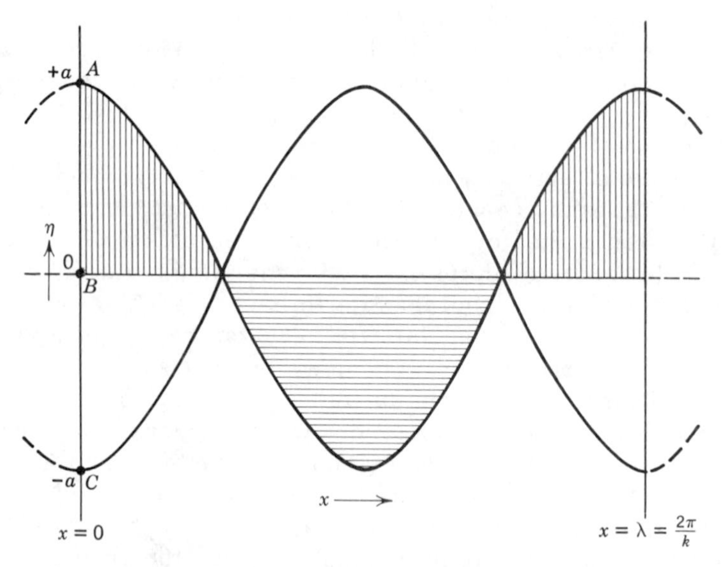


**Figure 5.21.** One wavelength of a standing ocean wave with successive positions *ABCBA*. Shaded areas represent the displacement of water from which the oscillation in potential energy is calculated.

energies between states $A$ and $B$ is

$$V=\int_0^{\lambda=2\pi/k}\frac{\eta}{2}g\rho\eta b\,dx=\tfrac{1}{2}g\rho b\int_0^{\lambda}\eta^2\,dx$$
$$=\tfrac{1}{4}\rho g\lambda ba^2\cos^2\omega t=\tfrac{1}{8}\rho g\lambda ba^2(1+\cos 2\omega t) \tag{5.38}$$

The potential energy oscillates with a frequency $2\omega$, having two maxima per cycle of the wave, corresponding to positions $A$ and $C$ in Figure 5.21. The corresponding average vertical force acting on the area considered may be written in terms of the vertical accelerations of all of the elementary water masses $dm$:

$$F=\int\frac{d^2z}{dt^2}\,dm=\frac{1}{g}\frac{d^2V}{dt^2} \tag{5.39}$$

where $z$ is the vertical position of a drop and $gz\,dm$ is its potential energy in the gravity field $g$. Differentiating Equation 5.38, we obtain the oscillatory pressure, or force per unit area:

$$P=\frac{F}{b\lambda}=-\tfrac{1}{2}\rho a^2\omega^2\cos 2\omega t \tag{5.40}$$

This is the average pressure acting over the whole area and is therefore propagated downward to indefinitely great depths and is apparent as a pressure fluctuation on the ocean floor, generating microseisms. Its frequency is twice that of the surface wave and the strength of the pressure oscillation depends on the square of the wave amplitude, that is, it is a second-order effect.

We note that there is also a first-order pressure fluctuation having the same frequency as the wave but that, unlike the second-order effect, it is attenuated with depth as rapidly as the particle motion, which hardly extends to depths greater than one wavelength. The second-order oscillation does not invoke any particle motion at depth, except that arising from compression of the water, as considered by Longuet-Higgins (1950), and yielding of the bottom.

The ideal situation of a standing wave extending to infinity is of course quite unlike the actual ocean surface. However, the Longuet-Higgins mechanism will operate whenever two wave trains with similar frequency components, traveling in approximately opposite directions, interfere over an area whose dimensions are comparable to or larger than the depth of water; in other words, the extent of the coherence of the interfering waves must be comparable to the depth of the water. The fact that ocean waves are trochoidal,* not sinusoidal, does not materially affect the argument.

The association between occurrences of large-amplitude microseisms and cyclonic disturbances at sea has been noted by numerous authors and reviewed by Deacon (1947). Whereas large microseisms always appear to be associated with storms (which may be well removed from coastlines), the reverse is not always true because the necessary condition of oppositely traveling wave trains may not occur. In particular, cyclonic disturbances generate microseisms but monsoon winds, which maintain an approximately constant direction, do not. However, the essential feature of the standing wave theory, which is confirmed by the observations, is the frequency doubling. The dominant period of ocean swell is generally 10 to 20 sec and for microseisms 5 to 10 sec; moreover, variations in both have been observed to correspond.

Some years ago meteorologists took interest in microseisms as possible indicators of hurricanes at sea. Tripartite stations were established to explore techniques of storm location, but the results were generally discouraging. Now that meteorological satellites are in operation, interest in microseisms is virtually restricted to the signal-to-noise problem of reducing their effects on seismic records. Since they are principally Rayleigh waves (vertically polarized surface waves) there

*An example of a trochoid is the curve traced out by any point on a wheel rolling along a flat surface.

may be some advantage in using seismometers in deep bore holes, where surface noise is less, but holes can hardly be deep enough to give much relief from noise with periods of order 5 sec. Another approach is to use arrays of seismometers coupled either electrically or in the analysis of records.

## 5.5 TSUNAMIS

Sea waves of long wavelengths are commonly generated by submarine earthquakes and are simply due to sudden subsidence or uplift of large areas of sea floor, or possibly combinations of both subsidence and uplift in adjacent areas. The waves sweep across open ocean at high speeds and have caused severe damage to coastal areas thousands of miles from the earthquakes which generated them. The Pacific ocean is particularly affected because so much of its perimeter is seismically active. These waves are commonly referred to as "tidal waves" but are quite unrelated to tides and the Japanese word *tsunami* is in universal scientific usage. Van Dorn (1965) has given a comprehensive review of the phenomenon. The Chilean earthquake of May 1960 was particularly effective in tsunami generation, producing waves with amplitudes of several meters all around the Pacific (e.g., Fig. 5.22). However, as in other similar cases, the amplitude was variable and by no means directly related to distance from the shock.

To understand the apparently fickle manner of tsunami propagation and the selection of certain sections of coastline for waves of destructive amplitudes it is necessary to recognize the depth dependence of wave velocity. This is a feature of shallow water waves, that is, those with wavelengths much greater than the depth of water. The velocity of this class of wave may be derived by assuming equipartition of the potential and kinetic energies of the wave motion (see also Proudman, 1953, pp. 247–251). The potential energy of a single wavelength, for a "thickness" $b$ measured along the crests, is, as in Section 5.4 (Eq. 5.38):

$$V = \tfrac{1}{4}\rho g \lambda b a^2 \tag{5.41}$$

the cosine term being now equated to unity because we are considering a traveling wave of constant shape. The condition that wavelength be very much greater than depth ensures that vertical motion of the water is slight compared with the horizontal motion. The kinetic energy is therefore calculable in terms of the horizontal velocities of the water particle, which must be such that in half of one wave period $\tau$, the horizontal motion replaces troughs by crests and vice versa. The volume of water transferred between adjacent quarter wavelengths of the path

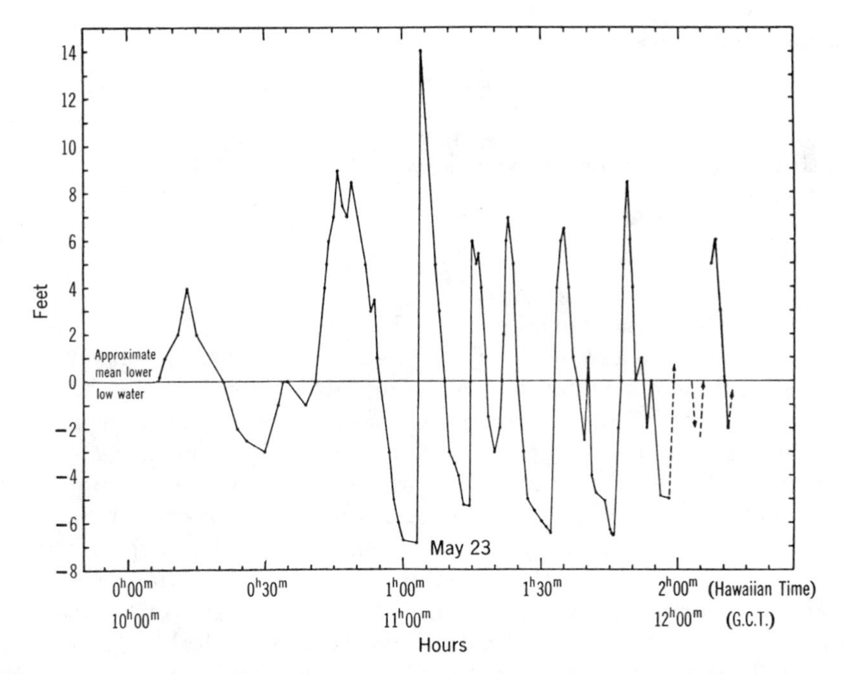


**Figure 5.22.** Tsunami at Hawaii from the Chilean earthquake of May 1960, from a sequence of observations at Wailuku River bridge by U.S. Geological Survey. Reproduced, by permission, from Eaton et al. (1961).

in time $\tau/2$ is $(a\lambda b/\pi)$, and, since the cross-sectional area of moving water is $hb$ for depth $h$ of water, the required mean water velocity during the half cycle of the wave is

$$\bar{v} = \frac{2}{\pi}\frac{a\lambda}{h\tau} \tag{5.42}$$

Since the velocity is sinusoidal the peak value is

$$v_0 = \frac{\pi}{2}\bar{v} = \frac{a\lambda}{h\tau} \tag{5.43}$$

and the mean square velocity is therefore

$$\overline{v^2} = \tfrac{1}{2}v_0^2 = \frac{1}{2}\left(\frac{a\lambda}{h\tau}\right)^2 \tag{5.44}$$

Since the total mass of water in one wavelength is

$$m = \rho\lambda hb \tag{5.45}$$

its kinetic energy is

$$\tfrac{1}{2}m\overline{v^2} = \frac{1}{4}\frac{\rho\lambda^3 ba^2}{h\tau^2} \tag{5.46}$$

Equating kinetic and potential energies (Eqs. 5.41 and 5.46), we obtain the wave velocity

$$c = \frac{\lambda}{\tau} = (gh)^{1/2} \tag{5.47}$$

Equation 5.47 is a special case of the equation given in the footnote on p. 139, since $c = \omega/k$ and $kh \ll 1$ is assumed. With this assumption, the velocity depends on depth only, being 0.22 km $\text{sec}^{-1}$ (790 km $\text{hour}^{-1}$) in ocean 5 km deep. The periods of tsunamis are 10–40 min; thus the wavelengths in deep water are hundreds of kilometers long and are quite unnoticeable at sea, but as they approach a coastline the depth and consequently the velocity decrease, and the waves increase in height as the wavelength contracts. Furthermore, the waves are refracted by the submarine topography so that contours of the ocean floor focus the waves on to particular sections of the shore. Also, oscillations (seiches) can be excited in harbors and estuaries that have suitable dimensions. Eaton et al. (1961) considered that the higher-frequency fluctuations that followed the onset of the tsunami after about an hour in the record shown in Figure 5.22 were due to the excitation of local oscillations, but a very similar record of the Chilean tsunami was obtained by the author from an instrument on Norfolk Island, which is only a small perturbation of the ocean floor, so that the recording was virtually the (smaller amplitude) open ocean wave. Apparently the propagation is dispersive, longer wavelengths arriving first, and the simple analysis which leads to Equation 5.47 is incomplete; the important neglected factor is the slope of the ocean floor, which is presumed to be responsible for the dispersion.

# 6

# Seismic Waves and the Internal Structure of the Earth

"Unwary readers should take warning that ordinary language undergoes modification to a high pressure form when applied to the interior of the Earth . . ."

BIRCH, 1952, p. 234

## 6.1 ELASTIC WAVES AND SEISMIC RAYS

Virtually all of our direct information about the interior of the Earth has been derived from observations of the propagation of the elastic waves generated by earthquakes. Certain classes, such as the surface waves, are guided by the density and velocity layering, particularly at and near the surface, and are important in the elucidation of crustal and upper mantle structures, but of greater general interest and more easily understood are the body waves that propagate throughout the Earth. Since much of the Earth is an elastic solid, two kinds of body wave can be recognized, compressional or $P$ waves and shear or $S$ waves. The designations $P$ and $S$ refer to primary and secondary arrivals; the compressional waves, being faster, arrive first at a recording station. Since both $P$ and $S$ are elastic waves, their velocities $V_P$ and $V_S$ are given in terms of the ratio of the elastic modulus, appropriate to the particular deformation associated with each wave, to density $\rho$:

$$V_P = \sqrt{\frac{m}{\rho}} \tag{6.1}$$

$$V_S = \sqrt{\frac{\mu}{\rho}} \tag{6.2}$$

The passage of an $S$ wave involves a pure shear of the medium, so that $\mu$ is the ridigity or shear modulus of the medium. The elastic modulus appropriate to $P$ waves is less obvious; we refer to it as the modulus of simple longitudinal extension because the material undergoing successive compressions and dilations in a $P$ wave is subject to lateral constraint. It cannot expand and contract laterally in response to the longitudinal compressions and dilations as can a rod transmitting compressional waves, that are long compared with the rod diameter. $m$ is readily related to the more familiar elastic constants—Young's modulus $q$, bulk modulus $K$, rigidity $\mu$, and Poisson's ratio $\nu$—by equations for deformation of elastic bodies given, for example, by Pearson (1959).

The variations in elastic properties that result from the very high compressions of the Earth's interior must be described in terms of a *finite strain theory*, that is, one in which elastic moduli may not be regarded as even approximately constant, as considered in Section 6.4. Nevertheless the elastic moduli remain perfectly valid concepts under these conditions, being defined in terms of the stresses required to produce incremental strains (in particular those in seismic waves) that are *infinitesimal*, that is, so small that conventional elasticity theory applies to them, even though a high compression is superimposed.

We may consider a body to be subjected to three mutually perpendicular, normal stresses $\sigma_1$, $\sigma_2$, and $\sigma_3$. Dealing first with the response to $\sigma_1$ only, the material is not laterally constrained and so its longitudinal strain $\epsilon_1$ is simply related to $\sigma_1$ by Young's modulus:

$$\epsilon_1 = \frac{\sigma_1}{q} \tag{6.3}$$

and the lateral strain is given by Poisson's ratio:

$$\epsilon_2 = \epsilon_3 = -\nu\epsilon_1 = -\frac{\nu\sigma_1}{q} \tag{6.4}$$

Thus by superimposing the strains for all three stresses, we have

$$\left.\begin{aligned} \epsilon_1 &= \frac{1}{q}[\sigma_1 - \nu(\sigma_2 + \sigma_3)] \\ \epsilon_2 &= \frac{1}{q}[\sigma_2 - \nu(\sigma_3 + \sigma_1)] \\ \epsilon_3 &= \frac{1}{q}[\sigma_3 - \nu(\sigma_1 + \sigma_2)] \end{aligned}\right\} \tag{6.5}$$

Now in the constrained medium $\sigma_2$ and $\sigma_3$ are self-adjusted to make $\epsilon_2 = \epsilon_3 = 0$, so that

$$\sigma_2 = \sigma_3 = \frac{\nu\sigma_1}{1-\nu} \tag{6.6}$$

and

$$\epsilon_1 = \frac{\sigma_1}{q}\left[1 - \frac{2\nu^2}{(1-\nu)}\right] \tag{6.7}$$

Thus

$$m = \frac{\sigma_1}{\epsilon_1} = \frac{q(1-\nu)}{(1-2\nu)(1+\nu)} \tag{6.8}$$

We can now give the more familiar expression for $m$ in terms of $K$ and $\mu$ from the relationships between elastic constants:

$$K = \frac{q}{3(1-2\nu)} \tag{6.9}$$

$$\mu = \frac{q}{2(1+\nu)} \tag{6.10}$$

and

$$m = K + \tfrac{4}{3}\mu \tag{6.11}$$

Then, writing Equations 6.1 and 6.2 in the usual form

$$V_P = \sqrt{\frac{K + \frac{4}{3}\mu}{\rho}} \tag{6.12}$$

$$V_S = \sqrt{\frac{\mu}{\rho}} \tag{6.13}$$

A more rigorous application of elasticity theory to the derivation of Equations 6.12 and 6.13 is given by Bullen (1963). The derivation assumes perfect elasticity, which is a reasonable approximation to the truth; anelastic response causes attenuation, as discussed in Chapter 10, but this is sufficiently slight for moderate earthquakes to be readily observable with seismometers on the opposite side of the Earth. Strictly, adiabatic elastic moduli must be used because thermal diffusion is too slow to dissipate the temperature changes associated with compressions and rarefactions in a seismic wave, but the adiabatic and isothermal rigidities are equal and the two bulk moduli differ only by a few percent, even in the deep interior of the Earth where the high temperature increases the difference relative to laboratory conditions.* This becomes significant when we consider the finer details of the composition and properties of the deep interior (Section 6.4). Other assumptions made here are that the elastic medium is homogeneous and isotropic. Seismic anisotropy (directional dependence of seismic velocity) results from preferred alignment of noncubic minerals or from alignment of grain elongations. Anisotropies of several percent have been observed in certain parts of the crust and upper mantle (Raitt, 1969; Raitt et al., 1971; Keen and Tramontini, 1970) and are possibly quite widespread in the Earth, affecting both surface waves and body waves. However, for an overall picture of gross Earth structure (meaning the radial variations in properties) it suffices to neglect both anisotropy and lateral inhomogeneity.

Three types of radial variation in property are recognized:

1. A gradual change of density and elastic constants with depth, due to effects of pressure and temperature on chemically homogeneous material.
2. A sharp boundary between chemically or physically distinct media. "Sharp" here means relative to the wavelengths of seismic waves of interest.
3. A chemical or phase transition which, although not "sharp" in the sense of category (2), gives a much stronger progressive change in properties than (1).

All three types cause refraction of seismic waves and sharp boundaries cause also reflections and partial conversion of $P$ waves to $S$ waves or $S$ to $P$. The laws of reflection and refraction follow from seismic ray

*Adiabatic and isothermal incompressibilities at temperature $T$ are related by

$$\frac{K_S}{K_T} = \frac{C_P}{C_V} = (1 + \gamma\alpha T)$$

or

$$K_S = K_T + \gamma^2 \rho C_V T$$

where $\alpha$ is volume expansion coefficient, $\gamma$ is the thermodynamic Grüneisen parameter (Appendix E), $C_P$ and $C_V$ are the specific heats at constant pressure and constant volume, respectively, and $\rho$ is density.

theory, in close analogy with geometrical optics. Central to this theory is Fermat's principle of stationary time, according to which the actual path of light, or of a seismic wave, between two points is generally shorter (but for some well-observed seismic arrivals is longer) in time (Jeffreys, 1970, pp. 126–127) than any immediately neighboring hypothetical path. This does not imply that there is necessarily only one path; in general, there may be several alternative paths, but each requires a minimum (or maximum) of transit time relative to small deviations in the path. In practice this means that we can describe wave propagation in terms of rays (optical or seismic), which are everywhere normal to the wavefronts of the waves. In terms of Huygens' construction, illustrated in Figure 6.1 for the case of a wave refracted at a sharp boundary, each point on a wavefront acts as a source of new wavelets, whose envelope represents the wavefront at a later time. Constructive interference of wavelets occurs only at the envelope and propagation is, therefore, necessarily normal to the wavefronts.

Motion of the medium in compressional or *P* waves is longitudinal so that there is no polarization of a *P* wave. However, *S* waves, being transverse, are polarized and it is necessary to distinguish vertical and horizontal polarizations, termed *SV* and *SH* waves,* which behave differently at sharp horizontal boundaries. The horizontal motion of an

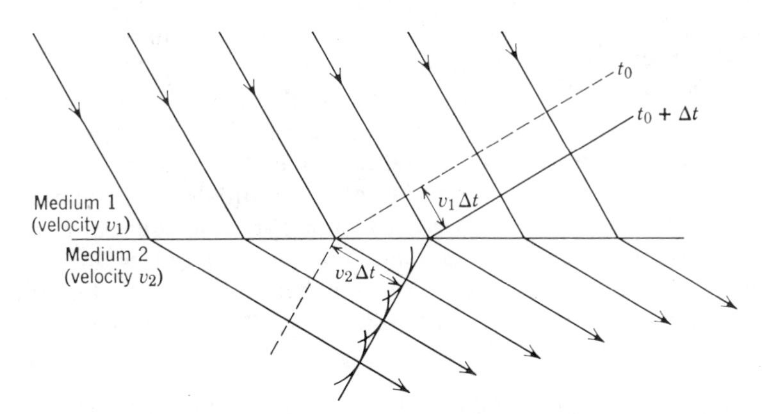


**Figure 6.1.** Huygens' construction for refraction of a seismic wave at a sharp boundary. Each point in the boundary may be regarded as a source of wavelets whose envelope is the refracted wavefront. The time interval is $\Delta t$ between the positions of a wavefront indicated by broken and solid lines.

*Particle motion in an *SH* wave is horizontal, and remains so as the wave propagates, so long as the refracting structure is horizontally layered. In an *SV* wave particles oscillate in the vertical plane, perpendicular to the path.

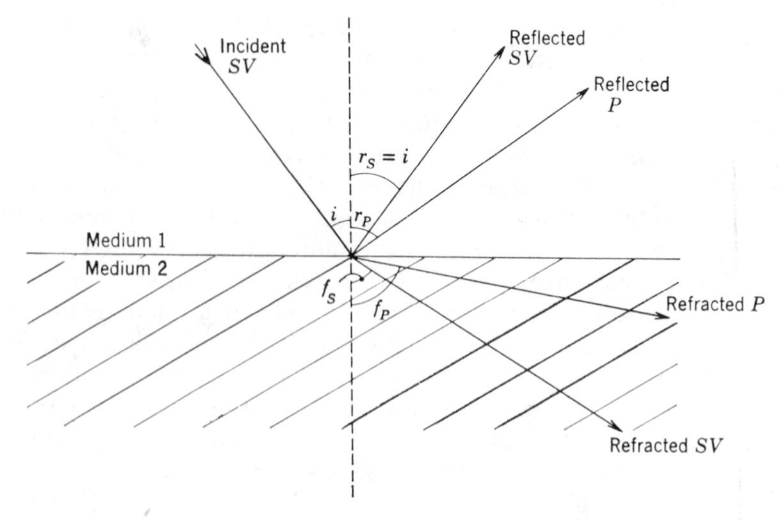


**Figure 6.2.** Reflected and refracted rays derived from an *SV* ray incident on a plane boundary. The boundary is assumed to be "welded," that is, there is a continuity of solid material across it.

*SH* wave is transferred across a boundary as a simple refraction with some reflection, but an *SV* wave, which has a component of motion normal to the boundary, generates *P* waves in addition to the refracted and reflected *SV* waves, as illustrated in Figure 6.2. The division of energy at a boundary is a function of the angle of incidence; the relevant equations are given by Bullen (1963). In general an *S* wave may have both *SV* and *SH* components, in which case the polarization becomes nearer to *SH*, since the *SV* component is diminished in generating *P*. Conversely, *P* waves generate *SV* but not *SH* at a boundary.

Snell's law is applicable to seismic waves, as in optics. In Figure 6.2, the angles of reflection and refraction are related to the angle of incidence by the wave velocities $V_P$ and $V_S$ of *P* and *S* waves in media 1 and 2:

$$\frac{\sin i}{V_{S_1}} = \frac{\sin r_S}{V_{S_1}} = \frac{\sin r_P}{V_{P_1}} = \frac{\sin f_s}{V_{S_2}} = \frac{\sin f_P}{V_{P_2}} \tag{6.14}$$

Similar equations can be written by inspection for reflections and refractions of an incident *P* ray or for an *SH* ray (for which only reflected and refracted *SH* occur). In the problem considered in Section 6.2, in which there is a monotonic increase in velocity with depth, Snell's law is applied in differential form. In this context Snell's law is a consequence of the requirement that, for a wave that crosses a boundary, the phase velocity parallel to the boundary must be the same on both sides.

A liquid medium, such as the Earth's core, constitutes a special case in which $\mu = 0$ and there is therefore no *S*-wave propagation. *S* waves are, however, reflected from the core boundary and *SV* can generate *P*, which propagates through the core, where it is designated by the letter *K*.

A limitation of ray theory is that it applies only to wavefronts or obstacles large compared with the wavelengths. The effects of small obstacles or sharply curved wavefronts must be described in terms of diffraction. Small-scale inhomogeneities cause scattering. This is important in the crust and appears significant also at the bottom of the mantle near to the core-mantle boundary (Haddon and Cleary, 1974).

In addition to the body waves, *P* and *S*, there are surface waves of two kinds,* which propagate around the boundaries between layers of different materials or in strong velocity and density gradients near the surface of the Earth. Rayleigh waves may be guided by a single boundary, the free surface, or in a surface wave guide in which velocity and density increase downward. In Rayleigh wave motion the paths of the particles of the medium are ellipses whose major axes are normally vertical and minor axes are in the direction of propagation of the wave. The sense of the particle motion in a normally layered Earth is usually retrograde (as a wheel rolling back toward the source). Love waves are *SH* waves (particle motion horizontal and normal to the direction of propagation) that require a wave guide for propagation. Commonly one boundary of the wave guide is the Earth's free surface and the other is the normal increase in shear wave velocity, $V_S$, downward in the Earth. However, a low velocity channel internal to the Earth, such as the widespread, perhaps worldwide, upper mantle minimum in shear wave velocity at depths of 100 to 200 km, can also act as a guide for Love waves.

Rayleigh waves, which propagate at the free surface of a homogeneous medium, are nondispersive, with a velocity of about 0.92 $V_S$ in a medium with Poisson's ratio 0.25, but in the Earth, with velocity layering, Rayleigh waves are dispersive and Love waves are always dispersive because velocity layering or gradients are necessary for their propagation. Surface wave velocity normally increases with wavelength, because $V_S$ increases with depth in the wave guide and the longer wavelengths penetrate deeper. The shear wave velocities in the adjacent media $V_{S\,max}$, $V_{S\,min}$, set the limits on surface wave velocities.

*A comprehensive treatment of surface waves is given by Ewing, Jardetsky and Press (1957).

For Love waves $V_{S\,max} > V_L > V_{S\,min}$ and for Rayleigh waves $0.92\ V_{S\,max} > V_R > 0.92\ V_{S\,min}$ approximately. Love waves characteristically have a minimum in group velocity at a particular frequency, which arises at the end of a wave train as a rather clean sinusoidal *Airy phase*. Summaries of observed surface wave dispersions are by Oliver (1962) and Dorman (1969).

The body waves, *P* and *S*, from an earthquake both arrive before the surface waves. This is fortunate because surface waves from distant shallow earthquakes normally arrive with much larger amplitudes than the body waves; the energy of a surface wave is spread along an expanding line around the Earth, whereas the wavefront of body waves is an expanding surface and the amplitude thus diminishes more rapidly with distance. Figure 6.3 is reproduced from a seismogram showing arrivals of *P*, *S*, and surface waves at Charters Towers, Queensland, from an earthquake off Northern Sumatra. This is a record from a long period instrument which emphasizes the greater amplitudes of surface waves (which have longer periods). The recognition of the several different arrivals is a skill acquired by practice. An important recent

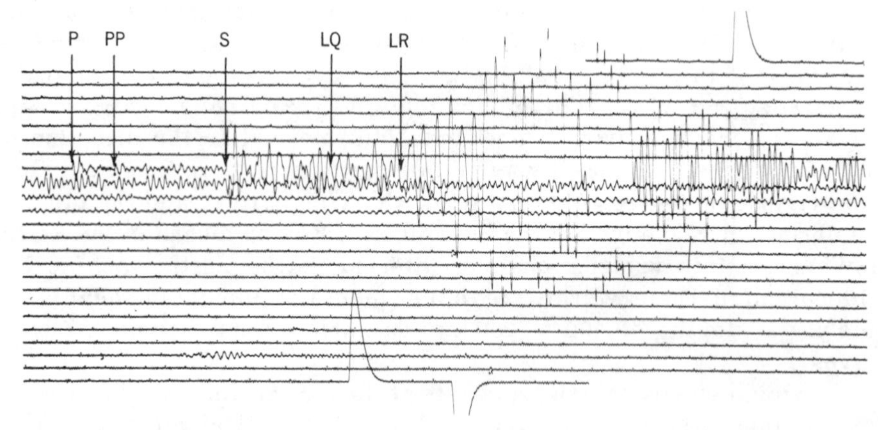


**Figure 6.3.** Seismogram obtained at Charters Towers, Queensland (Station CTA) showing arrivals of *P*, *PP*, *S* and surface waves *LQ*, *LR*, from a magnitude 5.9 earthquake off the west coast of Northern Sumatra (distance 6100 km, $\Delta = 54.9°$) on August 21, 1967. The successive lines are parts of a continuous helical trace on a seismogram that was unwrapped from a recording drum. This recording is from a long period East-West instrument; an upward deflection of the trace represents an eastward movement of the ground, the maximum peak-to-peak ground movement here being about 200 $\mu$m. Minute marks superimposed on the trace show that almost 8 minutes elapsed between the *P* and *S* arrivals. The top and bottom traces show calibration pulses. Figure courtesy of Dr. J. P. Webb and Mr. P. Gaffy.

development is the use of arrays of seismometers, whose outputs can be phased to select arrivals from particular directions and thus to diminish greatly the complications in seismic wave trains. This can make the several different arrivals on a record much more distinct.

The distinctions between several classes of elastic wave are necessarily blurred. Thus we can represent Love waves as $S$ waves totally internally reflected by the boundaries of a layer and so propagating along it as a superposition of two oblique (upward and downward propagating) waves.* This makes clear the physical basis of the dispersion of surface waves since the development of a standing wave depends on the angle of incidence and wavelength of component body waves relative to the thickness of a layer. Similarly, Brune (1964) has pointed out an equivalence of free oscillations to multiply reflected body waves. As is pointed out in Section 6.3, free oscillations are readily recognized as standing surface waves and the normal mode spectra merge into the surface wave dispersion curves. Surface waves of wavelength $\lambda$ effectively penetrate the Earth to a depth of order $\lambda$, so that very long wavelengths, corresponding to the lower modes of free oscillation, effectively sample the whole Earth.

## 6.2 TRAVEL TIMES OF BODY WAVES AND THE VELOCITY STRUCTURE OF THE EARTH

If the velocities of elastic waves were uniform within the Earth, then seismic rays would be straight lines, following chords as in Figure 6.4. The travel time from a surface focus to an observatory at an angular

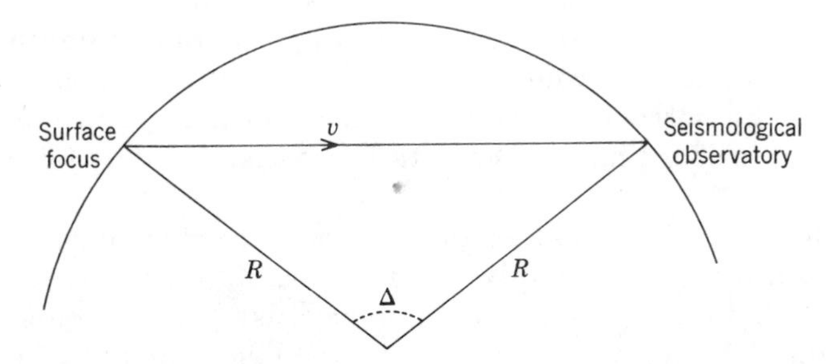


**Figure 6.4.** Path of a seismic ray through a hypothetical uniform Earth from a surface earthquake to an observatory. In seismology the distance between two surface points is commonly given as the angle $\Delta$ between radii from the center to the surface points.

*Propagation occurs as "leaking modes" if the internal reflections are not total.

distance $\Delta$ from the focus would be

$$T = 2\frac{R}{V}\sin\frac{\Delta}{2} \tag{6.15}$$

for waves of velocity $V$. A vital feature of the travel times is that they increase less strongly with distance than is indicated by Equation 6.15. Observed $T-\Delta$ curves (Fig. 6.5) are more curved than this equation would suggest. The velocity at depth is thus greater than at the surface and seismic rays are refracted, as in Figures 6.6 and 6.7. Elucidation of the details depends on accurate travel time data, which have been developed by successive refinements and improvements over many years. Data on many earthquakes from numerous seismological observatories have been used, allowing compensation for crustal and upper mantle heterogeneity in terms of "station corrections," although the sampling is biased by two limitations: earthquakes occur only in seismic zones, which are not necessarily representative of the whole Earth, and seismic stations are restricted to continental areas. Improved source control has been possible with large nuclear explosions whose locations, origin times, and depths of focus (zero) are known and so do not have to be deduced from the observed seismic arrivals, but local, station anomalies must still be taken into account. A comprehensive tabulation of travel times, which has been used as the reference standard for many years, is given by Jeffreys and Bullen (1940), whose results are represented in Figure 6.5. A more recent tabulation of $P$-wave times has been presented by Herrin (1968) and coauthors. The multiplicity of the curves arises because, in addition to refractions, there are reflections and partial $P$ to $S$ and $S$ to $P$ conversions at the core-mantle boundary and at the surface of the Earth, as shown by the ray paths of Figure 6.6. In precise travel-time studies, the tables require correction for the ellipticity of the Earth, which is a decreasing function of depth, as noted generally in Section 3.1 and with particular reference to the core in Section 3.4. The corrections are minimized by using geocentric rather than geographic coordinates. The details are discussed by Jeffreys (1970) and Bullen (1963).

Seismic rays from a near earthquake or an explosion have generally been represented rather simply, as in Figure 6.7, in terms of a layered structure of the outer part of the Earth or crust. The principal layers of the continental crust (total thickness about 40 km), each of which may be present in varying thickness or absent, are, in order of both depth and increasing seismic velocity, (1) sediments, with very low velocities and high attenuation, (2) an upper igneous layer, commonly of acid (granitic) composition, and (3) an "intermediate" layer or layers, probably grading into more basic composition (see Section 7.1). Underlying the inter-

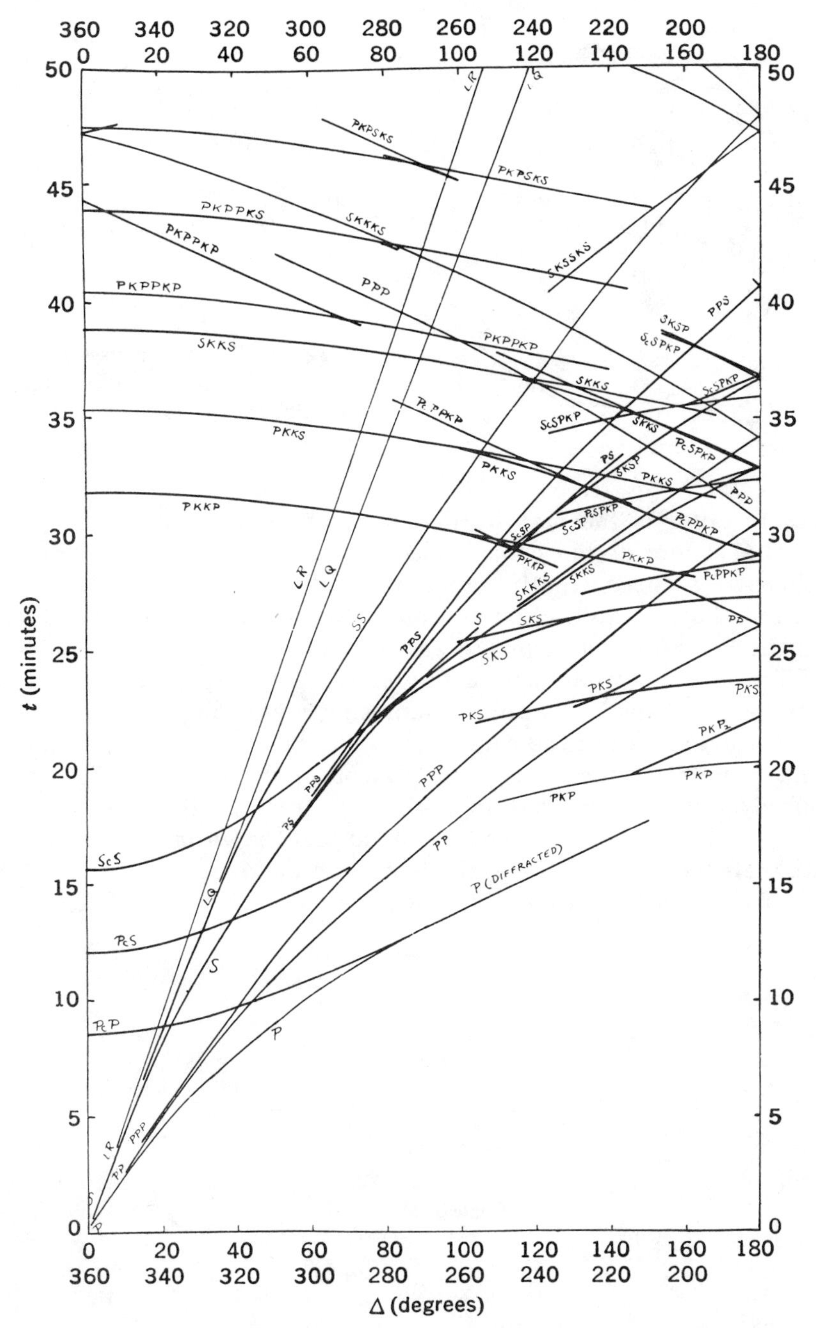


Figure 6.5. The Jeffreys-Bullen (J-B) travel time curves. Reproduced, by permission, from Jeffreys and Bullen (1940).

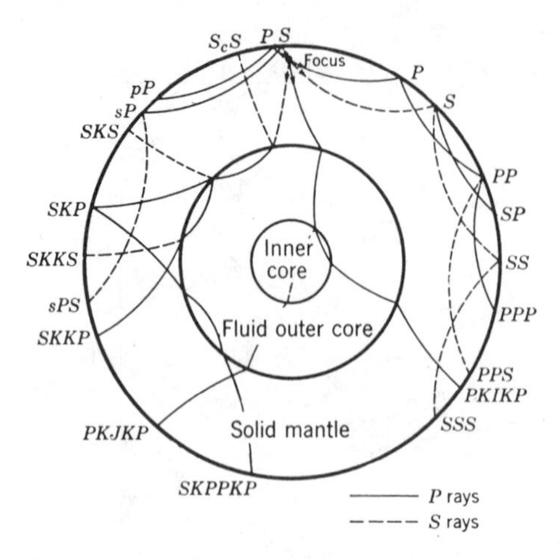


**Figure 6.6.** Seismic rays whose travel times are represented in Figure 6.5. Reproduced, by permission, from Bullen (1954).

mediate layer is the mantle, the boundary being marked by a sharp and almost worldwide discontinuity in seismic velocities. This is the Mohorovičić discontinuity, generally abbreviated to "M" layer or "Moho." The oceanic crust is much thinner (of order 5 km), being a basaltic layer with an overlying veneer of sediments.

Travel times from near earthquakes can be represented in terms of "families" of rays, a "family" comprising rays whose points of deepest penetration are all in the same layer. The family represented in Figure 6.7 has penetrated the Mohorovičič discontinuity to the top of the mantle. The refraction is such that all of the rays in a family have very similar paths in all layers except the lowest, so that the variation of travel time with distance is due to the variable path in the deepest layer penetrated and, if the velocity there is $V$, we have

$$V = \frac{dL}{dT} = a\frac{d\Delta}{dT} \tag{6.16}$$

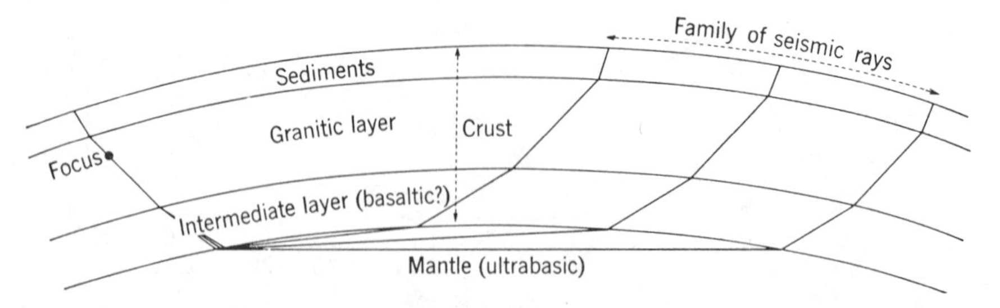


**Figure 6.7.** Seismic ray from a near earthquake in a layered Earth, showing refraction through characteristic layers of the crust.

where $L = a\Delta$ is distance from the epicenter, related to angular distance $\Delta$ by the radius of the Earth $a$. If a sufficient array of seismometers is used, as in seismic exploration, the velocities in a sequence of layers within which velocity increases downward can be determined, although the problem is generally complicated by variable thicknesses of the layers and by heterogeneity, particularly in the upper crust.

Following closely the treatment of Bullen (1963), we may also derive the $T - \Delta$ relationship for remote earthquakes (teleseisms), which give information about the deeper parts of the Earth. In near earthquake studies the Earth may be assumed flat, but in studying teleseisms a spherically layered Earth is considered. The layering need not be discrete (as in an onion), but it is convenient to envisage it that way initially. Figure 6.8 shows the geometry of a ray in a three-layer Earth in which the velocities increase inward from the surface. Applying Snell's law to each of the boundaries, $A$, $B$ in Figure 6.8:

$$\frac{\sin i_1}{V_1} = \frac{\sin f_1}{V_2} \tag{6.17}$$

$$\frac{\sin i_2}{V_2} = \frac{\sin f_2}{V_3} \tag{6.18}$$

but from the two triangles

$$q = r_1 \sin f_1 = r_2 \sin i_2 \tag{6.19}$$

so that

$$\frac{r_1 \sin i_1}{V_1} = \frac{r_1 \sin f_1}{V_2} = \frac{r_2 \sin i_2}{V_2} = \frac{r_2 \sin f_2}{V_3} \tag{6.20}$$

Equation 6.20 could be extended to refractions at any number of boundaries or to gradual refraction in a layer of progressively increasing velocity. Thus for the particular ray considered

$$\frac{r \sin i}{V} = \text{constant} = p \tag{6.21}$$

where $i$ is now quite generally the angle between the ray and the radius at any point. $p$ is termed the parameter of the ray, a geometrical constant for all points along it. By determining the parameter of a ray we obtain a value of $r/V$ at its deepest point of penetration, where $\sin i = 1$.

To proceed further with the direct analysis of travel times in terms of the velocity structure, we must impose a limitation on the possible radial variations of velocity, such that the travel time curve $(T - \Delta)$ is everywhere continuous and single valued. This restriction means that the analysis applies only to a single "family" of rays, (as in Fig. 6.7), all of which have their deepest points of penetration in a particular layer. It

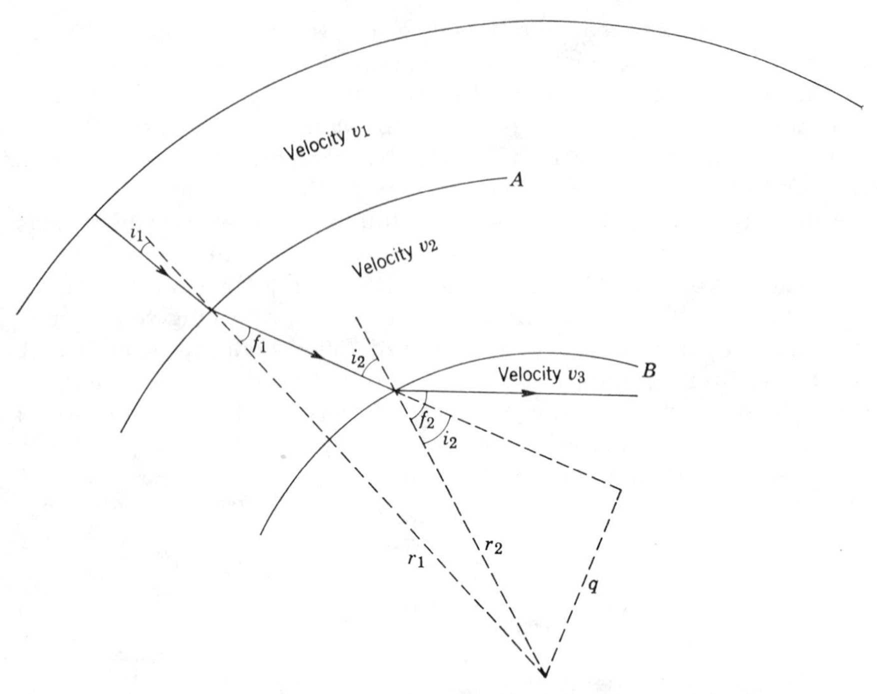


**Figure 6.8.** Teleseismic ray in a three-layer Earth, with constructions used to show the geometrical significance of the parameter of a seismic ray.

requires that the downward curvature of a ray in a layer of velocity *decreasing inward* should be less than the curvature of a level surface ($r^{-1}$). This means that

$$\frac{dV}{dr} < \frac{V}{r} \tag{6.22}$$

$dV/dr$ being positive in this case, that is $V$ *increasing outward*. Violation of Equation 6.22 occurs in limited ranges in the Earth, notably at the core-mantle boundary (2900 km depth), where it causes a shadow zone within which no arrivals are observed, as in Figure 6.9, that is, there is a break in the travel time curve. The other exception arises in a depth range having a rapid increase in velocity with depth. Then increasingly sharp refraction of rays making greater penetrations causes them to travel shorter distances, as in Figure 6.10*a* resulting in *triplication* of the travel-time curve as in Figure 6.10*b*. Since, by Equation 6.21, increasing initial steepness of a ray is represented by decreasing $p$, the condition imposed is that decreasing $p$ means increasing $\Delta$, that is,

$$\frac{d\Delta}{dp} < 0 \tag{6.23}$$

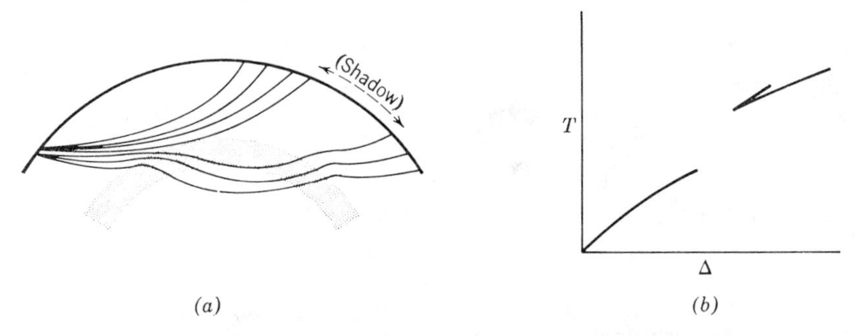


**Figure 6.9.** (*a*) Effect on seismic rays of a layer in which velocity decreases with depth more strongly than is permitted by Equation 6.22. (*b*) Corresponding travel time curve.

This condition is violated by rays that have their deepest penetrations in the transition zone of the upper mantle or at the boundary between the fluid outer core and solid inner core. A more detailed discussion of these arguments is given by Bullen (1963).

Accepting the conditions imposed by Equations 6.22 and 6.23, we can obtain a simple relationship between $p$ and the travel-time curve by considering the geometry of two infinitesimally different rays $PP'$ and $QQ'$, as in Figure 6.11. $PN$ is a normal from $P$ to $QQ'$, which means that it is a wavefront, and the difference in travel time between $PP'$ and $QQ'$ is thus

$$dT = 2\frac{QN}{V_0} \qquad (6.24)$$

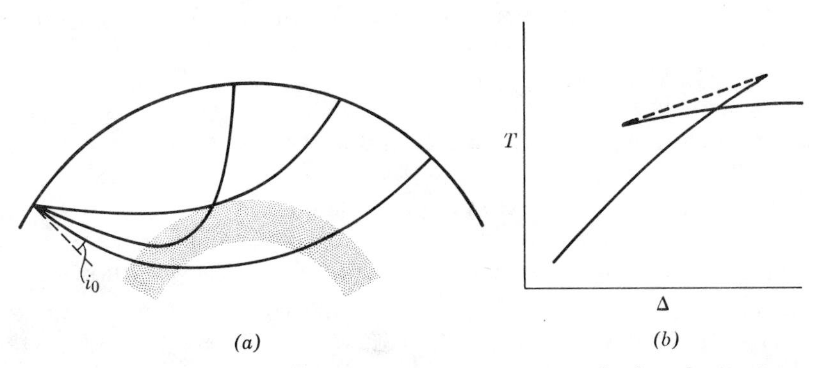


**Figure 6.10.** (*a*) Effect on seismic rays of a layer in which velocity increases rapidly with depth in an otherwise "normal" Earth. (*b*) Corresponding features of the travel-time curve. If the velocity increase is a sharp transition, then the section of the curve represented by the broken line is missing.

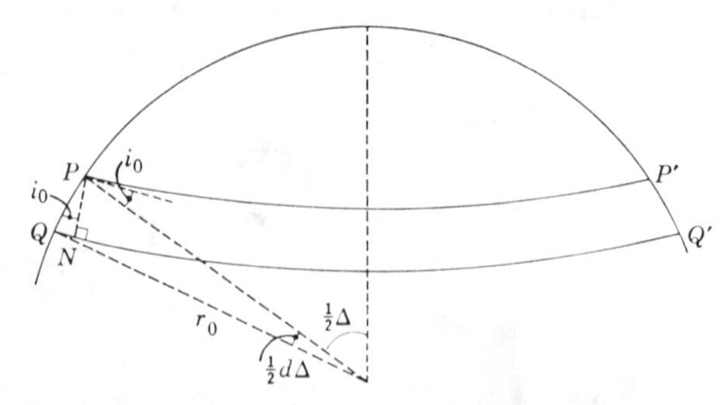


**Figure 6.11.** Adjacent teleseismic rays with geometrical construction used to derive Equation 6.26.

if $V_0$ is the seismic velocity at the surface. But

$$QN = PQ \sin i_0 = \tfrac{1}{2} r_0 \, d\Delta \sin i_0 \qquad (6.25)$$

Therefore

$$\frac{dT}{d\Delta} = \frac{r_0 \sin i_0}{V_0} = p \qquad (6.26)$$

Values of $T$, $\Delta$ are tabulated from observations and over limited ranges can be fitted by continuous analytical functions of which $dT/d\Delta$ is the gradient* so that $p$ is a known function of the angular distance $\Delta$ traversed by seismic rays.

The angular distance can be written as an integral. We first note that

$$p = \frac{r \sin i}{V} = \frac{r}{V} \cdot \frac{r \, d\theta}{ds} \qquad (6.27)$$

where $\theta$ is the partial angular distance, as in Figure 6.12, and $s$ is the actual distance measured around the ray path to the point on the path specified by $\theta$. Also,

$$(ds)^2 = (dr)^2 + (r \, d\theta)^2 \qquad (6.28)$$

Eliminating $ds$ from Equations 6.27 and 6.28,

$$\left(\frac{r^2 \, d\theta}{Vp}\right)^2 = (dr)^2 + (r \, d\theta)^2 \qquad (6.29)$$

and introducing, for convenience, a factor $\eta = r/V$, we obtain

$$\frac{d\theta}{dr} = \frac{p}{r(\eta^2 - p^2)^{1/2}} \qquad (6.30)$$

*Use of a spaced array of seismometers gives directly $dT/d\Delta$, which is the inverse of the apparent phase velocity of a wave across the array.

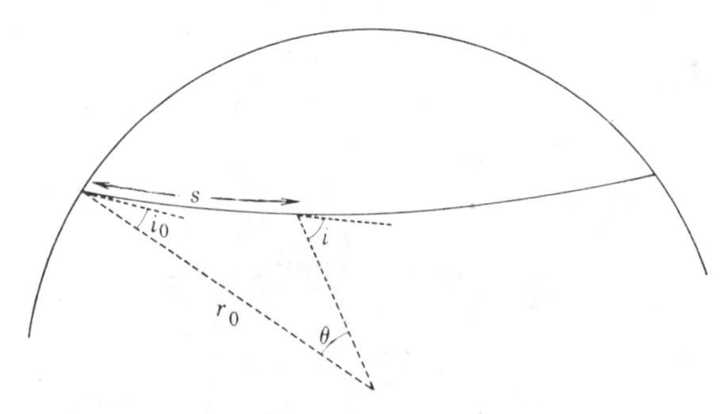


**Figure 6.12.** Geometry used to obtain the integral expression for $\Delta$.

Integrating from the deepest point of the ray $(r')$ to the surface $(r_0)$:

$$\tfrac{1}{2}\Delta = \int_{r'}^{r_0} \frac{p\,dr}{r(\eta^2 - p^2)^{1/2}} \tag{6.31}$$

Since $\Delta, p$ are measured quantities, Equation 6.31 is an integral equation giving $\eta$ (and hence $V$) as a function of $r$. A method of solution reported by Jeffreys (1970) from a communication by G. Rasch and using simplifications due to E. Wiechert, L. Geiger, and others is given in Appendix D. The result is

$$\int_0^{\Delta_1} \cosh^{-1}\left(\frac{p}{p_1}\right) d\Delta = \pi \ln\left(\frac{r_0}{r_1}\right) \tag{6.32}$$

Equation 6.32 is a convenient form for numerical integration from a tabulation of travel times in equal intervals of $\Delta$, because $p$ is a known function of $\Delta$ by Equation 6.26 and $p_1$ is the value of $p$ at $\Delta = \Delta_1$. Equation 6.32 thus determines $r_1$ corresponding to $\Delta_1$ and therefore to $\eta_1 = r_1/V_1$, and so gives $V(r)$ within the range of $r$ down to the maximum penetration of a particular type, or family, of seismic ray.

The foregoing treatment is applicable to most of the Earth, since the normal behavior is a slow increase in velocity with depth, by virtue of the more rapid increase in elastic moduli than density with pressure.* Breaks or multiplication of arrivals correspond to transitions between regions of the Earth that are distinguished by compositional or phase differences. A scale drawing of the rays and wavefronts within the Earth is given in Figure 6.13. Ambiguities in interpretation of travel times are clarified by the effective duplication of data resulting from the

*In the upper mantle the decrease in elasticity with temperature may outweigh the increase with pressure over a limited depth range.

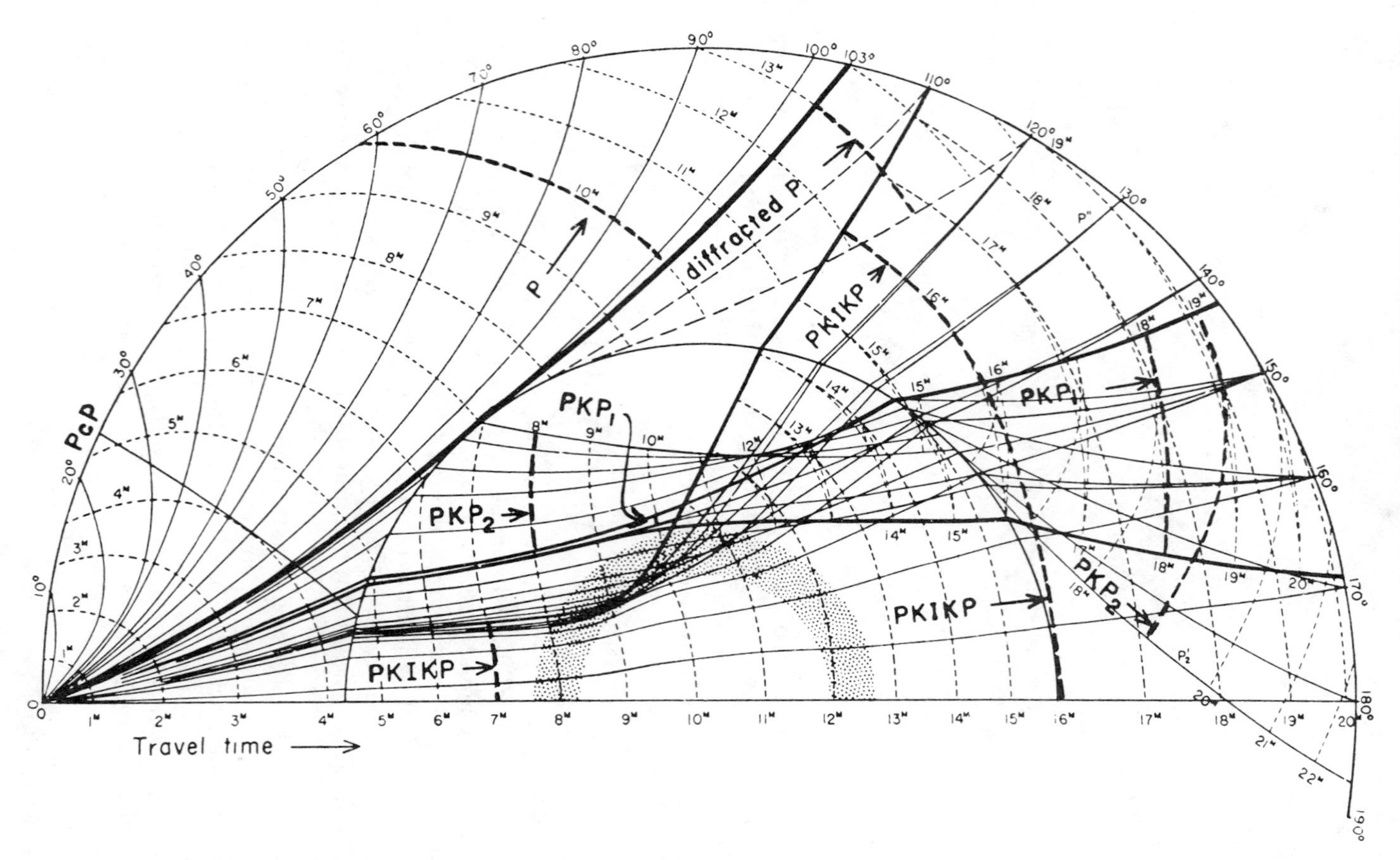


**Figure 6.13.** Rays and wavefronts for compressional waves in the Earth. Reproduced, by permission, from Gutenberg (1959). (Copyright Academic Press.)

multiplicity of rays; for example, core velocities must be the same from both *PKP* and *SKS* travel times. Furthermore, constraints on Earth models are imposed by observations of free oscillations (Section 6.3) and surface waves (especially for upper mantle structure), so that recent Earth models differ from one another in only minor ways. The seismic velocity profile for the Earth as a whole is given in Figure 6.14.

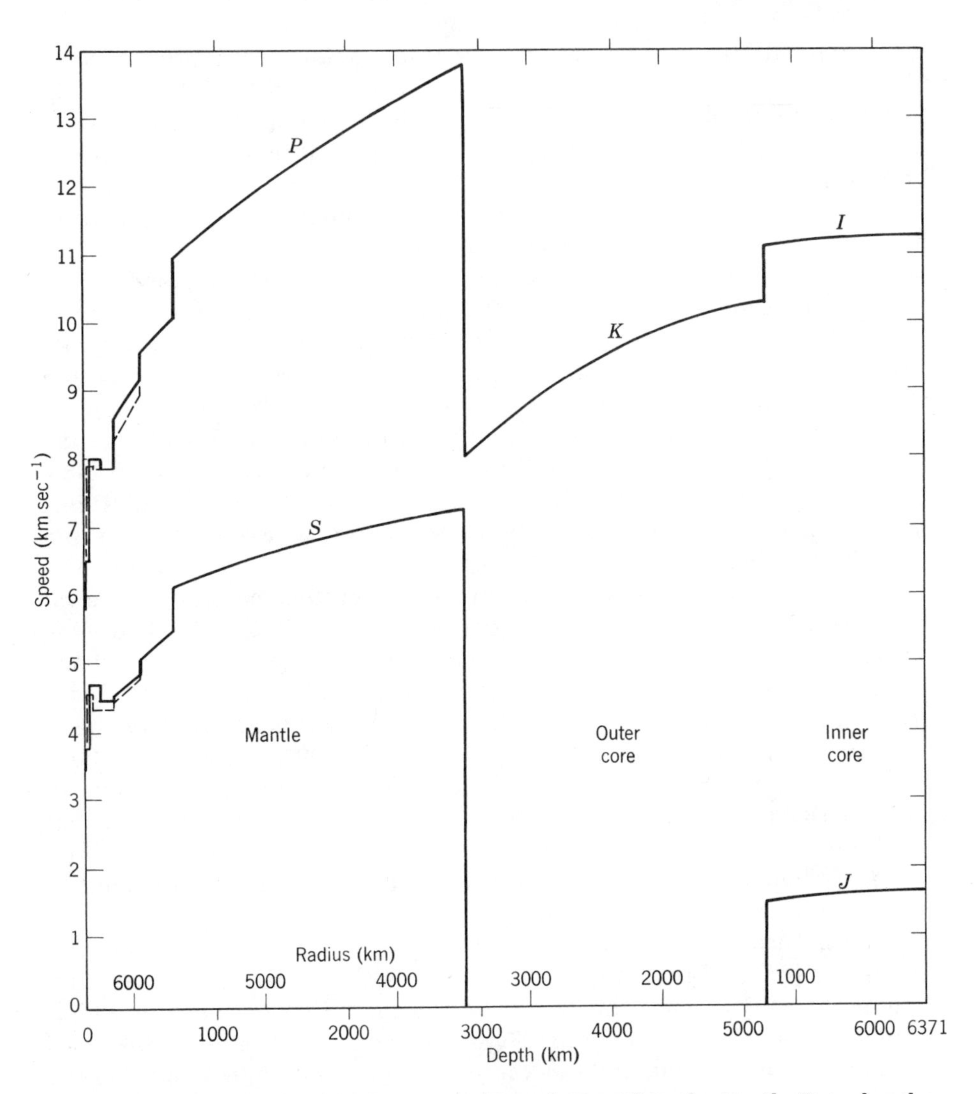


**Figure 6.14.** Velocities of body waves (*P* and *S*) within the Earth. Data for the earth model by Dziewonski et al. (1975)—see Appendix G.

## 6.3 FREE OSCILLATIONS

The excitation of free oscillations of the Earth by large earthquakes has been compared to the ringing of a bell. Recognition that the Earth as a whole could, if suitably excited, undergo free vibrations in an indefinite number of modes extends back 100 years. It was a product of the development of the theory of elasticity by Lord Kelvin, H. Lamb, A. E. H. Love, Lord Rayleigh, and others. The early work, which is mainly concerned with the oscillations of a hypothetical uniform Earth, has been reviewed by Stoneley (1961). Calculations in terms of more realistic Earth models were hardly possible without electronic computers and were in any case of limited interest until it became apparent that free oscillations were observable.

Interest in the subject was renewed by H. Benioff's development of instruments to observe very long-period seismic waves. From an examination of his records following the Kamchatka earthquake of 1952, Benioff tentatively identified a 57-min period as a fundamental mode of free oscillation. In the following few years instrumental improvements were directed specifically to the observation of free oscillations and Alterman et al. (1959) calculated the periods of oscillation for realistic Earth models. When the next really large earthquake, capable of exciting oscillations of observable amplitude, occurred in Chile in May 1960, several geophysical laboratories were able to record the oscillations. Bullen (1963, p. 260) has described the scientific excitement at the Helsinki meeting of the International Union of Geodesy and Geophysics, later in 1960, when representatives of several groups met and compared their preliminary results. This meeting effectively established free oscillations as a new branch of seismology. Subsequent observations, especially following the Alaskan earthquake of March 1964, have confirmed and amplified the 1960 results. The observational data have been summarized by Slichter (1967) and S. W. Smith (1967) and a comprehensive tabulation and identification of modes with observed periods is given by Dziewonski and Gilbert (1972, 1973) and Gilbert and Dziewonski (1975). Since the oscillations die away in a few days (even less for the shorter periods), it is remarkable how much has been accomplished in the few brief periods of recording.

The deformations of the Earth in free oscillations are indicated in Figure 6.15 for the fundamental modes of the three different types. The simplest to envisage are the *radial* oscillations, that is, those in which particle motion is purely radial (Figure 6.15*a*). The fundamental is an alternating compression and rarefaction of the whole Earth and there is an infinite series of overtones for which spherical nodal surfaces occur within the Earth. The radial oscillations are a special case of spheroidal

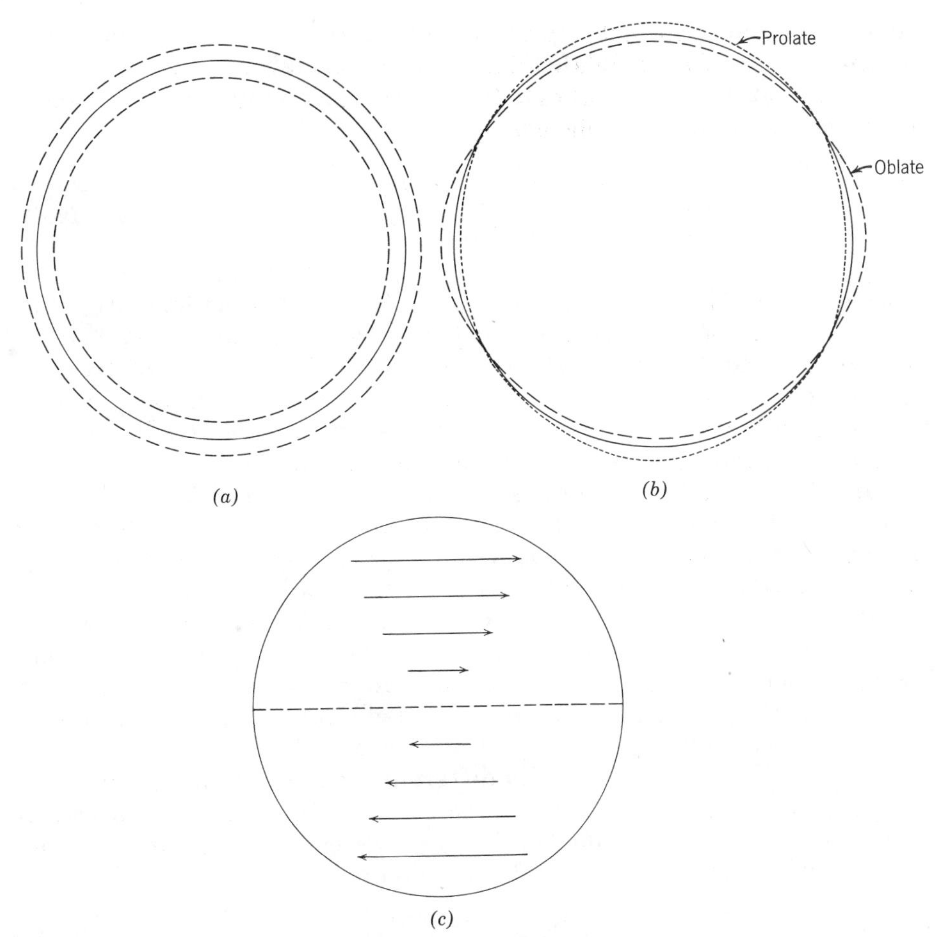


**Figure 6.15.** The simplest representative modes of free oscillation. (*a*) Radial oscillation ${}_0S_0$. (*b*) Spheroidal "football" mode ${}_0S_2$. (*c*) Instantaneous motion in toroidal mode ${}_0T_2$.

oscillations, in which particle motion has both radial and tangential components and the Earth's surface is deformed in the manner of the spherical harmonic functions $[P_l^m(\cos\theta)\cos m\lambda]$ (see Appendix C). The tesseral harmonics, in which $m > 0$, have not been distinguished observationally from the zonal harmonics $P_l(\cos\theta)$, with which their periods are degenerate and the superscript $m$ is dropped. Then $l = 0$ [i.e., $P_0(\cos\theta) = 1$] gives the case of radial oscillations; $l = 1$ [i.e., $P_1(\cos\theta) = \cos\theta$] is precluded by the fact that it represents a net translation of the surface and therefore of the center of gravity; $l = 2$

represents the "football" mode, in which the Earth deformation is alternately prolate and oblate (Fig. 6.15), and all higher $l$ occur. The form of the motion in the general case (for $m = 0$) can be represented by the $r$ and $\theta$ components of displacement, $u$ and $v$:

$$u = U(r)P_l(\cos\theta)\sin\omega t$$
$$v = V(r)\frac{\partial P_l(\cos\theta)}{\partial\theta}\sin\omega t \tag{6.33}$$

where $U$ and $V$ are related radial distributions of the motion within the Earth and $\sin(\omega t)$ indicates an oscillation whose frequency $(\omega/2\pi)$ is related also to $U$, $V$ and determined by the internal elasticity and density structure. The "colatitude" $\theta$ refers not to geographic coordinates but to a pole determined by the geometry of the exciting earthquake. The radial functions $U$ and $V$ can take an indefinite number of distinct forms, each representing a fundamental oscillation of order $l$ or one of a series of overtones in which spherical surfaces within the Earth are nodes of the motion. The number of nodal surfaces is represented by a prefix $n$, so that the general representation of a spheroidal oscillation is ${}_nS_l^m$, or, since interest is restricted to zonal harmonics with $m = 0$, simply as ${}_nS_l$. The fundamental radial oscillation is ${}_0S_0$ and the fundamental "football" mode is ${}_0S_2$; spheroidal modes ${}_0S_3$, ${}_0S_4$, . . . , are motions with increasingly subdivided zonal distributions and each of these modes has overtones with internal nodal surfaces.

Oscillations of an essentially different type, in which the motion is entirely circumferential, with no radial component, are known as torsional or toroidal oscillations. In this case there are no displacements in the $r$, $\theta$ directions but only in the $\lambda$ direction:

$$w = \frac{W(r)}{\sin\theta}\frac{\partial P_l(\cos\theta)}{\partial\theta}\sin\omega t \tag{6.34}$$

where, as before, $\theta$ is measured from a "pole" of the motion that may be anywhere on the Earth. The simplest of these modes, ${}_0T_2$, is a twist between two hemispheres that oscillate in a rotational sense with a nodal plane between them (Fig. 6.15$c$).* Higher modes are due to a finer subdivision of the Earth into 3, 4, etc., zones with opposite motions and, as in the spheroidal case, there are overtones with internal nodal surfaces whose number is represented by the prefix $n$. The general

*An independent ${}_0T_1$ mode cannot exist because it implies an oscillation in the rate of rotation of the whole Earth and is incompatible with conservation of angular momentum. However, to preserve constant $I\omega$, such oscillations in angular velocity, $\omega$, must accompany (and coincide in period with) all spheroidal modes that cause an oscillation in axial moment of inertia $I$.

toroidal oscillation is ${}_nT_l^m$, with $m = 0$ in most cases of interest. Power spectra of Earth strains due to the Chilean 1960 and Alaskan 1964 earthquakes, recorded at Isabella, California, are reproduced in Figure 6.16, showing both spheroidal and toroidal periods. The relative strain amplitudes were not the same because the observatories were differently situated with respect to the nodes of the motions in the two cases.

There is a direct relationship between the higher-order free oscillations and long-period surface waves. The oscillations are simply standing waves (surface waves in the more familiar cases) and may be regarded as a superposition of oppositely traveling surface waves, Rayleigh waves (vertical polarization) in the case of the spheroidal modes and Love waves (horizontal polarization) for toroidal modes. This connection emphasizes an important feature of the oscillations: the lower modes, with periods up to 53 min, stress the whole Earth, but successively higher-frequency oscillations are increasingly concentrated in the outer part of the Earth, so that those with periods of a few minutes effectively stress only the upper mantle. The periods of the free oscillations are thus determined by the densities and elasticities of the internal layers with different weightings for the layers according to the mode of oscillation. It is this feature that makes free oscillations very useful in clarifying the details of internal structure.

Calculation of free oscillation periods for any particular Earth model is mathematically heavy work. Except for the special case of radial oscillations, the toroidal periods are easier to compute than the spheroidal ones, because toroidal modes involve a pure shear with no dilatation and therefore no change of density. Gravity does not contribute to the restoring forces on displaced particles as it does in the case of the spheroidal modes, in which there are radial displacements and density changes. Calculation of the free periods may be reduced to the solution of a second-order ordinary differential equation in the cases of radial or toroidal oscillations, to one of the fourth order for spheroidal oscillations if the effect of gravity is neglected, and one of sixth order if gravity must be allowed for. However, even the simplest case requires a lengthy numerical integration of the equations of motion through the assumed density and elasticity structure and the use of a digital computer is necessary if the computation is to proceed very far. For the development of the equations in suitable form, and a discussion of the method, reference should be made to Alterman et al. (1959). It turns out that earth models are more strongly constrained by periods of toroidal modes than of spheroidal modes.

Comparison of the periods computed for a particular Earth model with those of identified modes in the harmonic analysis of a free oscillation record, as in Figure 6.16, allows the model to be refined by a

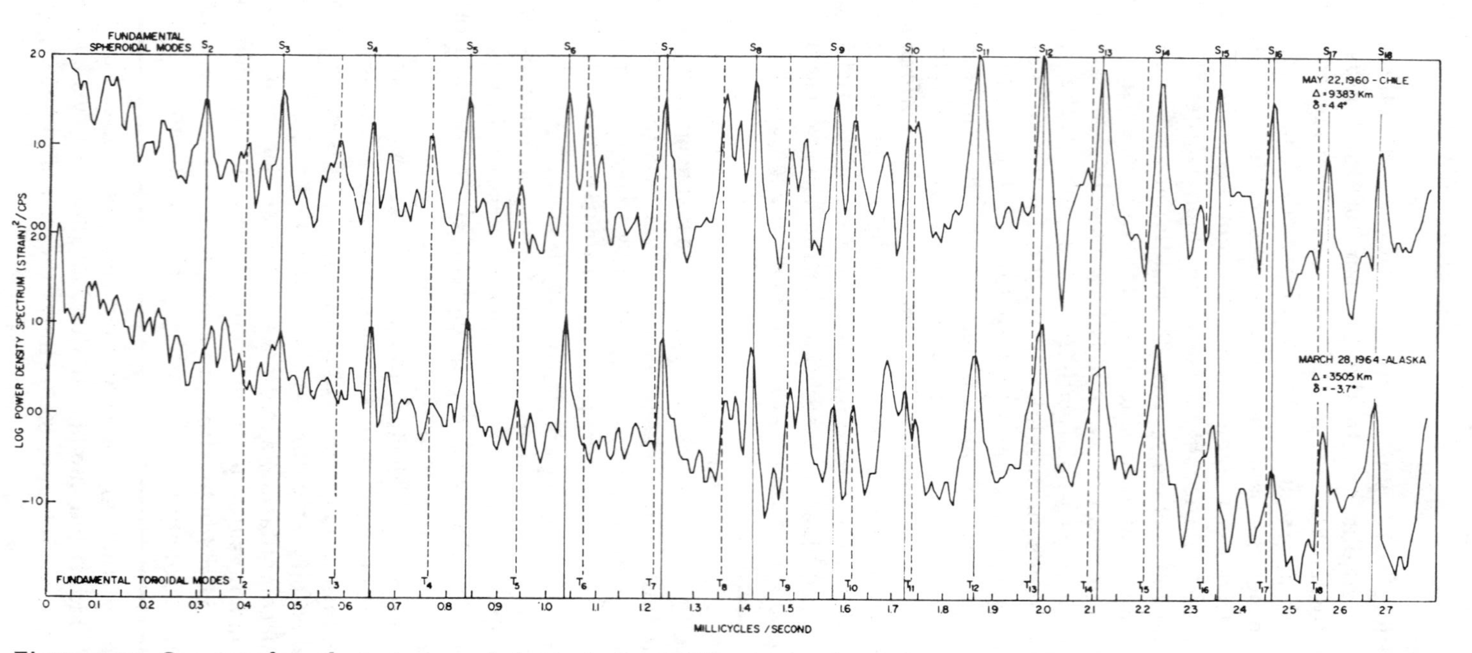


**Figure 6.16.** Spectra of earth strain recorded at Isabella, California, for the Chilean 1960 and Alaskan 1964 earthquakes. $\delta$ is the angle between the strain seismometer axis and the great circle path to the epicenter. Reproduced, by permission, from S. W. Smith (1967).

succession of minor adjustments that bring its periods into closer agreement with those observed. Bolt (1964) reviewed the situation after the Chilean earthquake records had been digested and more recent work has been reported by Bullen and Haddon (1967a, 1976b), Derr (1969), Press (1970), Jordan and Anderson (1974), Dziewonski et al. (1975), and Gilbert and Dziewonski (1975). The internal structure is sufficiently well established to allow unambiguous identification of the well-separated frequencies of the lower modes. Identification is aided, particularly for the higher modes, by the observational distinction between spheroidal and toroidal modes: strain and pendulum seismometers record both, but recording gravity meters see only the spheroidal oscillations because the toroidal modes involve no density changes or vertical accelerations. Also, certain modes may be poorly recorded by particular observatories which happen to be near to nodal lines for those modes. The term "terrestrial spectroscopy" was coined by Pekeris et al. (1961) for the study of free oscillation spectra; as they pointed out, the similarity to atomic spectroscopy includes the mechanical analogue of the Zeeman effect, an observed splitting of certain lines by the Earth's rotation and ellipticity.

Damping of free oscillations gives useful evidence of anelasticity in the deep interior. Assuming frequency independence of the intrinsic $Q$ of mantle material, the higher $Q$s observed for the lower modes reflect the high $Q$s for the deeper parts of the mantle (see Section 10.4). Of particular interest is the extraordinarily high $Q$ for the ${}_0S_0$ mode, reported by Slichter et al. (1966) to exceed 25,000 following the Alaskan earthquake. Since this mode is purely compressive, it may be inferred that damping of stress waves is primarily due to the shear component of stress.

## 6.4 INTERNAL DENSITY AND COMPOSITION

The seismic velocities $V_P$ and $V_S$, plotted in Figure 6.14 from the model tabulated in Appendix G, allow $K/\rho$ and $\mu/\rho$ to be determined throughout the Earth by Equations 6.12 and 6.13. Until free oscillation data became available, the development of a model Earth—with $K$, $\mu$, $\rho$ tabulated independently—required an appeal to other data or plausibility arguments. Constraints were imposed by the total mass and moment of inertia of the Earth (Eq. 3.27) and by the condition of hydrostatic stability, that is, density cannot decrease significantly inward without demanding implausibly high strength. In each of the recognized zones of the Earth, within which the velocity profile is smooth, it is reasonable to assume chemical and phase homogeneity. This allows the variation of density $\rho$ due to increasing pressure $P$ with depth $z$ to be calculated by the method of L. H. Adams and E. D. Williamson. We consider a spherically symmetric Earth and write

$$\frac{d\rho}{dz} = -\frac{d\rho}{dr} = -\frac{d\rho}{dP}\cdot\frac{dP}{dr} = \frac{\rho}{K}\cdot(g\rho) = \frac{Gm(r)\rho^2}{Kr^2} \tag{6.35}$$

where

$$m(r) = \int_0^r 4\pi r^2\rho(r)\,dr \tag{6.36}$$

is the mass inside radius $r$, which is responsible for the gravity $g$ at $r$ and $G$ is the gravitational constant. The quantity $K/\rho$ is determined at all depths from the velocities, since

$$V_P^2 - \tfrac{4}{3}V_S^2 = \frac{K}{\rho} = \phi \tag{6.37}$$

Thus for such a homogeneous zone the density gradient is determined by the density distribution itself and the problem is simply that of obtaining a self-consistent model. Birch (1952) pointed out that since the seismologically determined incompressibility was the adiabatic value $K_S$, adiabatic compression (i.e., an adiabatic temperature gradient) is assumed in the Adams-Williamson equation (6.35). This appears to be an adequate approximation through most of the Earth, but for application to a depth range in which the temperature gradient exceeds the adiabatic value by $\tau$,* Birch (1952) added a correction term:

$$\frac{d\rho}{dr} = -\frac{g\rho}{\phi} + \alpha\rho\tau \tag{6.38}$$

where $\alpha$ is volume expansion coefficient. The pioneering work of K. E. Bullen in the 1930s (comprehensively discussed in his book—Bullen, 1963) established Earth models using these principles. Subsequent improvements have been remarkably slight.

With the availability of free oscillation data, independent checks on the density and elasticity profile became possible. The number of identified modes with measured periods is now such that detailed models have been calculated without resort to the Adams-Williamson equation. Nevertheless it is evident that this equation accurately describes large regions of the Earth. Jordan and Anderson (1974) and Dziewonski et al. (1975) concluded from their model work that, within the

*The adiabatic gradient is positive downward, that is, $(dT/dz)_{\text{Adiabatic}} > 0$. $\tau$ is positive when the actual temperature gradient is steeper than the adiabat, since $\tau = (dT/dz) - (dT/dz)_{\text{Adiabatic}}$. The adiabatic gradient is considered in Chapter 7; see especially Equations 7.10 and 7.22.

precision of observations, the lower mantle and the outer core were each homogeneous and obeyed the Adams-Williamson equation (see also Davies and Dziewonski, 1975). The density profile of one model is shown in Figure 6.17.

Inferences concerning the compositions of the interior zones of the Earth require extrapolation of the model densities to zero pressure.

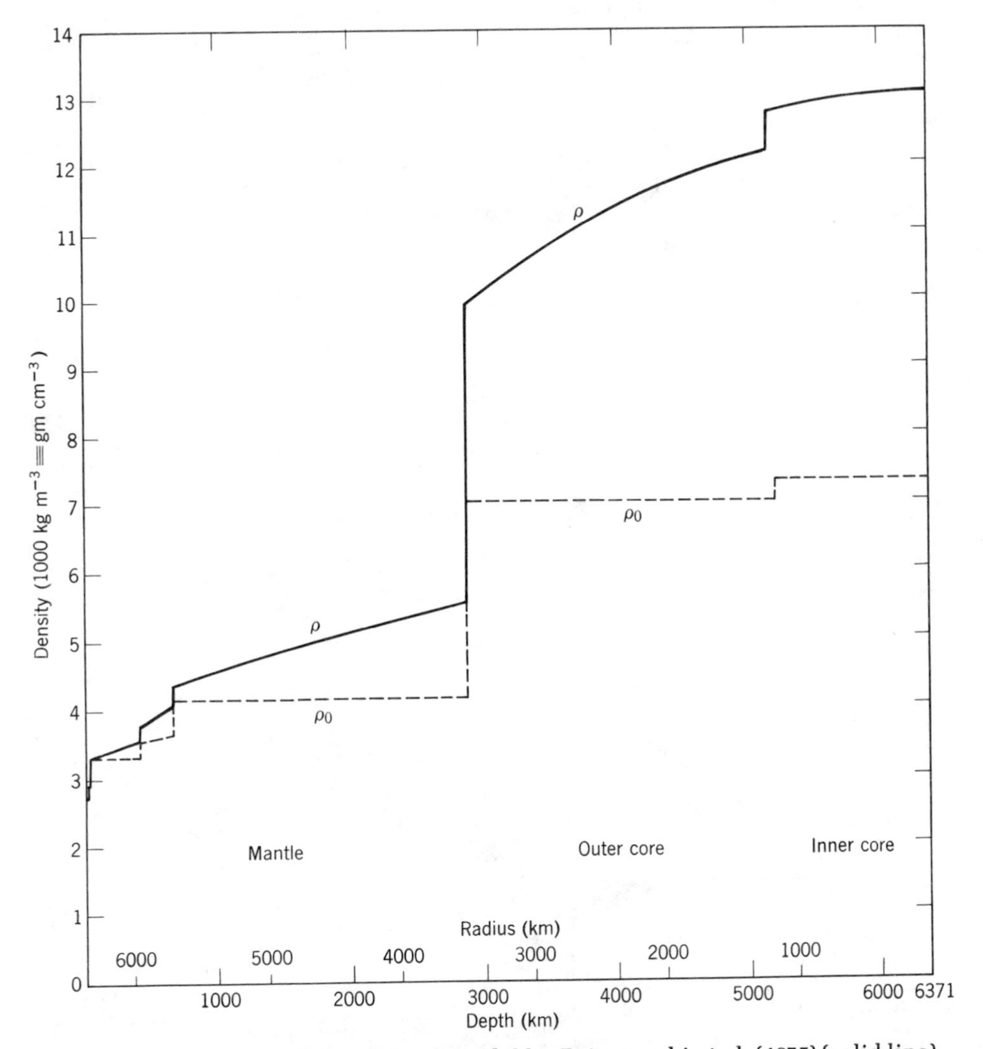


**Figure 6.17.** Density profile of Earth model by Dziewonski et al. (1975) (solid line) with corresponding extrapolated zero pressure (and room temperature) density (broken line).

Conventional elasticity theory, which assumes infinitesimal strain, is completely inadequate for this purpose because the elastic constants change dramatically under pressures comparable in magnitude to the elastic moduli. There are first-order changes in density, which can be described only by a finite strain theory. Figure 6.18 is a plot of the elastic moduli within the Earth as functions of pressure. This shows that the variations with pressure are sufficiently regular for simple linear relationships to represent reasonably well the behavior over moderate pressure ranges:

$$K = K_0 + K_0' P \tag{6.39}$$

$$\mu = \mu_0 + \mu_0' P \tag{6.40}$$

where subscripts 0 represent zero pressure values. In particular, for the upper part of the lower mantle (from 670 km to about 1500 km according

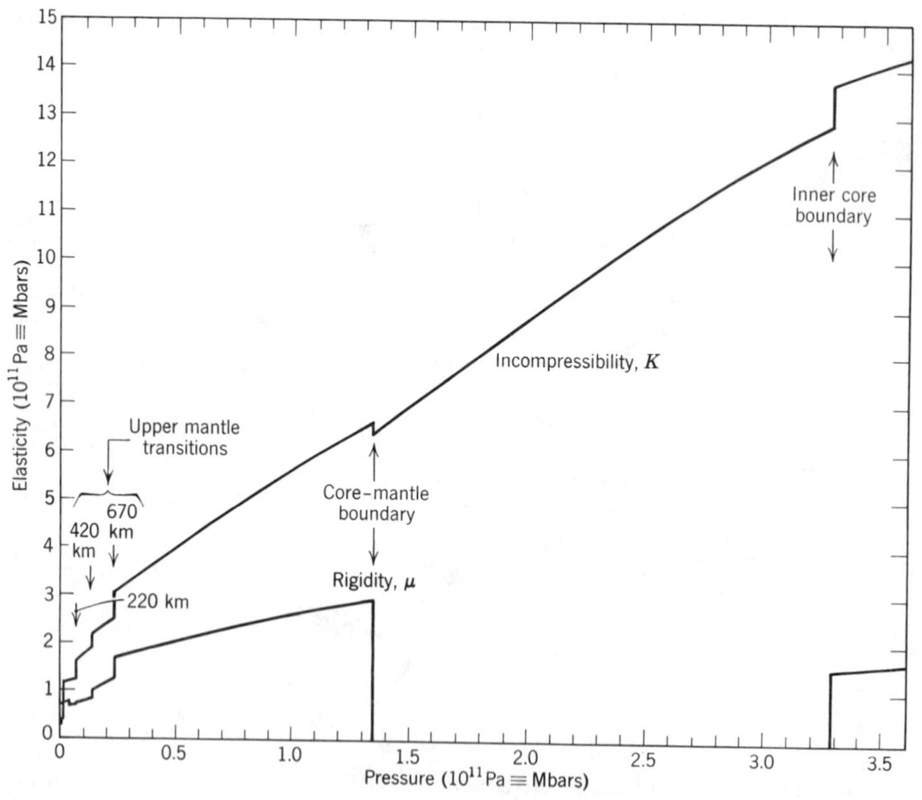


**Figure 6.18.** Variations of elastic constants with pressure in the interior of the Earth (model data by Dziewonski et al. 1975—Appendix G).

to the model data of Fig. 6.18):

$$K = 2.25 \times 10^{11}\,\text{Pa} + 3.35\,P \pm 2\% \tag{6.41}$$

$$\mu = 1.3 \times 10^{11}\,\text{Pa} + 1.4\,P \pm 5\% \tag{6.42}$$

Equation 6.39 was proposed by Murnaghan (1951) following an extended discussion of more elaborate hypotheses, and is frequently referred to as Murnaghan's equation.

Equation 6.39 allows the direct integration of a density profile, since

$$K = \rho \frac{dP}{d\rho} = K_0 + K_0' P \tag{6.43}$$

which integrates to

$$\frac{\rho}{\rho_0} = \left(1 + \frac{K_0'}{K_0} P\right)^{1/K_0'} = \left(\frac{K}{K_0}\right)^{1/K_0'} \tag{6.44}$$

The density profile of the lower mantle, as tabulated in Appendix G, fits this relationship well, with the values of $K_0, K_0'$ given in Equation 6.41, and yields a value for uncompressed density, $\rho_0 = 4.0 \times 10^3\,\text{kg m}^{-3}$. Since the adiabatic incompressibility is here used directly from the seismological model, the estimated zero pressure density represents adiabatic decompression from lower mantle conditions (see Chapter 7). Further extrapolation to room temperature gives $\rho_0 = 4.1_5 \times 10^3\,\text{kg m}^{-3}$. It is believed that this density corresponds to high-pressure phases of iron-magnesium-silicon oxides, as discussed later, and that there is no major discontinuity in chemical composition between the upper and lower parts of the mantle (Watt et al., 1975; Davies, 1974a; Davies and Dziewonski, 1975).

Linear extrapolation of the core data to zero pressure is less satisfactory primarily because it depends crucially on the values of $K_0, K_0'$, which are not well constrained, but also because the linear relationship, Equation 6.39, is not very satisfactory over this pressure range. Stacey (1972) endeavored to retain the simple, empirical approach by appealing to an estimate of the adiabatic incompressibility of laboratory iron at the melting point from the measured thermal expansion coefficient and Grüneisen's ratio (see Appendix E). He smoothly extrapolated the core data to the assumed zero pressure values ($K_{S_0} = 0.956 \times 10^{11}\,\text{Pa}$, $K_0' = 4.6$) and numerically integrated the expression

$$\ln\left(\frac{\rho}{\rho_0}\right) = \int_0^P \frac{dP}{K_S} \tag{6.45}$$

to obtain $\rho_0 = 6300\,\text{kg m}^{-3}$ (for $\rho = 9900\,\text{kg m}^{-3}$ at the top of the core). This estimate may err slightly on the low side but the important point is that there is no possibility of bringing it into line with the density of pure iron at the melting point (solid 7270 kg $m^{-3}$; liquid 7015 kg $m^{-3}$, Kirshenbaum and Cahill, 1962), which appears to be a reasonable approximation to the temperature of adiabatically decompressed core material (see Chapter 7). Thus the outer core must contain a substantial fraction of an element or elements much lighter than iron; as mentioned in Section 1.5 (see Table 1.2), and strongly advocated by Murthy and Hall (1970), sulfur must be rated as the main contender.

Numerous finite strain relationships more sophisticated than Equation 6.39 have been proposed. It is generally recognized that $dK/dP$ decreases at high pressures (a slight downward curvature is apparent in Fig. 6.18), so that $d^2K/dP^2$ is negative. But it is unsatisfying to truncate the series expansion with a negative term $-\frac{1}{2}K_0''P^2$, which eventually becomes dominant at sufficiently high pressures and leads to diminishing $K$. We need a more general expression, of which Equation 6.39 happens to be a good approximation for moderate pressures. The most satisfying approach is to appeal to an interatomic potential function that can be matched to an observed $K(P)$ relationship. The easiest to manipulate is the widely used Born-Mie potential:

$$E\left(\frac{r}{a}\right) = -A\left(\frac{a}{r}\right)^m + B\left(\frac{a}{r}\right)^n \tag{6.46}$$

in which generally $m = 1$, $n \approx 7$ and, for the condition that the atomic spacing, $r$, should have the equilibrium value, $a$, at zero pressure. $B = mA/n$. This may be most easily justified for ionic crystals. Probably more appropriate for materials with covalent or metallic bonds is a potential using an exponential repulsion term:

$$E(r) = -A \cdot \frac{a}{r} + B\left(\frac{a}{r}\right)^2 + C\exp(-r/R) \tag{6.47}$$

in which coefficient $B$ is probably necessary only in representing metals. The general form of an atomic potential function is shown in Figure E.1 (Appendix E).

The relationship between any potential function and the corresponding $K-P-\rho$ relations is obtained from the following equations. The total interaction energy* of $N$ atoms in a lattice is

$$\varepsilon = 3Nf_1E(r) \tag{6.48}$$

where $f_1$ is a structure factor, equal to unity for a simple cubic lattice, and the factor 3 arises because in a simple cubic lattice there are three bonds

*The interactions are not, in fact, independent but affect one another. The function $E(r)$, as used here, could not be observed with two independent atoms but is the average bond energy for uniform dilation or compression of a whole lattice. The number of bonds per atom (three for a simple cubic crystal) is half of the number of bonded neighbors to each atom.

per atom. These atoms occupy a volume

$$V = Nf_2r^3 \tag{6.49}$$

where $f_2$ is another structure factor, equal to unity in the simple cubic case. The ambient pressure is

$$P = -\frac{d\varepsilon}{dV} = -\frac{f_1}{f_2} \cdot \frac{1}{r^2}\frac{dE}{dr} \tag{6.50}$$

The low temperature isothermal incompressibility is

$$K(T = 0) = -V\frac{dP}{dV} = \frac{f_1}{3f_2}\left\{\frac{1}{r}\frac{d^2E}{dr^2} - \frac{2}{r^2}\frac{dE}{dr}\right\} \tag{6.51}$$

and the pressure dependence of $K$ is

$$\frac{dK}{dP} = \frac{dK/dr}{dP/dr} = \frac{\dfrac{d^3E}{dr^3} - \dfrac{3}{r}\dfrac{d^2E}{dr^2} + \dfrac{4}{r^2}\dfrac{dE}{dr}}{3\left(\dfrac{2}{r^2}\dfrac{dE}{dr} - \dfrac{1}{r}\dfrac{d^2E}{dr^2}\right)} \tag{6.52}$$

Of particular value in comparing these equations with observations, especially seismological data on the deep interior, is the relationship between dimensionless quantities $dK/dP$ and $P/K$, which do not include the structure factors. Thus at $P = 0$, Equation 6.46 gives

$$\left(\frac{dK}{dP}\right)_0 = K_0' = \tfrac{1}{3}(m + n + 6) \tag{6.53}$$

If $m = 2$, $n = 4$, Equation 6.46 yields the special case, for $\xi = 0$, of the well-known Birch-Murnaghan equation (Birch, 1952; Keane, 1954)

$$P = \tfrac{3}{2}K_0\left[\left(\frac{\rho}{\rho_0}\right)^{7/3} - \left(\frac{\rho}{\rho_0}\right)^{5/3}\right]\left\{1 - \xi\left[\left(\frac{\rho}{\rho_0}\right)^{2/3} - 1\right] + \cdots\right\} \tag{6.54}$$

and to include nonzero $\xi = \frac{3}{4}(4 - K_0')$ an additional repulsion term $+C(a/r)^6$ must be added to Equation 6.46. Other relationships that have been used are (1) the Morse potential (Slater, 1939, p. 452):

$$E(r) = A\left[e^{2c(a-r)} - 2e^{c(a-r)}\right] \tag{6.55}$$

(2) Bardeen's (1938) potential function:

$$E(r) = -A\left(\frac{a}{r}\right) + B\left(\frac{a}{r}\right)^2 + C\left(\frac{a}{r}\right)^3 \tag{6.56}$$

and (3) an empirical equation deduced from observations on metals by Grover et al. (1973):

$$\log K = \log K_0 + \alpha\frac{|\Delta V|}{V_0} \tag{6.57}$$

where $\alpha$ is a constant and $|\Delta V|/V_0$ is the fractional change in volume under pressure. Close examination shows that these equations are much more nearly alike than is immediately apparent. Further work is needed to narrow the range, but for most purposes Equation 6.47 (with or without the term involving $B$) should be favored.

Pressures in the deepest parts of the Earth are beyond the range of static, laboratory experiments but explosively generated, transient shock-waves have been used to measure compressions up to several megabars ($10^{11}$ Pa). The method is described by Rice et al. (1958), Al'tshuler (1965), and Keeler and Royce (1971) and applied to geophysical problems by McQueen and Marsh (1966) and McQueen et al. (1967). The general method is to fire a high-speed projectile at a target containing the material under examination, generating a shock wave that propagates through it at speed $v$, faster than the following material velocity, $u$ (or the normal sound speed). Taking the initial material or particle velocity to be zero, conservation of mass through the shock front gives

$$v\rho_0 = \rho(v - u) \tag{6.58}$$

since the shock moves at speed $v$ into material of uncompressed density $\rho_0$, but only at speed $(v - u)$ relative to the following, compressed material, density $\rho$. Thus the volume compression or density increase produced by the shock is

$$\frac{\rho}{\rho_0} = \left(1 - \frac{u}{v}\right)^{-1} \tag{6.59}$$

The pressure, $P$, in the shocked material is directly related to the rate of change of momentum per unit area of shock front. Taking momentum and pressure to be zero in the unshocked material, a mass per unit area $\rho_0 v$ (per second) is given velocity $u$, so that

$$P = \rho_0 v \cdot u \tag{6.60}$$

Thus, knowing $\rho_0$, observations of $u$ and $v$ for a shock give $\rho$ and $P$ by Equations 6.59 and 6.60 and a series of measurements with shocks of different intensities gives a $P - \rho$ curve or equation of state.

Such observations indicate the pressure-density relationship of a material at extremely high pressures, but the shocks produce intense heating for which a correction must be applied to obtain either isothermal or adiabatic equations of state. Shock compression is not simply adiabatic because the compressed material acquires kinetic energy, which is not allowed for in the standard thermodynamic equations representing adiabatic changes. The heating effect can be expressed in terms of conservation of energy through a shock front, across which there is a pressure difference, $P$, accelerating material to speed $u$. The power per unit area transmitted across a face in the shocked material (parallel to the shock front) is $Pu$ and this is equated to the rate at which kinetic energy plus internal energy of the material are increased:

$$Pu = \rho_0 v(\tfrac{1}{2}u^2 + E - E_0) \tag{6.61}$$

since the mass of shocked material increases at a rate $(\rho_0 v)$ per unit area, $E$ and $E_0$ being the internal energies (per unit mass) in the shocked and unshocked conditions. With substitutions from Equations 6.58 and 6.60 this gives

$$E - E_0 = \tfrac{1}{2}P\left(\frac{1}{\rho_0} - \frac{1}{\rho}\right) \tag{6.62}$$

An equation of state describing the behavior of a material in which the internal energy changes in the manner of Equation 6.62 is known as a Hugoniot, after one of the original authors of this equation. The Mie-Grüneisen equation (Eq. E.3—see Appendix E) relates the difference in pressure at fixed density between the initial low temperature* and a high temperature state of specified thermal energy since

$$E - E_0 = C_V(T - T_0) - \int_{V_0}^{V} (P - \alpha K_T T_0)\,dV \tag{6.63}$$

so that Equation E.3 gives

$$P_{T=0} = P - \gamma\rho C_V(T - T_0) = P\left[1 - \frac{\gamma}{2}\left(\frac{\rho}{\rho_0} - 1\right)\right] - \int_{V_0}^{V} (P - \gamma\rho C_V T_0)\,dV \tag{6.64}$$

where the integral refers to the shock compression curve.

Equation 6.64 corrects the pressure, as given by Equation 6.60, to that corresponding to the isothermal equation of state. For megabar pressures the correction is a substantial one, so that a reliable value of the Grüneisen parameter, $\gamma$, at high pressures is necessary for accurate determination of the equation of state. In Appendix E it is shown that $\gamma$ is related directly to the pressure dependence of incompressibility and can therefore be estimated from the shock-wave $P-\rho$ curve itself, but uncertainty in this correction has been one of the limitations in applying shock wave measurements (another being, the technical difficulties of the measurements). Nevertheless, shock wave data have given valuable confirmation of two important conclusions about the composition of the Earth's interior—the lower mantle is explicable in terms of close-packed high-pressure phases with an overall composition similar to that of the upper mantle (McQueen et al., 1967) and the core is about 8% less dense than pure iron (McQueen and Marsh, 1966).

*The argument used here is the classical one, which neglects the vanishing of specific heat at low temperatures, but since expansion coefficient vanishes similarly, keeping $\gamma$ approximately constant, this is an adequate assumption.

Birch (1961) introduced another approach to the problem of identifying the composition of the mantle from seismic data by reporting empirical evidence of a systematic relationship between seismic (or acoustic) velocities, densities, and mean atomic weights $\bar{m}$ of laboratory samples of oxides and silicates. He expressed his results in a simple linear form:

$$V_P = a(\bar{m}) + b\rho \tag{6.65}$$

which he suggested could accommodate variations due to pressure and temperature as well as phase and composition. It is now generally referred to as Birch's law. Subsequent workers (see, for example, O. L. Anderson and Nafe, 1965; D. L. Anderson, 1967; Shankland and Chung, 1974) have agreed that this linear relationship is a special case, for a limited range of values, of a more general power law, usually applied to the "bulk sound velocity,"

$$V_\phi = \phi^{1/2} = (V_P^2 - \tfrac{4}{3}V_S^2)^{1/2} \tag{6.66}$$

The trend represented by Equation 6.65 is generally agreed, but not without exceptions, difficulties and variations, so that no single law of simple form will explain all of the data (Liebermann and Ringwood, 1973; Liu, 1974a; Mao, 1974; Davies, 1974b). In particular the value of $b$ for phase transitions involving no change in anion-cation coordination is higher than for transitions with coordination changes. But the existence of an approximate systematic relationship between velocity and density for a wide range of oxides and silicates implies that the form of the interatomic potential energy function $E(r/a)$ is approximately a universal one, normalized to the equilibrium spacing, $a$.

A simple starting point for "derivation" of the general power law form of the velocity-density relationship, as advocated by Shankland (1972), is the integrated form of Murnaghan's equation (Eq. 6.44).

$$\frac{K}{K_0} = \left(\frac{\rho}{\rho_0}\right)^{K_0'} \tag{6.67}$$

which refers to the compression of a material at constant phase. But similarity of the form of the potential function for different materials means that for greater generality we should, in this equation, refer not to the ratio of densities but to the ratio of reciprocal atomic volumes $(\rho/\bar{m})$ where $\bar{m}$ is the average atomic weight, which is now allowed to be variable, constant composition or phase being no longer a requirement. Then we can put

$$K = A^2\left(\frac{\rho}{\bar{m}}\right)^{K_0'} \tag{6.68}$$

where $A$ is a constant. Thus the "bulk sound velocity" is given by

$$V_\phi = \left(\frac{K}{\rho}\right)^{1/2} = A \cdot \frac{\rho^{\frac{1}{2}(K_0'-1)}}{(\bar{m})^{\frac{1}{2}K_0'}} = A\frac{\rho^\lambda}{\bar{m}^{(\lambda+\frac{1}{2})}} \tag{6.69}$$

where $\lambda = \frac{1}{2}(K_0' - 1)$ is approximately a constant. Shankland (1972) added a factor $c$ to account for effects of different ionic sizes, especially $Ca$, which Simmons (1964) showed to require special treatment:

$$V_\phi = A\frac{\rho^\lambda}{\bar{m}^{[\lambda(1-c)+\frac{1}{2}]}} \tag{6.70}$$

Then, conventionally, the equation is normalized to a standard value of $\bar{m}$, $m_0$. With numerical values favoured by Shankland and Chung (1974), who put $m_0 = 20.2$:

$$V_\phi = 1.42\rho^\lambda(20.2/\bar{m})^{[\lambda(1-c)+\frac{1}{2}]} \tag{6.71}$$

for $V_\phi$ in km sec$^{-1}$, $\rho$ in $10^3$ kg m$^{-3}$ (gm cm$^{-3}$).

For the lower mantle $\lambda = \frac{1}{2}(K_0' - 1) = 1.2$, by Equation 6.41. Neglecting $c$ (which introduces no error if $\bar{m}/m_0 \approx 1$) and assuming values of $V_\phi$ and $\rho$ below the phase transition at 670 km in the model data in Appendix G, we obtain $\bar{m} = 20.2$ for the lower mantle. On this basis it appears that, within the uncertainties of the present calculations, the lower mantle has a composition indistinguishable from the upper mantle, in accord with the conclusions of several authors (e.g., Mao, 1974), although there have been suggestions for more complex lower mantle structures with iron content increasing with depth (e.g., Anderson and Jordan, 1970) and discontinuities or sharp gradients at depths greater than 670 km.

In reviewing the composition of the Earth it is usually claimed that homogeneity and spherical (or, more correctly, elliptical) symmetry improves with depth. This is reflected in the model in Appendix G, which recognizes that the structures of continents and oceans extend to several hundred kilometers at least but treats the Earth at greater depths as having no lateral heterogeneities. While it makes sense to suppose that lateral heterogeneities are consequences of tectonic processes (Section 10.1) and that these are at least much stronger in the upper mantle than at greater depths, it appears incautious to suppose that this is anything but a result of the inadequacy of geophysical methods to distinguish fine details at great depths. The depth and manner of mantle convection are still subject to doubt and disagreement, but the evidence for lower mantle convective "plumes" (see also Section 2.3) is becoming stronger and, as considered in Chapter 7, it is difficult to devise a mechanism for extracting core heat without allowing lower mantle convection. We note that at the base of the mantle, where seismic observations are sensitive to heterogeneities, evidence of them is in fact found (Haddon and Cleary, 1974). Further, Bolt and Nuttli (1966) found that wavefronts of seismic waves traveling across an array of seismic stations in California departed in a regular way from normals to the great circles through the epicenters from which they originated. The deviation was sensitive to the direction of the epicentre, the maximum value being 11°, which implies substantial mantle inhomogeneity. Thus the conventionally assumed "smoothness" of the lower mantle is probably merely our broad scale (i.e., distant) view of it.

The strongly bimodal distribution of crustal structures into characteristically continental and oceanic types is referred to in Section 4.3. The continents protrude above the general level of the ocean basins by virtue of their thicker crust (30 to 40 km compared with about 5 km), which is less dense than the underlying mantle by about 500 kg $m^{-3}$ (0.5 gm $cm^{-3}$). The boundary (Mohorovičić discontinuity or "Moho") is almost worldwide, but may be absent under ocean ridges where new basaltic crust is forming. As shown both by thermodynamic arguments (Bullard and Griggs, 1961), and by high pressure experiments on basic rocks (Ringwood and Green, 1966), the density contrast must be attributed to a difference in composition between crust and mantle materials. The crust grades from acid (silica rich) rocks, such as granite (density about 2700 kg $m^{-3}$), which is characteristic of the upper parts of the continents, to basic (low silica) rocks, such as basalt (density about 2900 kg $m^{-3}$), which forms the thin oceanic crust and probably also underlies the continental acid rocks. The mantle is termed ultrabasic, that is, very low in silica and correspondingly rich in magnesium and iron, with uncompressed density about 3300 kg $m^{-3}$.

Lateral inhomogeneity in the upper mantle is related to the presence or absence of a low velocity layer. That such a layer is expected in a chemically homogeneous region of the upper mantle can be seen directly from the data on the pressure and temperature dependences of the wave velocities $V_p$ and $V_s$, as summarized by Anderson et al. (1968). Taking forsterite ($Mg_2SiO_4$) as more representative of the mantle than the other minerals they consider, the relevant values are as follows:

For compressional waves:

$$\left(\frac{\partial V_p}{\partial p}\right)_T = 10.3 \times 10^{-3}\,\mathrm{km\ sec^{-1}\ kb^{-1}} \tag{6.72}$$

$$\left(\frac{\partial V_p}{\partial T}\right)_p = -4.1 \times 10^{-4}\,\mathrm{km\ sec^{-1}\ {}^{\circ}C^{-1}} \tag{6.73}$$

For shear waves:

$$\left(\frac{\partial V_s}{\partial p}\right)_T = 2.45 \times 10^{-3}\,\mathrm{km\ sec^{-1}\ kb^{-1}} \tag{6.74}$$

$$\left(\frac{\partial V_s}{\partial T}\right)_p = -2.9 \times 10^{-4}\,\text{km sec}^{-1}\,{}^\circ\text{C}^{-1} \tag{6.75}$$

A low velocity layer occurs if $(dV/dz)$ is negative, where $z$ is depth measured positive downward, and since

$$\frac{dV}{dz} = \left(\frac{\partial V}{\partial p}\right)_T \frac{dp}{dz} + \left(\frac{\partial V}{\partial T}\right)_p \frac{dT}{dz} \tag{6.76}$$

and $dp/dz = 0.33$ kb km$^{-1}$ in the upper mantle, the temperature gradients which must be exceeded to form a low velocity layer are 8.3°C km$^{-1}$ for compressional waves and 2.8°C km$^{-1}$ for shear waves. These values refer only to a particular mineral* but the conclusion is a more general one. The upper mantle temperature gradient is almost inevitably steep enough to develop a velocity minimum for $S$ waves, but the gradient at, say, 150 km depth can hardly suffice to give a minimum for $P$ waves without invoking melting. The velocity minimum is therefore usually interpreted as a consequence of partial melting at about that depth, but with the proportion of fluid not exceeding a few percent.

As implied earlier in this section, the transitions in density and elasticity in the upper mantle, at least two of which appear to be more or less sharp, are attributed to phase transitions of iron-magnesium silicates to denser forms that are stable at high pressures. The major discontinuities that are explained in this way are at depths of about 400 km and 650 to 700 km (420 km and 670 km in the model in Appendix G and Figs. 6.17 and 6.18). The representative upper mantle mineral is olivine, $(Fe, Mg)_2\,SiO_4$. At pressures of about 140 kbar ($1.4 \times 10^{10}$ Pa), corresponding to the first of these transitions, olivine transforms to a spinel structure with a density increment of about 10%† (Ringwood and Major, 1970). At pressures of order 250 kbar ($2.5 \times 10^{10}$ Pa), corresponding to the 670-km transition, a separation into independent oxides is observed (Ming and Bassett, 1975), in the case of quartz the rutile structure (stishovite) being favored. However silicates of perovskite structure that are slightly denser than the corresponding mixed oxides are also both expected and observed (Ringwood 1970b; Liu, 1974b) and these are undoubtedly important in the lower mantle, even if separate oxides also occur.

Interpretation of the core composition in terms of its density has been subject to some doubt about the phase of solid iron under core

*Critical gradients for a wider range of minerals are given by Liebermann and Schreiber (1969).

†Forsterite (Mg-rich) end members of the olivine ($[Mg, Fe]_2\,SiO_2$) solid solution series are first converted to an intermediate (beta) phase with a 7% density increment and to a spinel form at somewhat higher pressures.

conditions. At room temperature iron transforms to the $\epsilon$ (hexagonal close packed) phase at a pressure of about 120 kbar ($1.2 \times 10^{10}$ Pa) (Clendenen and Drickamer, 1964; Takahashi and Bassett, 1964) with a density increment, extrapolated to zero pressure, of about 5%. Properties of $\epsilon$ iron are presumed to be similar to those of cobalt, which has the h.c.p. structure at low pressure and higher values of $K_0$ and $K_0'$ than $\alpha$ or $\gamma$ iron. However, the pressure at which the transition occurs increases with temperature and equation of state studies using core data on $dK/dP$ versus $P/K$ by R. D. Irvine and the author clearly favor $\gamma$ iron as the equilibrium solid phase corresponding to the outer core. The density decrement on melting of $\gamma$ iron at 1 bar is 4.3% [3.5% for $\alpha$-iron (Kirshenbaum and Cahill, 1962) plus 0.8% to account for the $\alpha \rightarrow \gamma$ transition] and at core pressures comparison of the Clausius-Clapeyron and Lindemann melting laws (Eqs. 7.24, 7.25) with the assumption of constant entropy of melting gives a density change of 2.5% but virtually no difference between the values of $K$ for the solid and liquid phases. The least well-known contribution to estimated density arises from the effect of electron pressure on the equation of state (see, for example, Appendix J, Problem 6.11) but it probably amounts only to 1%. More important is the problem of thermal expansion (or thermal pressure—see Appendix E). Assuming core temperatures as in Section 7.3, the core material is found to be about 7% less dense than pure iron; to this figure should be added 1.5% for every 1000° underestimation of the temperature. This is in excellent agreement with shock wave measurements (McQueen and Marsh, 1966) which led to an estimated 8% difference. If the lighter component is primarily sulfur (Murthy and Hall, 1970; King and Ahrens, 1973) then a mass amounting to 10 to 15% of the outer core appears to be required.

Interpretation of inner core data is more problematical. Solidity is indicated by the appearance of rigidity, but the rigidity is abnormally low relative to incompressibility and the very high value of Poisson's ratio ($\nu = 0.45$) is difficult to reconcile with a close-packed crystalline solid. A composition of solid nickel-iron, or of (Fe, Ni)-S eutectic, is generally favored. The relevant phase of solid iron is the $\gamma$ (face-centered cubic) form. Although iron transforms to the $\epsilon$ (hexagonal close packed) form at $1.35 \times 10^7$ Pa (135 kbar) at room temperature, the $\gamma \rightarrow \epsilon$ transition is governed by an equation of the form (7.25) and so the transition cannot intersect the melting point at any pressure. The possible relevance of an electronic phase transition in iron (Elsasser and Isenberg, 1949) is now discounted (Bukowinski, 1976).

# 7

# The Earth's Internal Heat

"Conclusions and hypotheses concerning the earth's interior are in a state of flux."

GUTENBERG, 1959, p.v.

## 7.1 THE GEOTHERMAL FLUX

The outward flow of heat through the crust is the only directly measurable feature of the Earth's internal heat. Determinations of heat flow require measurements of temperature gradient in the crust and of the thermal conductivities of the rocks in which the gradient is measured. In principle these measurements are straightforward but in practice the necessary precautions, such as the avoidance of ventilation in mines and of rock formations in which flow of ground water could upset measurements, require the exercise of considerable care. Most of the useful data are very recent and since it is now easier to obtain data at sea, most of the values are of heat flow through the ocean floor.

The data used here are from the review by Lee (1970), who summarised over 3500 measurements. New data are being acquired continuously but large areas of the Earth are still unrepresented in heat flow tabulations. However, the general global pattern and relationships to geological features are clear. A collection of reviews, including measurement techniques, has been edited by Lee (1965).

Following the conclusion of Birch (1948), it has generally been assumed that global climatic changes (especially ice ages) have only a slight effect on the heat flux deduced from measurements of temperature gradient in the upper 1 km or so of crust, but Crain (1968) and Beck (1970) questioned this assumption. A thermal wave imposed on the surface penetrates the crust*, modulating the steady geothermal gradient, but attenuating with depth. Under the least favorable conditions the temperature increment introduced to the lower half of a borehole used for thermal measurements is no more than 17% of the peak to peak wave amplitude at the surface. The error is of course minimized by taking measurements over the greatest possible depth range and it appears that boreholes not less than 1 km deep are required for reliable observations. Thus marine measurements, which are made with probes penetrating the sea floor sediment by only a few meters, depend even more critically on constancy of the ambient (sea water) temperature. Fortunately it does appear that, so long as the Earth has frozen polar caps, melt water flows along the ocean bottom to all deep parts, maintaining a constant, low temperature. Von Herzen and Maxwell (1964) found that the temperature gradient down a deep-ocean drill hole gave the same value of heat flow as simple probes nearby, so that it is now improbable that a significant fraction of the ocean floor heat flow is due to residual (stored) heat from a climatic warm period with no polar caps. For most areas it appears that Birch's (1948) conclusion—that calculated values of heat flux are subject to an error less than 10%—is valid, but special consideration may be required for areas that were heavily glaciated in the Pleistocene period.

A general correlation of heat flow with geology is apparent in Table 7.1. The more recent the volcanic origin (orogeny) of an area, the higher its heat flow is likely to be. Although this generalization is presumed to be significant, it must be noted that heat flow can be locally very variable, especially over features such as the mid-Atlantic ridge, where there are apparently localized sources of heat within a few kilometers of the surface. Possibly these are merely equivalent to volcanic or thermal areas on land and do not need to be too highly regarded in the overall pattern of heat flow, to which their contribution is small. It is easy to understand why the heat flux from a recently volcanic area should be high, while the ancient, pre-Cambrian shields, which have been inactive for billions of years and from which a considerable mass of acidic rock (with greater than average radioactive content) has been eroded, are characterized by low heat flux. But the data in Table 7.1 really show

*A thermal oscillation $T_0 \sin \omega t$ imposed on the plane surface ($z = 0$) of a semi-infinite medium of diffusivity $\eta$, penetrates as an attenuating wave represented by

$$T(z) = T_0 \exp\left(-\sqrt{\frac{\omega}{2\eta}}\, z\right) \sin\left(\omega t - \sqrt{\frac{\omega}{2\eta}}\, z\right)$$

having speed $\sqrt{2\eta\omega}$ and attenuation length $z^* = \sqrt{2\eta/\omega}$. For $\eta = 1.2 \times 10^{-6}$ m$^2$ sec$^{-1}$, representative of igneous rock, and $\omega = 2 \times 10^{-11}$ sec$^{-1}$, corresponding to a period of $10^4$ years, $z^* = 350$ m. (See also footnote, p. 188.)

**Table 7.1** Average Values of Heat Flow from Each of Several Tectonically Different Types of Crust[a]

| | | |
|---|---|---|
| Continental areas | | |
| Pre-Cambrian shields | 4.10 ± 1.00 | (214) |
| Post pre-Cambrian, nonorogenic areas | 6.24 ± 1.72 | (96) |
| Paleozoic orogenic areas | 5.99 ± 1.67 | (88) |
| Mesozoic-Cenozoic orogenic areas | 7.37 ± 2.43 | (159) |
| Continents—grid average[b] | 6.11 ± 1.93 | (95) |
| Oceanic areas | | |
| Ocean basins | 5.32 ± 2.22 | (683) |
| Midocean ridges | 7.95 ± 6.20 | (1065) |
| Ocean trenches | 4.86 ± 2.93 | (78) |
| Continental margins | 7.53 ± 3.89 | (642) |
| Oceans—grid average[b] | 6.15 ± 3.27 | (591) |
| Whole Earth—grid average[b] | 6.15 ± 3.10 | (673) |
| Whole Earth—mean heat flow | 6.15 ± 0.34 | |

[a]Data from Lee (1970), *converted to units of* $10^{-2}$ $W\ m^{-2}$, with standard deviations. Figures in parentheses give numbers of data points averaged.

[b]Each of the values contributing to a grid average is itself an average over a 300 × 300 nautical mile square (5° × 5° at the equator). This reduces the bias of heavily sampled areas but may overemphasize isolated measurements.

remarkably little variability, bearing in mind the great differences between crustal types represented. Most striking is the equality of the continental and oceanic averages. This must be considered in the light of the differences in the radioactive contents of the different types of crust. Sclater and Francheteau (1970) reviewed the variability of heat flux over the Earth's surface in the light of current tectonic ideas and reached the important conclusion that the oceanic heat flux is largely explained by the cooling of fresh lithosphere, as considered semiquantitatively later in this section.

Several authors have sought a correlation between heat flux and gravity, or more specifically geoid height, supposing that the hot, rising limb of a convective cell would be lighter than average, by virtue of thermal expansion, giving a correspondence between high heat flow and geoidal lows. Higbie and Stacey (1971) examined the problem by comparing heat flows from the 5° × 5° areas of Lee (1970) with the geoid heights at the centers of these areas, as determined from the Gaposchkin and Lambeck (1971) spherical harmonic coefficients of the geoid. For both continental and oceanic areas the correlation coefficients were positive, rather than negative as anticipated, but were so small that the proper conclusion is that there is no correlation. The randomness of the data is apparent in Figure 7.1.

The thermally significant radioactive elements are U, Th and K. Rb is next most important but, as far as we are aware, it can be neglected in geothermal considerations. Heat productions by these elements are given in Table H3 (Appendix H) and the element data are combined with average concentrations in representative rocks (and meteorites) in Table 7.2. As was first noticed by Strutt (Lord Rayleigh) (1906), typical crustal igneous rocks have far higher concentrations of radioactivity than required for the Earth as a whole to explain the geothermal flux. In fact a 22-km layer of granite with the average radioactive content in Table 7.2 would produce the entire continental heat flux. Clearly not even the continental crust could be granite with this composition. Taking a more probable continental crustal structure, with intermediate (i.e., more basic) rocks at depth, we still find that at least two thirds of the continental heat flux originates in the crust itself, with not more than 0.015 to 0.02 $\mathrm{Wm^{-2}}$ coming from the mantle. On the other hand the oceanic crust, considered to be simply 5 km of tholeitic basalt, produces only $4 \times 10^{-4}\ \mathrm{Wm^{-2}}$, implying that virtually the entire oceanic heat flux is from the mantle. The observed variations in ocean floor heat flux must be due to differences in the heat flux from the mantle. This observation contrasts with the conclusion of Roy et al. (1968) that variability of the

**Table 7.2** Approximate Average Concentrations of Radioactive Elements and Corresponding Heat Productions in Different Geological Materials.

| | Material | Concentration (parts per million by weight) U | Th | K | Heat Production ($10^{-12}$ W $\mathrm{kg^{-1}}$) |
|---|---|---|---|---|---|
| Igneous crust | granites | 4.6 | 18 | 33000 | 1050 |
| | alkali basalts | 0.75 | 2.5 | 12000 | 180 |
| | tholeitic basalts | 0.11 | 0.4 | 1500 | 27 |
| Upper mantle | ecologites | 0.035 | 0.15 | 500 | 9.2 |
| | peridotites, dunites | 0.006 | 0.02 | 100 | 1.5 |
| Meteorites | carbonaceous chondrites | 0.020 | 0.070 | 400 | 5.23 |
| | ordinary chondrites | 0.015 | 0.046 | 900 | 5.85 |
| | iron meteorites | — | — | — | $<3 \times 10^{-4}$ |

Values of heat production are to be compared with the geothermal flux per unit mass of the Earth, $5.2_6 \times 10^{-12}$ W $\mathrm{kg^{-1}}$. Values of heat production for elements are given in Table H3 (Appendix H).

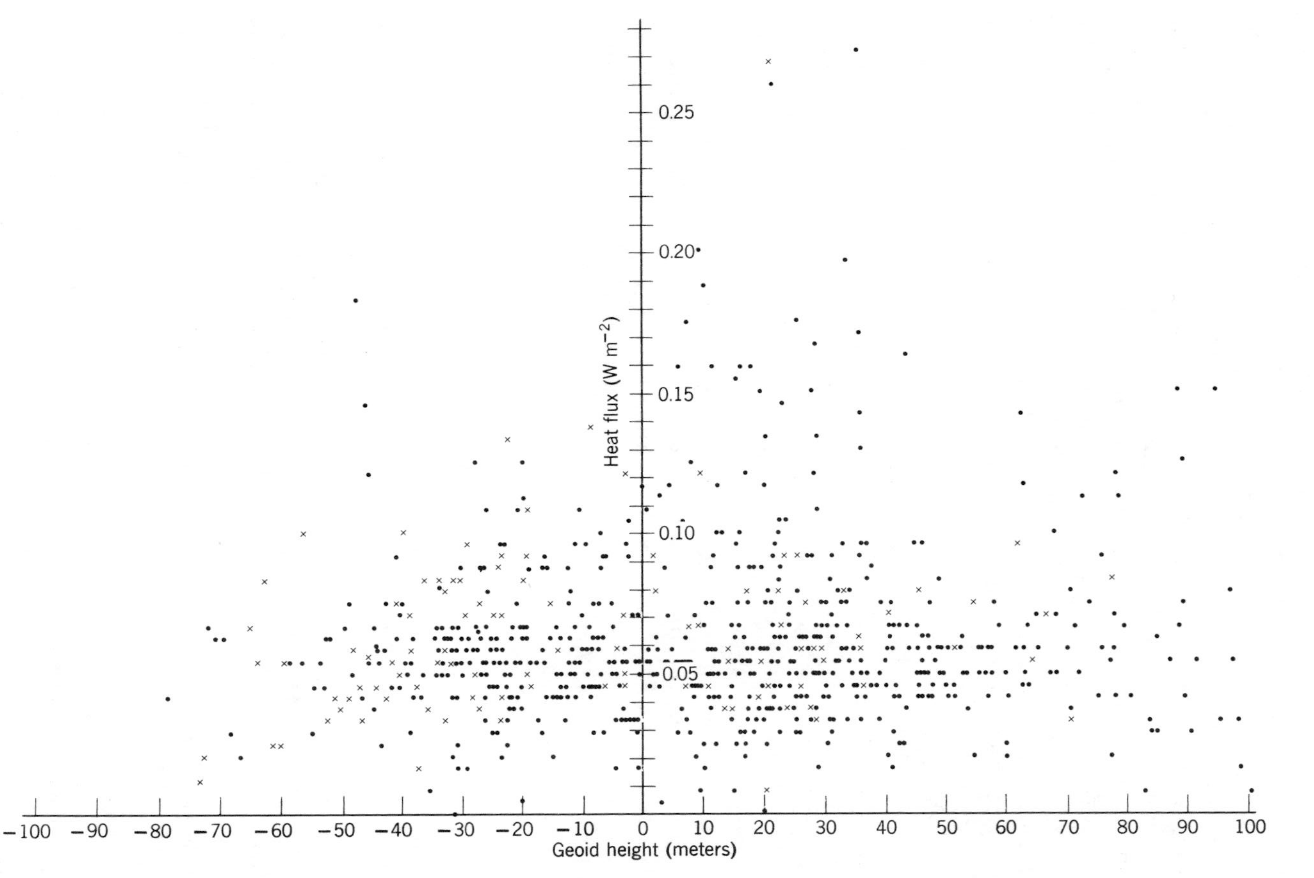


**Figure 7.1.** Comparison of average heat flows for 5° × 5° areas from the data of Lee (1970) with geoid heights at the centers of the areas from geoid coefficients of Gaposchkin and Lambeck (1971). Crosses represent land data and circles, marine data. Based on a plot by Higbie and Stacey (1971).

heat flux within a limited continental area (heat flow "province") was explicable entirely in terms of the radioactive contents of different crustal rocks, with a constant mantle flux.

Equality of the continental and oceanic heat fluxes has conventionally been explained as a consequence of differentiation whereby most of the radioactive elements of the subcontinental mantle have been segregated into the acidic rocks that have risen to form the continental crust, but leaving the suboceanic mantle essentially undepleted, so that the total radioactive heat generation in any vertical column is roughly constant. But this explanation is too simplistic. Both continents and ocean floors are moving in the process of plate tectonics (Section 10.1). Although the ocean floors apparently move more readily, we cannot regard the continents as stationary features overlying immobilized, depleted regions of the mantle. Instead, the heat flux equality must be regarded as a consequence of the tectonic process, whether fortuitously, as argued by Sclater and Francheteau (1970) or as a necessary consequence of a dynamic balance, as Elsasser (1967) suggested. For this purpose we should not regard the crust as a separate entity but refer to the *lithosphere*, the essentially rigid layer, 70 to 200 km thick, that comprises the tectonic plates overlying the softer mantle. The crust moves merely as part of the lithosphere. If we take the oceanic lithosphere to have a thickness $D = 70$ km (rather thinner than the continental lithosphere) and to be comprised of igneous rock with a thermal diffusivity* $\eta = 1.2 \times 10^{-6}\,\mathrm{m^2\,sec^{-1}}$, which is representative of laboratory samples, then the thermal time constant for cooling of the layer is

$$\tau_{\mathrm{Thermal}} \approx \frac{D^2}{\eta} = 1.3 \times 10^8 \text{ years} \tag{7.1}$$

But this is approximately its residence time at the surface between formation at an ocean ridge and return to the deeper mantle at a subduction zone. Thus there is not time for heat flux from the deeper part of the mantle to become established through the oceanic lithosphere, which essentially gives up its residual heat (plus a small radiogenic contribution) and then disappears again. This conclusion is

*Thermal diffusivity is defined in terms of conductivity $K$, density $\rho$, and specific heat (per unit mass) $C$:

$$\eta = \frac{K}{\rho C}$$

being the ratio of conductivity to heat capacity per unit volume. It expresses the ability of a body to lose its internal heat by conduction. That the time required varies as the square of size may be verified readily for cases of simple geometry since a doubled size doubles the heat capacity but halves the temperature gradient for fixed temperature drop.

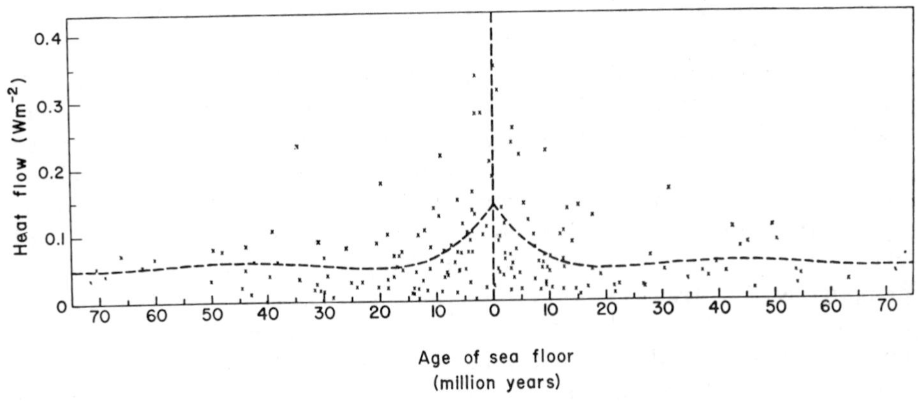


**Figure 7.2.** Heat flow versus age of the Atlantic sea floor as determined by distance from the mid-Atlantic ridge. Adapted from Lee (1970).

reinforced by the data in Figure 7.2. The oceanic heat flux must therefore be regarded as convectively transported heat from the deep interior, not a consequence of strong radioactivity of the suboceanic mantle, but R.A. Hart (1973) argues that a slow exothermic reaction between sea water and basalt contributes appreciably to the ocean floor heat flux.

We cannot apply the same elementary argument directly to the continental lithosphere. Since the heat flux from the subcontinental mantle is only a third or quarter, on average, of the flux from the suboceanic mantle, it is reasonable to expect the continental lithosphere to be thicker, giving a slower transfer of heat. But we do not see evidence that it is reabsorbed into the mantle. In this case it is more reasonable to suppose that the lower boundary (perhaps rather ill-defined) is maintained at the temperature of softening, even to the point partial melting, by convection of the lower more fluid region and that heat is conducted to the surface in a quasi-equilibrium state. A reasonable estimate for the temperature of this boundary would be 1300°C. The temperature at the base of the crust (depth about 35 km) is estimated to be about 600°C, requiring a temperature drop, $\Delta T$, of about 700°C in the subcrustal continental lithosphere. The mantle-derived heat flux through this layer is $(1/A)(dQ/dt) \approx 0.015$ W m$^{-2}$ and with a conductivity, $K = 2.5$ W m$^{-1}$ deg$^{-1}$, characteristic of igneous rock, we can estimate the required mantle thickness $\Delta z$ from the equation for thermal conduction

$$\frac{1}{A}\frac{dQ}{dt} = K\frac{\Delta T}{\Delta z} \tag{7.2}$$

giving $\Delta z \approx 117$ km. Added to the assumed crustal thickness (35 km), this gives a total thickness of about 150 km for the continental lithosphere, in reasonable accord with other estimates.

Another apparently significant coincidence, noted by Birch (1958) and documented in Table 7.2, is the close equality between the geothermal flux and the heat generated by radioactivity in a corresponding mass of average chondritic composition. Both for ordinary chondrites and for carbonaceous chondrites the agreement is still well within the variability of the observations. However, as Gast (1960) pointed out and Wasserburg et al. (1964) emphasized, the relative abundances of the principal radioactive elements in terrestrial rocks and in meteorites are consistently different, so that one or more processes of differentiation must have occurred and we must enquire whether this numerical agreement is fortuitous. The differences are conveniently displayed as a plot of K/U ratio as a function of potassium content (Fig. 7.3). Th is generally found associated with U, being about 3.5 times as abundant.

One possibility is that the Earth as a whole has the U, Th, and K abundances of carbonaceous chondrites and that the consistently lower K/U ratio of terrestrial rocks is compensated by an excess of K at depth. Several authors (Goles, 1969; Lewis, 1971; Hall and Murthy, 1971; Goettel, 1972) have argued that there are geochemical reasons for assigning the excess to the outer core, but this suggestion lacks laboratory support (Oversby and Ringwood, 1972; Seitz and Kushiro, 1974; Ganguly and Kennedy, 1977). The main theme of the argument has been over the question, does sulfur in the fluid outer core attract potassium? Laboratory evidence on the partitioning of potassium between Fe-S and silicate melts indicates that it does not but there is another possibility. At sufficiently high pressures potassium is expected to undergo an electronic phase transition similar to that observed at 45 kbar ($4.5 \times 10^9$ Pa) in cesium (Sternheimer, 1950). Such a transition would compress the $4s$ (valence) electrons into the empty $3d$ shells, converting potassium electronically to an element of the first transition series (which includes iron and nickel). In this state potassium would become less active chemically and may quite reasonably have an affinity for metallic iron rather than oxides, independently of the presence of sulfur. This becomes geophysically significant if the transition pressure for K is, say, 260 kbar ($2.6 \times 10^{10}$ Pa), as suggested by Drickamer and Frank (1973, pp. 112–114) from low temperature observations.

Independently of geochemical arguments, the geophysical requirement for a substantial heat source in the core, sufficient to maintain it in a liquid state against the conducted heat loss down an adiabatic temperature gradient, is compelling (Section 7.3). We should therefore examine the consequences of taking at face value the abundance data of Figure 7.3. Assuming the carbonaceous chondritic abundances ($K/U = 2 \times 10^4$,

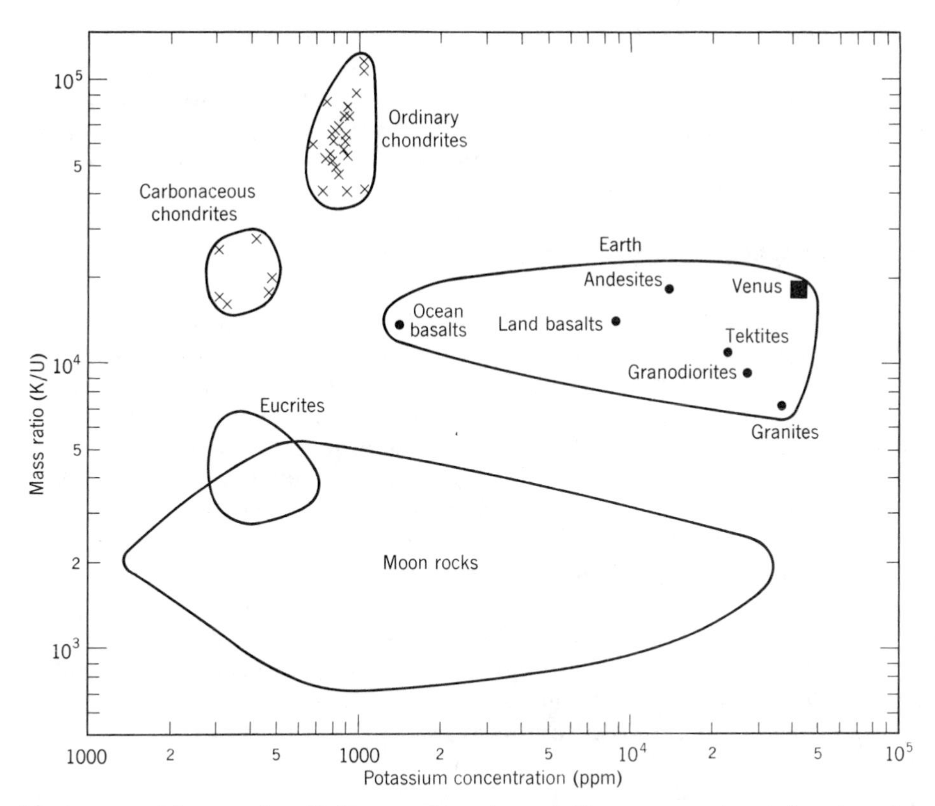


**Figure 7.3.** Mass ratios K/U as a function of K concentrations showing the independent groupings for representative crustal rocks, ordinary chondrites, carbonaceous chondrites, eucrites (a class of achondrite), moon rocks and a Venus sample. Most of the data are from a figure by Eldridge et al. (1974). The single value for Venus is from the X-ray spectrometer on a soft-landed probe (Vinogradov et al., 1973).

Th/U = 3.5 with $4 \times 10^{-4}$ parts by mass of K) to represent the Earth as a whole with K/U = $1.3 \times 10^4$ in the mantle and crust, we assign 35% of the total potassium to the core. This amounts to $8.4 \times 10^{20}$ kg and produces $2.9 \times 10^{12}$ W of radiogenic heat, which still appears to be barely sufficient to satisfy the core heat flux requirement. The possibility that large regions of the mantle have suffered much stronger depletion of K is noted in Section 2.3, and if we accept this argument then we can reasonably assign $5 \times 10^{12}$ W to the core. However, K/U ratios for the Earth as a whole and for the mantle are still very uncertain.

Whatever mix is favored for the Earth, these considerations leave unanswered the basic question of the difference in K/U ratio between carbonaceous and ordinary chondrites.* Since such strong fractionation occurred in the solar system anyway, is it not reasonable to suppose that the Earth represents just another fraction? Evidently the geophysical requirement for a core heat source, considered further in the following sections, imposes a stronger constraint than the geochemical arguments.

## 7.2 THERMAL REGIME OF THE MANTLE

If the loss of heat by the Earth were due solely to conduction, and if the thermal conductivity K or diffusivity $\eta$ of the deeper parts are similar to the laboratory values for igneous rocks ($K \approx 2.5$ W m$^{-1}$ deg$^{-1}$, $\eta \approx 1.2 \times 10^{-6}$ m$^2$ sec$^{-1}$) then the interior of the Earth would not yet know that the outside is cold. By Equation 7.1 the dimension of a layer with a thermal time constant of $4.5 \times 10^9$ years is only 400 km and very little heat from greater depths would have escaped in the whole history of the Earth. The lattice (phonon) contribution to the thermal conductivity of a rock decreases with temperature (Roufosse and Klemens, 1974) but at high temperatures radiative transfer of heat may become significant, depending upon the opacities of the minerals (Clark, 1957). For a "grey" material, in which opacity $\epsilon$ (reciprocal of mean free path of radiation) is independent of wavelength and temperature, the radiative conductivity $K_R$ increases strongly with temperature $T$:

$$K_R = \frac{16}{3}\frac{n^2}{\epsilon}\sigma T^3 \tag{7.3}$$

where $n$ is the refractive index and $\sigma = 5.67 \times 10^{-8}$ W m$^{-2}$ deg$^{-4}$ is the Stefan-Boltzmann constant. However, it now appears that high temperatures cause spreading of the infrared absorption bands due to $Fe^{2+}$ ions, increasing $\epsilon$ sufficiently to discount the possibility of a really significant increase in conductivity in the deep interior (Fujisawa et al., 1968; Fukao et al., 1968; Shankland, 1970; Schatz and Simmons, 1972; Mao and Bell, 1972). This leads to the conclusion that thermal conduction is essentially irrelevant to the thermal state and history of the deep interior.

The same conclusion follows from an argument suggested by Tozer (1967, 1972) and pursued by Tolland (1974). An obvious lower bound to the rate of heat production in the Earth shortly after its formation is given in the last column of Table H3 (Appendix H). Short-lived isotopes, now extinct, undoubtedly increased the early heat generation several fold, so that, even if the Earth started completely cold, which is questionable in any case, the heat release would have sufficed to bring it close to

*The K/U ratios of lunar rocks are in the range 1300 to 2300—much lower even than in the Earth.

melting in no more than $10^8$ years. Convection would then have been established, maintaining a more or less steady internal temperature pattern independently of the heat flux. Diminishing heat would slow the rate of convection but not significantly affect the temperature profile which must satisfy only two conditions: (1) it must not be less than adiabatic, as discussed below, and (2) the material in at least most of the Earth must be hot enough, and therefore "soft" enough to permit convective flow. The situation is self-stabilizing because the strength or viscosity of the mantle is a strong function of temperature. Slowing of convection causes the Earth to heat up slightly, so becoming softer and allowing convection to accelerate again. If the heat flux at any level diminishes to the point that convection could no longer be maintained then cooling by conduction would take over, but this is so slow that the temperature profile would remain at the point of convective instability for billions of years.

The two conditions to be satisfied by the temperature profile, mentioned above, require examination. First, the necessity for a gradient not less than adiabatic may be seen by considering a homogeneous medium with a temperature gradient (temperature increasing downward) in which a particular element of material is given a small vertical displacement (say upward), without allowing any heat transfer to or from it by contact with its surroundings. By virtue of the adiabatic decompression (due to the lower pressure at greater height) its temperature falls. If it then finds itself at the same temperature as the new surroundings, we say that the temperature gradient of the medium is exactly adiabatic, that is it corresponds to the pressure gradient for adiabatic changes. If the gradient is less then the element is now cooler and denser than its surroundings and tends to sink again. The medium is then stable with no tendency to convect. On the other hand if the temperature gradient is steeper, the element is hotter and thus less dense than its new surroundings and the flotation force exerted on it will tend to make it rise further. The medium is then unstable and may convect spontaneously. The steeper the temperature gradient the stronger is the convection, which has the effect of transferring heat upward and so reducing the temperature gradient toward the adiabatic value (at which convection ceases).

The variation of temperature $T$ with pressure $P$ during adiabatic compression is conveniently obtained from one of Maxwell's thermodynamic relations:*

$$\left(\frac{\partial T}{\partial P}\right)_S = \left(\frac{\partial V}{\partial S}\right)_P = \left(\frac{\partial V}{\partial T}\right)_P \left(\frac{\partial T}{\partial S}\right)_P \tag{7.4}$$

*A more extended discussion of the principles of thermodynamics, as applied to geophysical problems, is given by Officer (1974).

which applies to material of volume $V$ and entropy $S$. This can be expressed in terms of conventional physical parameters, since volume expansion coefficient is given by

$$\alpha = \frac{1}{V}\left(\frac{\partial V}{\partial T}\right)_P \tag{7.5}$$

and entropy is defined by

$$dS = \frac{dQ}{T} = \frac{mC\,dT}{T} \tag{7.6}$$

where $dQ$ is the quantity of heat supplied to a mass $m$ of specific heat $C$ to raise its temperature by $dT$, so that

$$\left(\frac{\partial T}{\partial S}\right)_P = \frac{T}{mC_P} \tag{7.7}$$

Substituting Equations 7.5 and 7.7 in 7.4 and writing $m/V = \rho$, we obtain

$$\left(\frac{\partial T}{\partial P}\right)_S = \frac{\alpha T}{\rho C_P} = \frac{\gamma T}{K_S} \tag{7.8}$$

where $\gamma$ is Grüneisen's parameter, as defined by Equation E1 (Appendix E). This is the adiabatic increase in temperature with pressure, but we also know the increase in pressure with depth $z$, or radius $r$ in the Earth

$$-\frac{dP}{dr} = \frac{dP}{dz} = g\rho \tag{7.9}$$

where $g$ is gravitational acceleration. Thus

$$-\left(\frac{dT}{dr}\right)_{\text{Adiabatic}} = \left(\frac{dT}{dz}\right)_{\text{Adiabatic}} = \left(\frac{\partial T}{\partial P}\right)_S \frac{dP}{dr} = \frac{\alpha T g}{C_P} = \frac{\gamma\rho T g}{K_S} = \frac{\gamma T g}{\phi} \tag{7.10}$$

The most convenient equation to integrate to obtain temperature differences over a wide range is Equation 7.8 of which the integral form is

$$\ln\left(\frac{T_2}{T_1}\right) = \int_{P_1}^{P_2} \frac{\gamma}{K_S}\,dP \tag{7.11}$$

Following Irvine and Stacey (1975), it is suggested in Appendix E that the variation in $\gamma$ over each of several depth ranges in the Earth (upper mantle, lower mantle, core) is slight, so that, as a good approximation, $\gamma$ may be taken outside the integral in Equation 7.11. Then we may note that since the density change by adiabatic compression is given by

$$\left(\frac{\partial \rho}{\partial P}\right)_S = \frac{\rho}{K_S} \tag{7.12}$$

we have

$$\ln\left(\frac{\rho_2}{\rho_1}\right) = \int_{P_1}^{P_2} \frac{dP}{K_S} \tag{7.13}$$

and so

$$\frac{T_1}{T_2} \approx \left(\frac{\rho_1}{\rho_2}\right)^{\gamma} \tag{7.14}$$

Given an assumed fixed point on the temperature profile of the Earth, we can estimate the temperature variation through any adiabatic range (of uniform composition) simply from the density variation (as tabulated in Appendix G), assuming a knowledge of $\gamma$ (for which the author's preferred values are given in Appendix E).

In a very large, slowly convecting fluid system, in particular the outer core, it appears that the temperature gradient must be maintained very close to the adiabatic value. But in the case of the mantle we must also consider the second of the basic conditions which the temperature profile must satisfy. The material must be "soft" enough (because it is hot enough) to *creep* or deform slowly under moderate shear stress. As mentioned in discussion of this problem in Section 10.2, estimates of the required conditions involve considerable extrapolation from the slowest laboratory creep experiments ($\sim 10^{-8}\ \text{sec}^{-1}$) to the rate of deformation of the mantle ($10^{-15}$ to $10^{-16}\ \text{sec}^{-1}$). However, the obvious simple interpretation is that to creep easily enough, the mantle must be reasonably close to melting. But in a complex medium, such as the mantle, there is no single simple melting point but a range of several hundred degrees between the *solidus*, at which the first fluid component appears, and the *liquidus*, at which the last solid component disappears. It seems that at least a significant region of the upper mantle is at or even slightly above its solidus temperature and contains a small fraction of fluid. Unfortunately we do not have reliable information on the pressure dependence of the solidus for mantle material, but several geophysical observations suggest that (as for simpler materials) the solidus temperature increases more rapidly with pressure than does the temperature of adiabatically compressed material initially at its melting point, that is, that adiabatic compression causes hardening. Attenuation of seismic waves (Section 10.4), which is greatest in the "weak" layer of the upper mantle, decreases at greater depths; as discussed in Section 4.2, the lower order harmonics of the gravitational potential of the Earth appear to be supported by stresses at several hundred kilometers depth, whereas even obvious surface features, such as continents, are in very precise isostatic balance. Thus it appears that for the mantle to convect it must have a temperature profile close to its melting point, which is steeper than the adiabatic gradient. The mantle solidus curve of Kennedy and Higgins (1972) was used as a guide although the solidus in Figure 7.4 is slightly less steep than theirs.

We are now able to attempt a plausible guess at the temperature profile of the mantle [which may eventually be subject to petrological checks (Boyd, 1973; Boettcher, 1974; MacGregor, 1974; Mercier and Carter, 1975)]. The lithosphere, being rigid, transmits heat upward only by conduction but, as considered in Section 7.1, the temperature gradient

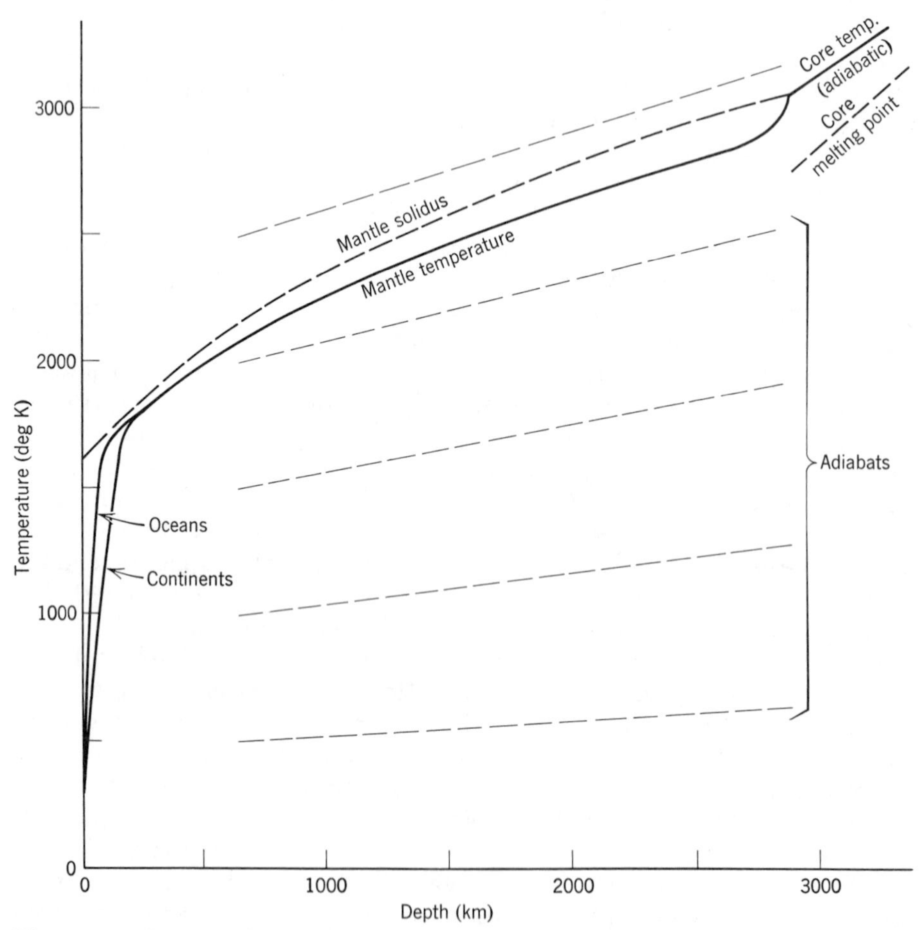


**Figure 7.4.** Suggested temperature profile of the mantle, a slight modification of the curve by Stacey (1975). The gradient in the lithosphere is steeper under the oceans than under continents for the reasons considered in Section 7.1, but the two profiles converge in the asthenosphere or strongly convecting weak layer that is at or close to the temperature of partial melting (solidus). At the base of the mantle the temperature gradient rises sharply, in response to core heat, to meet the mantle solidus at the boundary. This provides a source of "plume" lava by the mechanism discussed in Section 7.2. The lower mantle adiabats are plotted from Equation 7.14 with $\gamma = 1.0$. Numerical values should be regarded as illustrative, not definitive.

is much steeper under oceans where fresh, hot lithosphere is continuously appearing than under continents, which have reached diffusive equilibrium. The softening point (say 1300°C or 1600 K) is reached at an average depth of about 75 km under the oceanic lithosphere but only at 150 km or so under the continents. Below this the temperature gradient diminishes rapidly to a value slightly less than the solidus gradient but greater than the adiabatic gradient. The upper mantle convects strongly, the whole upper mantle being involved, but the increase in strength or viscosity at greater depths imposes a constraint on lower mantle convection. It now appears that the lower mantle must also convect on a broad scale, by virtue of its self-heat, perhaps as part of a pattern of whole mantle convection, but that the rate of flow is lower by a factor of order 10 (Andrews, 1975). Additionally, we must examine the concept of localized convective plumes. Consider a local region at the core-mantle boundary that is heated sufficiently to rise in the firm or viscous lower mantle. Once it begins to do so it readily accelerates because the lower mantle gradient exceeds the adiabatic gradient, as considered earlier in this section, but also because the gradient is less than the solidus gradient, so that as the material rises adiabatically it becomes increasingly molten, establishing a hot channel which guides the subsequent flow. It appears that "plumes" can develop only in response to a heat source in the core, as self-heat in the lower mantle gives a broader scale of convection.

Evidence that the lower mantle as well as the upper mantle is convecting, although in a quite different manner, implies that thermal time constants of the kind calculated by Equation 7.1 have no relevance to the Earth as a whole. The temperature profile of the Earth is virtually constant, with the surface heat flux in equilibrium with internal sources. This strengthens the significance of the coincidence of the geothermal flux with the heat generation in a chondritic Earth (Table 7.2) and therefore the arguments for a reservoir of potassium in the core (Section 7.3).

## 7.3 TEMPERATURE AND ENERGY BALANCE OF THE CORE

If, as considered in Section 3.4, we accept the elimination of the possibility that the geomagnetic dynamo is driven by precessional torques (Rochester, 1974; Rochester et al., 1975) then we are left with no alternative but outer core convection to maintain the dynamo. This means that the temperature gradient in at least the outermost part of the core must exceed the adiabatic value, but probably so slightly that we can take the adiabatic gradient as a sufficient approximation to it. The value of this gradient in the outermost part of the core is estimated from

Equation 7.10, with $g$, $\rho$, $K_S$ from Appendix G, $\gamma = 1.4$ from Appendix E and assuming $T \approx 3333$ K at the core-mantle boundary from Figure 7.4:

$$\left(\frac{dT}{dz}\right)_{z=2900\text{ km}} \approx 7.1_5 \times 10^{-4} \text{ deg m}^{-1} = 0.71_5 \text{ deg km}^{-1} \quad (7.15)$$

in substantial accord with an early estimate by Valle (1954).

By virtue of its electrical conductivity the core is a much better thermal conductor than is the mantle. Thermal conductivity $K$ is dominated by the electronic contribution $K_e$ and may be estimated from electrical conductivity by the Wiedemann-Franz law (see, for example, Kittel, 1971, p. 263):

$$K \approx K_e = LT\sigma_e = L\frac{T}{\rho_e} \approx 25 \text{ W m}^{-1} \text{ deg}^{-1} \quad (7.16)$$

where $\sigma_e$, $\rho_e$ are the electrical conductivity and resistivity, respectively, $L = 2.45 \times 10^{-8}$ W $\Omega$ deg$^{-2}$ is the Wiedemann-Franz constant and, for the purpose of obtaining a numerical value, $\rho_e = 3 \times 10^{-6}\ \Omega m$ is assumed from Section 8.3, with $T = 3333$ K as before. With the temperature gradient by Equation 7.15 and conductivity by Equation 7.16 we obtain the conducted (diffusive) heat flux from the core (surface area, $A = 1.5 \times 10^{14}$ m$^2$)

$$Q'_D = \left(\frac{dQ}{dt}\right)_{\text{Diffusive}} = KA\left(\frac{dT}{dz}\right) \approx 2.7 \times 10^{12} \text{ W} \quad (7.17)$$

This is the estimate of heat transmitted from the core into the mantle by conduction by virtue of the adiabatic temperature gradient. The heat transported up through the core by convection, which powers the dynamo as considered in Section 7.4, is additional to the flux represented by Equation 7.17 which can do no useful external work. We suppose that the total heat transport is perhaps twice this value, that is, about $5 \times 10^{12}$ W. Note that this estimate depends critically on the assumed value of $T$, since both $K$ and $(dT/dz)$ are proportional to $T$, but is probably not in error by a factor exceeding 2.

There is only one plausible source of this much heat: radioactive potassium, as considered in Section 7.1. We suppose either that this potassium is distributed uniformly through the mass of the core or of the outer core only. The latter is generally regarded as more plausible, assuming that potassium is rejected from the solid inner core,* so that we may write the heat flux through a level $r$ in terms of the total core

*A doubtful assumption if $K$ is in an electronically collapsed state, as suggested in Section 7.1.

heat $Q'_2$ and the fraction of the outer core mass, $m(r)$, inside $r$:

$$Q' = Q'_2 \frac{m(r) - m(r_1)}{m(r_2) - m(r_1)} \tag{7.18}$$

where the inner and outer radii of the outer core are $r_1$, $r_2$. To make use of tabulated quantities, it is convenient to write $m(r)$ in terms of $g$, the gravity at radius $r$, as in Equation 6.35:

$$m(r) = \frac{r^2 g}{G} \tag{7.19}$$

whence

$$Q' = Q'_2 \frac{r^2 g - r_1^2 g_1}{r_2^2 g_2 - r_1^2 g_1} \tag{7.20}$$

subscripts 1, 2 referring to values at $r_1$, $r_2$. The condition that the outer core be convecting at a particular level is that $Q'$ exceed the diffusive heat flux down the adiabatic gradient at that level,

$$Q' > 4\pi r^2 K \left(\frac{dT}{dz}\right)_{\text{Adiabatic}} \tag{7.21}$$

Combining Equations 7.10 and 7.14, $(dT/dz)_{\text{Adiabatic}}$ is also conveniently represented in terms of tabulated Earth model parameters

$$\left(\frac{dT}{dz}\right)_{\text{Adiabatic}} = \frac{\gamma g}{\phi} T = \frac{\gamma g}{\phi} T_2 \left(\frac{\rho}{\rho_2}\right)^{\gamma} \tag{7.22}$$

Equation 7.21 is plotted in Figure 7.5 from the model parameters of Appendix G with $\gamma = 1.4$. (Note that $\phi$ is given by Eq. 6.37.) As the figure shows, the critical total core heat required to initiate convection is virtually the same over most of the outer core, a conclusion reached by Metchnik et al. (1974) with the slightly different initial assumption that potassium is distributed uniformly with respect to outer core volume instead of mass. Except for a layer immediately above the inner core the diminution in heat flux per unit area with decreasing radius is almost exactly compensated by the decrease in $(dT/dz)_{\text{Adiabatic}}$ arising from the decrease in $g$ with depth. Thus we must take the temperature gradient in the outer core to be adiabatic except for a relatively thin inner range where a lower, diffusively controlled gradient prevails:

$$\left(\frac{dT}{dz}\right)_{\text{Diffusive}} = \frac{Q'(r)}{4\pi r^2 K} \tag{7.23}$$

This is shown in the outer-core temperature profile in Figure 7.6. If, however, the inner core contains potassium, even this small diffusive zone is eliminated.

The solidity of the inner core and the presumption that, like the

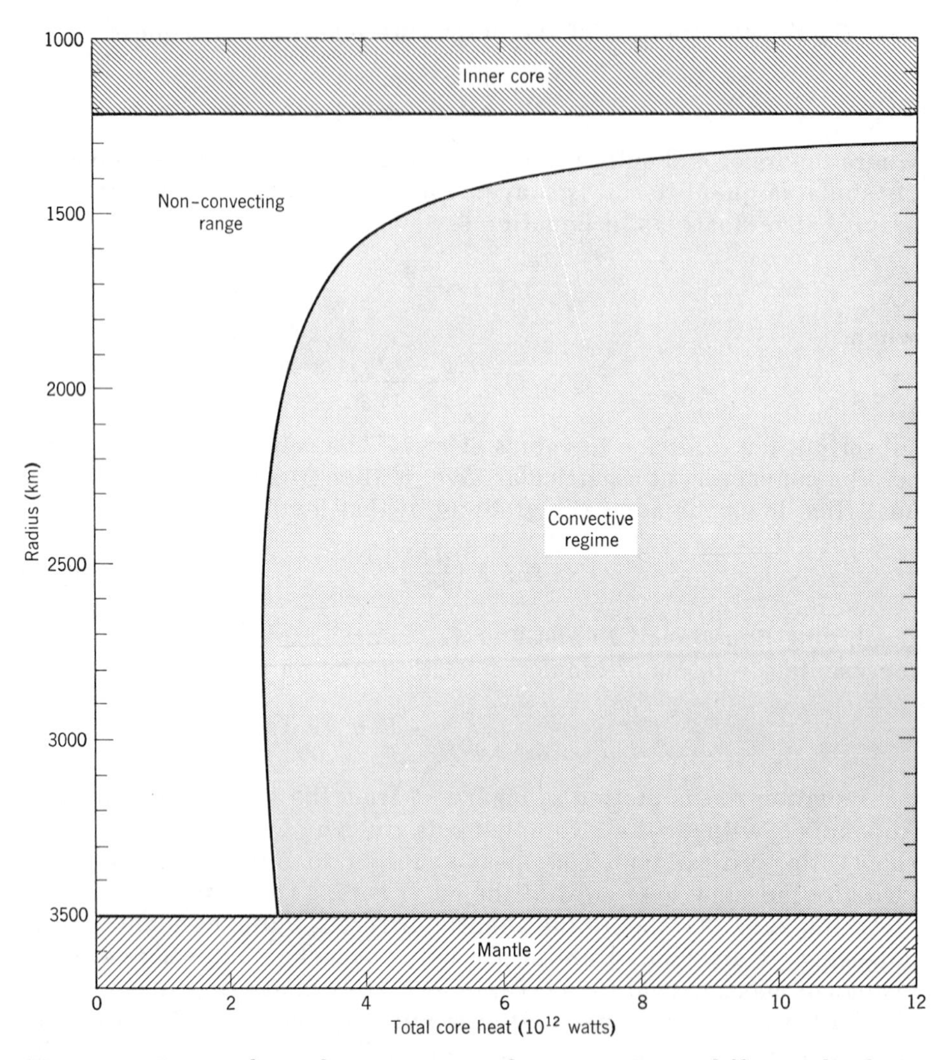


**Figure 7.5.** Required core heat generation for convection at different depths in the outer core, plotted from Equation 7.21. The existence of convection in the outer core at any depth implies that it occurs over virtually the entire depth.

outer core, it is composed largely of iron, led to several attempts to deduce the temperature at the inner-core boundary by extrapolating the solidus of iron, or of an iron-rich mix, to the appropriate pressure. However, melting is a very poorly understood phenomenon and the extrapolations of melting point-pressure curves to core pressures have led to widely disparate estimates (e.g., Higgins and Kennedy, 1971;

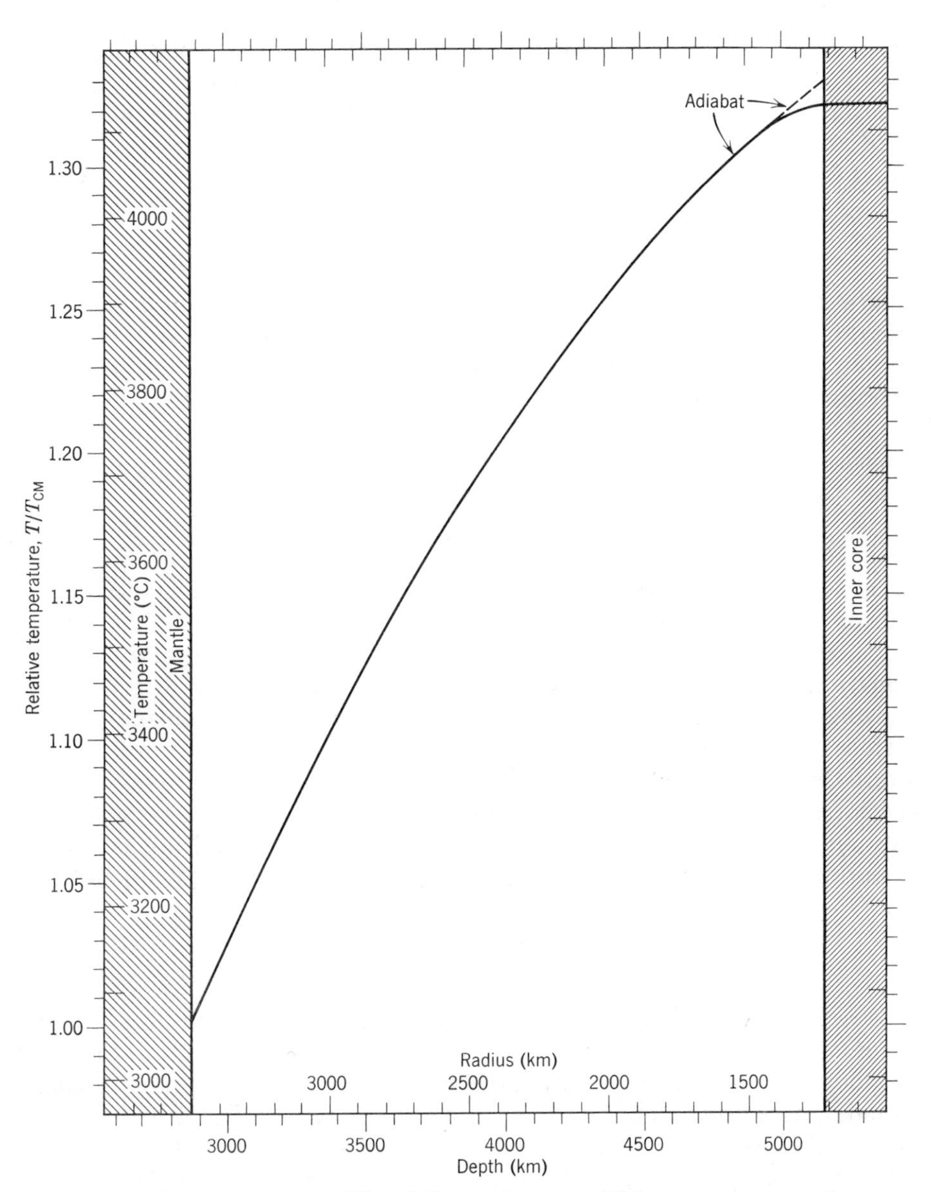


**Figure 7.6.** Temperature profile of the outer core. This assumes an outer core heat generation of $5 \times 10^{12}$ W, so that the gradient is adiabatic down to 4850 km, as in the data of Figure 7.5. Relative temperatures depend only upon the assumed value of Grüneisen's parameter, but absolute values are subject to greater uncertainty.

Leppaluoto, 1972; Boschi, 1974). The basic equation for melting of a simple material is the standard thermodynamic relationship of Clausius and Clapeyron, which relies on the fact that the Gibbs free energies of the solid and liquid states at the melting point $T_M$ and pressure $P$ must be equal (see, for example, Morse, 1969, p. 121):

$$\frac{1}{T_M}\frac{dT_M}{dP}=\frac{\Delta V}{L} \tag{7.24}$$

where $\Delta V$ and $L$ are the volume change and latent heat (both per unit mass) on melting. Both $\Delta V$ and $L$ are pressure dependent and, without knowing their pressure dependences, Equation 7.24 cannot be used for extrapolation. A plausible approximation is $\Delta S = L/T_M =$ constant, that is, the entropy of melting of a particular solid phase is independent of pressure (or of the melting point) and this is in any case a useful approximation, but $\Delta V$ is not as easily extrapolated.

In one significant respect the approach to a melting theory is probably simpler at very high pressures than at laboratory pressures, because at high pressures solid phases are close packed and liquid phases can for some purposes also be regarded as close-packed structures. The difference is simply that melting introduces crystal dislocations until the structure is saturated with freely mobile dislocations, melting point being simply the temperature at which the free energy of a dislocation vanishes. Then it is a straightforward matter to obtain an approximate expression for the ratio of volume increment to energy for the introduction of a dislocation to a structure and this ratio gives an explicit expression for the pressure dependence of melting point by Equation 7.24*. The result is very similar to Lindemann's (1910) melting law

$$\frac{1}{T_M}\frac{dT_M}{dP}\approx\frac{2(\gamma-1/3)}{K} \tag{7.25}$$

where $\gamma$ is Grüneisen's ratio (Appendix E) and $K$ is incompressibility. Lindemann's theory was originally inspired by the observation that the product of volume expansion coefficient and melting point is very close to being constant ($\alpha T_M \approx 0.07$), especially for materials with the same crystal structure, and was based on a supposition that melting occurs when a solid is hot enough for thermal vibrations to cause neighbouring atoms to "collide." It is generally now presented in terms of the hypothesis that melting occurs when atomic vibrations reach a fixed fraction of the atomic spacing (Gilvarry, 1956a), but follows more naturally from the assumption that a melting point is the temperature at which an edge dislocation has zero free energy.

*These arguments appear as thesis material by R. D. Irvine, still to be published.

The thermodynamic basis for Equation 7.25 is a simple adaption of the Mie-Grüneisen equation (Eq. E.2–Appendix E) which relates the pressure increment $\Delta P$ due to heating of a material in constant phase (and at constant volume) to the thermal energy $\Delta E$ applied to it. To a good approximation the thermal energy of a solid is equipartitioned between kinetic energy and potential energy of the atomic vibrations. Thus we can rewrite the Mie-Grüneisen equation in terms of the potential energy increment alone, $\Delta E_P = \Delta E/2$. With the restrictions considered, we can regard melting as a process of heating in which all of the thermal energy applied appears as potential energy and therefore approximately identify $\Delta E_P$ with $L$ to obtain the pressure increment due to melting at constant volume:

$$\Delta P = \frac{2\gamma \Delta E_P}{V} \approx \frac{2\gamma L}{V} \tag{7.26}$$

which (with the relationship between $\Delta P$ and $\Delta V$, $\Delta P \approx K\ \Delta V/V$) leads directly to the approximate result, Equation 7.25, but without the term 1/3.

Equation 7.25 was obtained by making certain assumptions which apply reasonably well to some materials at laboratory pressures but are clearly wrong for others. It is the melting of close-packed structures for which the equation is most satisfactory; the alkali metals are particularly badly represented, essentially because in such "loose" atomic structures the effective number of neighbors to each atom may change significantly during melting, so that Equation 7.26 cannot apply. But at high pressures the exceptions to Equation 7.25 must disappear and we can, therefore, apply it with some confidence to the problem of the core. It invalidates much recent discussion of the core problem based on an extrapolation of the melting curve of iron by Higgins and Kennedy (1971) who inferred, particularly from observations on the highly compressible alkali metals, that the melting points of metals increased linearly with volume compression of the solid phase.

It is however important that sulfur is now widely accepted as a constituent of the core and that it dramatically reduces the melting point of iron. At pressures up to about 55 kbar ($5.5 \times 10^9$ Pa) the eutectic temperature of the F-S system is almost constant at about 1265 K (990°C) as the sulfur content of the eutectic composition decreases to about 24 $\omega t$% (Usselman, 1975). At this point there is a clear break to nearly constant eutectic composition (close to $Fe_2S$) and a normal increase in melting point with pressure, that is, a variation represented rather well by Eq. 7.25. If we use this equation to extrapolate the eutectic temperature from 1265 K at 55 kbar to the pressures of the core-mantle boundary and to the inner core boundary we obtain temperatures of about 3090 K and 4250 K, respectively, which accord reasonably with core temperatures estimated by extrapolation from mantle conditions (Section 7.2). There is now little reason to doubt that the inner core is a normal solidification of outer core material. A suggestion that it is due to an electronic phase change (Elsasser and Isenberg, 1949) must be discounted (Bukowinski, 1976) and, as mentioned in Section 6.4, the existence of hexagonal close-packed ($\epsilon$) iron, noted by Birch (1972), is not relevant at melting temperatures.

## 7.4 CONVECTION AS A POWER SOURCE FOR THE GEODYNAMO AND FOR PLATE TECTONICS

Convection has become a vital concept in geophysics. In the previous sections it is argued that the upper and lower mantles and the core are all convecting, although in different manners and at different rates. Mantle convection is responsible for plate tectonics (Section 10.1) and continental drift and, in fact, virtually all geological activity; core convection is believed to provide the power source for the geomagnetic field (Section 8.4). Thus it is obvious that convection on so grand a scale is a very powerful process. Since the power is derived by the transfer of heat from the hot interior to the cooler exterior of the Earth, it is apparent that we are dealing with a straightforward thermodynamic, heat engine problem, the mechanical work, which is the output of the engine, being available to deform the mantle material against the constraint of its strength or viscosity or move the conducting core fluid, extending and amplifying the magnetic field. In this section a straightforward application of thermodynamic principles is used to derive the efficiency of a convective engine, so that the available power can be specified in terms of the known (or estimated) heat sources.

Generalizing and extending slightly an argument used by Stacey (1967) and Metchnik et al. (1974) we consider a layer in the Earth, with an arbitrary temperature profile $T(r)$, which transports heat convectively from a lower boundary at $T_1$ to an upper one at $T_2$, the temperature difference being not less than adiabatic. A component mass $m$ is taken around a convective cycle $ABCDA$, as in Figure 7.7, receiving heat $Q_1$ on the lower, isobaric limb $D \rightarrow A$ of the cycle and giving out heat $Q_2$ on the upper limb. It is supposed that the vertical movements $A \rightarrow B$ and $C \rightarrow D$ are accomplished adiabatically, regardless of the difference in temperature between the mass element and its surroundings ($T(r)$). On the limb $A \rightarrow B$ the mass element is hotter than its surroundings at the corresponding level, the temperature increment $\Delta T$ being an arbitrary function of height or radius, $r$, by virtue of the arbitrary profile of the

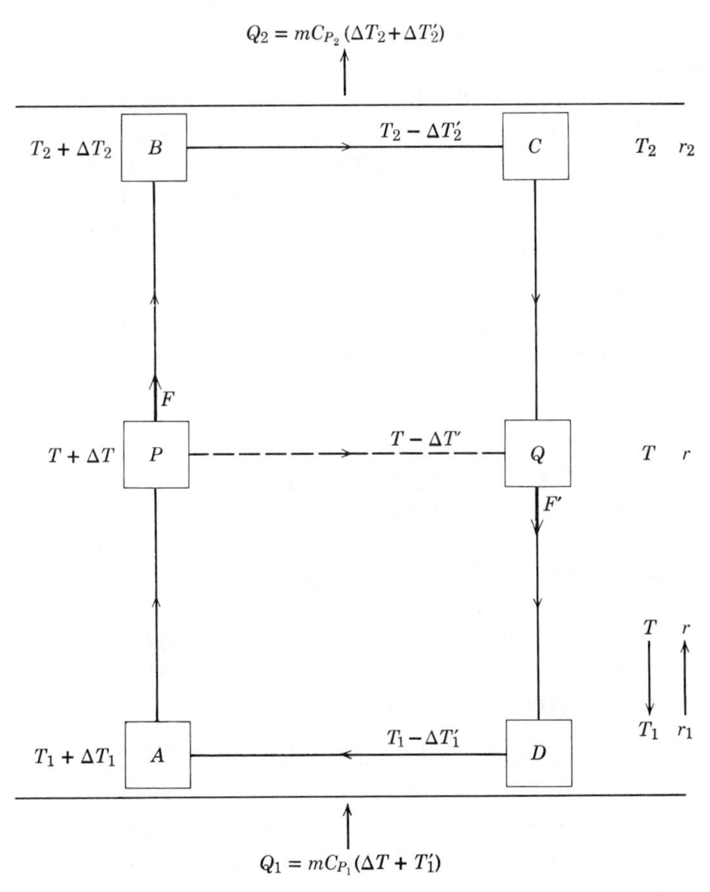


**Figure 7.7.** Convective cycle of a mass $m$ in a medium of arbitrary temperature gradient, $T(r)$, between a heat source at $r_1$ and a heat sink at $r_2$. The paths $A \to B$ and $C \to D$ represent adiabatic expansion and compression, respectively, and $B \to C$, $D \to A$ are isobars.

medium, but it is reasonable to assume that $\Delta T$ is not large compared with $T$. Similarly on the falling convective limb the temperature is lower than ambient by $\Delta T'$, but either $\Delta T$ or $\Delta T'$ may be zero or even locally negative without affecting the argument.

By virtue of the temperature contrast with the ambient profile of the medium, the mass element experiences of flotation force $F$ on the upward limb and a sinking force $F'$ on the downward limb, where

$$F = m\bar{\alpha}\Delta T g \tag{7.27}$$

$$F' = m\bar{\alpha}'\Delta T' g \tag{7.28}$$

Here $\bar{\alpha}$ is the mean volume expansion coefficient of the material over the temperature range $T$ to $(T+\Delta T)$ at the ambient pressure and $\bar{\alpha}'$ is the mean between $T$ and $(T-\Delta T')$ at the same pressure. We can reasonably put $\bar{\alpha}=\bar{\alpha}'=\alpha$ if $(\Delta T+\Delta T')$ is not too large, because expansion coefficients are only weakly dependent on temperature (at least over limited ranges in material of constant phase which is well above its Debye temperature), but they may be strongly pressure-dependent. Since upward and downward movements (of different mass elements) must occur simultaneously we can justify coupling $F$ and $F'$ in a single expression to find the work done by the total of the flotation and sinking forces in one complete cycle:

$$W=\int_{r_1}^{r_2}(F+F')\,dr=m\int_{r_1}^{r_2}(\Delta T+\Delta T')\,\alpha g\,dr \tag{7.29}$$

This is simply the total of the mechanical work done by the material on the vertical limbs of the cycle, no work being available from the horizontal limbs. It is presumed to be used to overcome the constraints to the motion (e.g., viscosity or magnetic field).

Since the vertical limbs are adiabats, the entropy difference between any two points on opposite limbs must be the same. Expressing this equality for transfer at an arbitrary level ($P\rightarrow Q$ in Fig. 7.7) and at the bottom of the cycle, $A\rightarrow B$, we have

$$\frac{mC_P(\Delta T+\Delta T')}{T^*}=\frac{mC_{P1}(\Delta T_1+\Delta T_1')}{T_1^*} \tag{7.30}$$

again presuming that the $\Delta T$s are small compared with the mean temperatures $T^*$ at each level. Substituting for $(\Delta T+\Delta T')$ from Equation 7.30 in Equation 7.29:

$$W=\frac{mC_{P1}(\Delta T_1+\Delta T_1')}{T_1^*}\int_{r_1}^{r_2}\frac{T^*\alpha g}{C_P}\,dr \tag{7.31}$$

Equation 7.31 now contains two familiar expressions, since the heat input at $r_1$, where the specific heat is $C_{P1}$, is

$$Q_1=mC_{P1}(\Delta T_1+\Delta T_1') \tag{7.32}$$

and the integrand is recognized as the adiabatic temperature gradient, by Equation 7.10. Thus

$$W=\frac{Q_1}{T_1^*}(T_1^*-T_{2s}^*)=Q_1\left(1-\frac{T_{2s}^*}{T_1^*}\right) \tag{7.33}$$

where $T_{2s}^*$ is the temperature at $r_2$ of material adiabatically expanded from temperature $T_1^*$ at $r_1$. Therefore the efficiency of the convective engine is simply the ideal thermodynamic (Carnot) efficiency of an engine working over the adiabatic temperature difference of the working substance between the pressures of the heat source and sink. For a homogeneous medium in constant phase, Equation 7.14 may be substituted to obtain an expression for efficiency that is convenient to use with tabulated earth model parameters

$$\eta = \frac{W}{Q_1} = 1 - \left(\frac{\rho_2}{\rho_1}\right)^{\gamma} \tag{7.34}$$

Equation 7.33 or 7.34 express a geophysically important, if intuitively well recognized, result. Convection in a relatively incompressible material (such as the core or mantle) can achieve significant thermodynamic efficiency only on a global scale, such that the pressure differences suffice to give first-order changes in temperature (or density) by adiabatic compression or expansion. As a following analysis shows, convective engine efficiency of order 10% may be obtained in a body with the size of the core. However, the power available from liquid convection on a laboratory or industrial scale is quite trivial. Gases offer more favorable conditions, being more compressible, so that density changes by a factor of order two are possible in atmospheric convection, which does therefore operate at a respectable thermodynamic efficiency and thus can generate a great deal of mechanical power. Temperature gradients or differences exceeding adiabatic values are, of course, necessary to generate the finite density differences that are required for convection actually to start but from the point of view of thermodynamic efficiency the excess is wasted.

The simplest application of the thermodynamic engine is to the outer core. Bullard and Gellman (1954) first appealed to the temperature difference between the inner and outer boundaries for an estimate of the efficiency and suggested 10%. On the basis of more recent estimates of core parameters, including the conducted heat loss, and assuming the core heat to originate uniformly through the volume of the outer core, Metchnik et al. (1974) obtained an upper limit of about 7%. This calculation is readily repeated here using the revised value of Grüneisen's ratio for the core ($\gamma = 1.4$—Appendix E) and assuming radioactive heat to be distributed uniformly with respect to outer core mass (rather than volume), as in Equations 7.18 and 7.20. Then, as Figure 7.5 shows, the conducted heat flux is virtually the same fraction of the total heat flux for all depths down to $r \approx 1600$ km, below which there is insufficient heat to cause convection. Then we can refer to a convective heat flux $Q'_c$, which also originates uniformly through the core (but only down to $r = 1600$ km) and is given by

$$Q'_c = Q'_1 - Q'_D \tag{7.35}$$

$Q'_1$ being the total heat flux at the outer boundary ($r_1$) and $Q'_D$ is the diffusive component. Dividing the outer core into layers, as in the tabulation of Appendix G, the convective heat flux originating between radii $r_n$ and $r_{n-1}$ is

$$\Delta Q'_c = (Q'_1 - Q'_D)\frac{r_n^2 g_n - r_{n-1}^2 g_{n-1}}{r_2^2 g_2 - r_1^2 g_1} \tag{7.36}$$

$r_2 = 3485.7$ km and $r_1 = 1217.1$ km, being the outer and inner radii of the whole outer core. The corresponding contribution to convective power is, by Equation 7.34,

$$\Delta P = \Delta Q'_c\left[1 - \left\langle\left(\frac{\rho_2}{\rho_n}\right)^\gamma\right\rangle\right] \tag{7.37}$$

where for $\langle(\rho_2/\rho_n)^\gamma\rangle$ the mean value over the range $r_n$ to $r_{n-1}$ is taken. Then, summing over all elements listed in Table G2, the efficiency is

$$\eta = \frac{P}{Q'_1 - Q'_D} = 0.11 \tag{7.38}$$

that is, relative to the *convected* core heat the thermodynamic efficiency is 11%.

A slight correction is required to take account of the fact that the power $P$ is itself dissipated mainly as ohmic heating within the outer core. Thus we should consider a corrected efficiency $\eta^*$, since

$$P = \eta(Q + P) \tag{7.39}$$

so that

$$\eta^* = \frac{\eta}{1-\eta} = 0.12 \tag{7.40}$$

Relative to the total core heat, the efficiency is smaller by the ratio of convected heat to total heat:

$$\eta_T = \eta^* \frac{Q'_1 - Q'_D}{A'_1} \tag{7.41}$$

This is the "total" efficiency, which is zero if $Q_1 = Q_D \approx 2.7 \times 10^{12}$ W but increases toward $\eta^* = 12\%$ as $Q'_1 \gg Q'_D$. We have no clear indication of the value of $Q'_1$ to choose. $Q'_1 = 5 \times 10^{12}$ W is suggested in Sections 7.1 and 7.3, giving $\eta_T = 5.5\%$ and $P = 2.75 \times 10^{11}$ W, but if $Q_1 = 10^{13}$ W, which is the upper limit of the plausible range, $\eta_T = 8.8\%$ and $P = 8.8 \times 10^{11}$ W.

Application of the theory of the convective engine to the mantle raises some additional questions. First, temperature differences between up-going and down-going material may not be as small as in the core. This determines the magnitude of convective stress, as considered

**Table 7.3** Thermodynamic Efficiency and Power of Core and Mantle Convection.

| Zone | Assumed Heat | Efficiency | Power |
|---|---|---|---|
| Outer core (self-heat) | $5 \times 10^{12}$ W | 5.5% | $0.27_5 \times 10^{12}$ W |
| Lower mantle | | | |
| Core heat | $5 \times 10^{12}$ W | 21.1% | $1.0_5 \times 10^{12}$ W |
| Self heat | $11 \times 10^{12}$ W | 10.6% | $1.2 \times 10^{12}$ W |
| Upper mantle | | | |
| Core heat | $5 \times 10^{12}$ W | 10.6% | $0.5 \times 10^{12}$ W |
| Lower mantle heat | $11 \times 10^{12}$ W | 10.6% | $1.2 \times 10^{12}$ W |
| Self-heat | $6 \times 10^{12}$ W | $6.0_6$% | $0.36 \times 10^{12}$ W |
| Crust | $9.5 \times 10^{12}$ W | — | — |

Values for the lower mantle were obtained by integrating equations of the form of 7.34 with $\gamma = 1.0$. For the upper mantle the compression was estimated from Equation 6.45, ignoring phase transitions, and $\gamma = 0.8$ was assumed, as in Appendix E.

in Section 10.3. However, it affects the present argument only to the extent that expansion coefficient and specific heat at any pressure are temperature dependent and does not introduce an error comparable to the uncertainty in distribution of heat sources. More serious are the effects of phase transitions in the upper mantle. Verhoogen (1965) considered this question and concluded that phase transitions do not necessarily inhibit convection. Their influence on the pattern of convection is considered in Chapter 10; we are here concerned with their effect on the efficiency of the convective engine, which depends only on the thermodynamic reversibility of the phase changes, that is, do they occur in both directions at the same pressure and temperature? If not, then the irreversibility appears as a temperature or pressure hysteresis that removes energy from the convective cycle, lowering its efficiency. It appears unlikely that this is an important effect.

Convective powers can be calculated for any assumed convective regime and distribution of heat sources, by the method used above for the core. A summary of results of calculations of this kind is given in Table 7.3, in which the various contributions are separated, so that the efficiencies can be applied readily to any assumed convective pattern. The core energy is referred to in connection with the dynamo in Section 8.4 and the implications of the mantle values are considered in Chapter 10.

# 8

# The Geomagnetic Field

"There are too many ways in which the core be made to convect to permit an unambiguous interpretation."

ELSASSER, 1963, PAGE 29

## 8.1 THE MAIN FIELD

In this section the geomagnetic field is considered as a static entity, whose features must be explicable in terms of any satisfactory theory of its origin. The variation of these features with time is discussed in Section 8.2.

To a useful first approximation the field may be represented by that of a magnetic dipole situated at the center of the Earth and having a dipole moment, $m = 7.94 \times 10^{22}$ A m$^2$*, with its axis inclined at about 11° to the Earth's geographic axis. The field of a dipole is conveniently represented in terms of the scalar magnetic potential $V_m$, which may be

*This is the conventional unit in the S.I. system, to which the present edition has been converted. The traditional units of geomagnetism are cgs electromagnetic units (emu), in which the Earth's magnetic moment has the value $7.94 \times 10^{25}$ G cm$^3$. The field is then measured in gauss (G) (= oersteds, in the absence of a magnetizable medium). The equations in this chapter may be used in emu if the bracketted factors involving $\mu_0$ and $4\pi$ are omitted (and assuming magnetizable media to be excluded). The S.I. unit of field intensity is the tesla (= $10^4$ G) and a commonly used subdivision is the nano tesla (nT) (= $10^{-5}$ G = 1 gamma). One tesla ≡ 1 weber m$^{-2}$ and in many texts is written as Wb m$^{-2}$.

differentiated to obtain any component of the field:

$$V_m = \frac{\mathbf{m} \cdot \mathbf{r}}{(4\pi)r^3} \equiv \frac{m \cos \theta}{(4\pi)r^2} \tag{8.1}$$

where $\theta$ is the angle between the dipole axis and the radius vector $\mathbf{r}$ from the dipole to the point considered. Then the field is

$$B = -(\mu_0) \text{ grad } V_m \dagger \tag{8.2}$$

and the horizontal (i.e., circumferential) and vertical (i.e., radial) components, $B_\theta$ and $B_r$ are

$$B_\theta = -(\mu_0) \frac{1}{r} \frac{\partial V_m}{\partial \theta} = \left(\frac{\mu_0}{4\pi}\right) \frac{m}{r^3} \sin \theta \tag{8.3}$$

$$B_r = -(\mu_0) \frac{\partial V_m}{\partial r} = \left(\frac{\mu_0}{4\pi}\right) \frac{2m}{r^3} \cos \theta \tag{8.4}$$

The variations in the dipole field over the surface of the Earth, taken to be a sphere of radius $a$, are

$$B_\theta = B_0 \sin \theta = -H \tag{8.5}$$

$$B_r = 2B_0 \cos \theta = -Z \tag{8.6}$$

where

$$B_0 = \left(\frac{\mu_0}{4\pi}\right) \frac{m}{r^3} = 3.07 \times 10^{-5} \text{ T} = 0.307 \text{ gauss} \tag{8.7}$$

is the equatorial strength of the nearest-fitting dipole field to the observed terrestrial field, and $H$, $Z$ are the conventional horizontal (northward) and vertical (downward) components (but see Equations 8.13, from which it is seen that $H^2 = X^2 + Y^2$).

†In mathematical terms this is justified by the fact that for points external to the Earth, with electric currents excluded, and disallowing time variations, Maxwell's equations give

$$\text{curl } \mathbf{H} = 0$$

so that $\mathbf{H}$ can be derived from a scalar potential $V_m$

$$\mathbf{H} = -\text{grad } V_m$$

and Equation 8.2 follows, since $\mathbf{B} = \mu_0 \mathbf{H}$, with $\mu_0$ constant. Also, another of Maxwell's equations,

$$\text{div } \mathbf{B} = 0$$

means that

$$\text{div grad } V_m = \nabla^2 V_m = 0$$

which is Laplace's equation, the starting point for spherical harmonic analysis (see Appendix C).

The total field strength is obtained by combining Equations 8.5 and 8.6:

$$B = (B_\theta^2 + B_r^2)^{1/2} = B_0(1 + 3\cos^2\theta)^{1/2} \tag{8.8}$$

and its inclination or angle to the horizontal is $I$, given by

$$\tan I = \frac{B_r}{B_\theta} = 2\cot\theta = 2\tan\phi \tag{8.9}$$

where $\theta$ is referred to as the magnetic colatitude, the angular distance from the magnetic pole, and $\phi = (90° - \theta)$ is magnetic latitude. Equation 8.9 is basic to the calculation of paleomagnetic pole positions (Chapter 9), since if the direction of magnetization of a rock records the ancient field direction and assuming that the field was dipolar, this equation gives the magnetic latitude at the time of its formation. Equation 8.9 is also the differential equation of a magnetic line of force:

$$\tan I = \frac{dr}{r\,d\theta} = 2\cot\theta \tag{8.10}$$

which integrates to

$$\frac{r}{a} = \frac{\sin^2\theta}{\sin^2\theta_a} \tag{8.11}$$

where $\theta_a$ is the colatitude at which the line of force crosses radius $a$. Most conveniently we take $\theta_a = 90°$, in which case $a$ is the distance from the dipole at which the line crosses the equatorial plane.

The equations for a dipole field apply equally to the field outside a uniformly magnetized sphere. That the Earth's field resembled that of a magnetized sphere was recognized by William Gilbert, physician to Queen Elizabeth I, whose contemporaries supposed the magnetic alignment of a lodestone, or natural compass, to be an extraterrestrial influence. Gilbert made the first quantitative study of geomagnetism* and his name has been given to one of the paleomagnetic polarity epochs, shown in Figure 9.13. However, we cannot conclude that the Earth is in fact a uniformly magnetized sphere; further evidence of the origin of the field is obtained by considering departures from the dipole character.

The potential of a geomagnetic field of arbitrary form can be represented as an infinite series of spherical harmonic functions, of which Equation 8.1 is the first term. A comprehensive discussion of this method of analyzing geomagnetic data is given by Chapman and Bartels

*Gilbert's treatise *De Magnete* . . . was published in 1600. An English translation (from Latin) by P.F. Mottelay appeared in 1893 and has been reprinted (1958) by Dover Publications.

(1940). The general expression for potential is obtained in Appendix C*.

$$V_m = \frac{1}{a} \sum_{l=1}^{\infty} \sum_{m=0}^{l} \left\{ \begin{matrix} \left[ C_l^m \left(\frac{a}{r}\right)^{l+1} + C_l'^m \left(\frac{r}{a}\right)^{l} \right] \cos m\lambda \\ + \left[ S_l^m \left(\frac{a}{r}\right)^{l+1} + S_l'^m \left(\frac{r}{a}\right)^{l} \right] \sin m\lambda \end{matrix} \right\} P_l^m (\cos \theta) \quad (8.12)$$

where $\theta$, $\lambda$ are the coordinates of magnetic colatitude and longitude and $a$ is the Earth radius. $V_m$ itself is not directly observable but components of the field northward ($X$), eastward ($Y$), and downward ($Z$) are measured over the Earth's surface ($r = a$). Then relationships

$$\left. \begin{aligned} X &= (\mu_0) \frac{1}{r} \left(\frac{\partial V_m}{\partial \theta}\right)_{r=a} \\ Y &= -(\mu_0) \left(\frac{1}{r \sin \theta} \frac{\partial V_m}{\partial \lambda}\right)_{r=a} \\ Z &= (\mu_0) \left(\frac{\partial V_m}{\partial r}\right)_{r=a} = -B_r \end{aligned} \right\} \quad (8.13)$$

(where $\theta$ now refers to the geographic axis rather than the magnetic axis as in Eq. 8.3), allow the coefficients in Equation 8.12 to be determined up to an order limited by the detail of the observations. Note that to determine separately the primed and unprimed coefficients in (8.12), we must obtain both $V_m$ and $(\partial V_m / \partial r)$. The vertical field $Z$ gives $(\partial V_m / \partial r)_{r=a}$ directly but $V_m$ is only indirectly obtainable from $X$ and $Y$, its horizontal derivatives. C.F. Gauss first applied this method to the analysis of the geomagnetic field and showed that the coefficients $C_l^m$ and $S_l^m$ described a field of internal origin, and that $C_l'^m$ and $S_l'^m$ represented a field of external origin, which he found to be nonexistent. We now know that the external field is not totally absent; a field amounting to perhaps 30 nT (about $10^{-3}$ of the total field) at the surface at geomagnetically quiet times, and several times as much during magnetic storms, is due to a "ring current" or drift of charged particles spiraling about geomagnetic field lines at several earth radii. However, Gauss's conclusion that the main field is of internal origin is valid and in the present context interest is restricted to the coefficients $C_l^m$ and $S_l^m$.

In geomagnetism it is convenient to use the Gauss coefficients $g_l^m$ and $h_l^m$, which are

$$\left. \begin{aligned} g_l^m &= (\mu_0) \frac{C_l^m}{a^2} \\ h_l^m &= (\mu_0) \frac{S_l^m}{a^2} \end{aligned} \right\} \quad (8.14)$$

*Note that there is no $l = 0$ term; this would correspond to a magnetic monopole within the Earth. The $l = 1$ terms correspond to a magnetic dipole.

The Gauss coefficients have the dimensions of magnetic field, whereas $C_l^m$ and $S_l^m$ have dimensions of magnetic pole strength (or mass in the case of gravitational potential), in particular for a dipole coaxial with the coordinate system, $g_1^0 = B_0$ by Equation 8.7. Equation 8.12 is then rewritten:

$$V_m = \frac{a}{(\mu_0)} \sum_{l=1}^{\infty} \left(\frac{a}{r}\right)^{l+1} \sum_{m=0}^{l} (g_l^m \cos m\lambda + h_l^m \sin m\lambda) P_l^m (\cos \theta) \qquad (8.15)$$

and Eqs. 8.13 apply as before.

Tabulations of the geomagnetic field and its change during the period 1905–1945, with harmonic analyses, are given by Vestine et al. (1947a,b) and more recent harmonic analyses are by Finch and Leaton (1957), Cain et al. (1967), IAGA Study Group (1976), and Hurwitz et al. (1974). The most interesting and perhaps the most significant aspect of these analyses is the separation of the best-fitting centered dipole field from the remaining (nondipole) part of the field. The form of the nondipole field is more apparent when the stronger dipole field is removed. This is emphasized by the work of Bullard et al. (1950), who used the Vestine data and produced contour maps of the nondipole field, one of which is reproduced in Figure 8.1. This is the field of the Earth, as represented by all spherical harmonic terms except $g_1^0$, $g_1^1$, and $h_1^1$, which correspond to components of the central dipole moment along the Earth's geographic axis [$g_1^0 = -0.3010\ (\times 10^{-4})$*], in the equatorial plane through the Greenwich meridian [$g_1^1 = -0.0202\ (\times 10^{-4})$*] and normal to both [$h_1^1 = 0.0568\ (\times 10^{-4})$*]. It should also be noted that the best-fitting dipole field is centered about 300 km off the Earth's center. This is referred to as the eccentric dipole field.

As many authors have recognized, the features of the geomagnetic field and its harmonic analysis can be used to infer the depths of the sources of the field. Of particular interest in this connection is the analysis by Lowes (1974), summarized in Figure 8.2. Lowes found that spherical harmonic coefficients of the geomagnetic field were reliable only up to order and degree about (8, 8) above which noise noticeably biassed spectral estimates. However, higher harmonic content can be inferred from a continuous line profile around the Earth obtained by Alldredge et al. (1963) and analyzed by Bullard (1967). Lowes (1974) grouped all spherical harmonic coefficients of the same degree to obtain the mean square value of the field represented by the terms of each harmonic degree:

$$R_l = (l+1) \sum_{m=0}^{l} [(g_l^m)^2 + (h_l^m)^2] \qquad (8.16)$$

*The factor $10^{-4}$ refers to values in tesla. Omit this factor for values in electromagnetic units (gauss). The total strength of the misaligned dipole is represented by $[(g_1^0)^2 + (g_1^1)^2 + (h_1^1)^2]^{1/2} = B_0$, as given by Equation 8.7, and the dipole moment is $(4\pi/\mu_0)B_0a^3$. Geomagnetic field coefficients for 1975 are given by Barraclough et al. (1975).

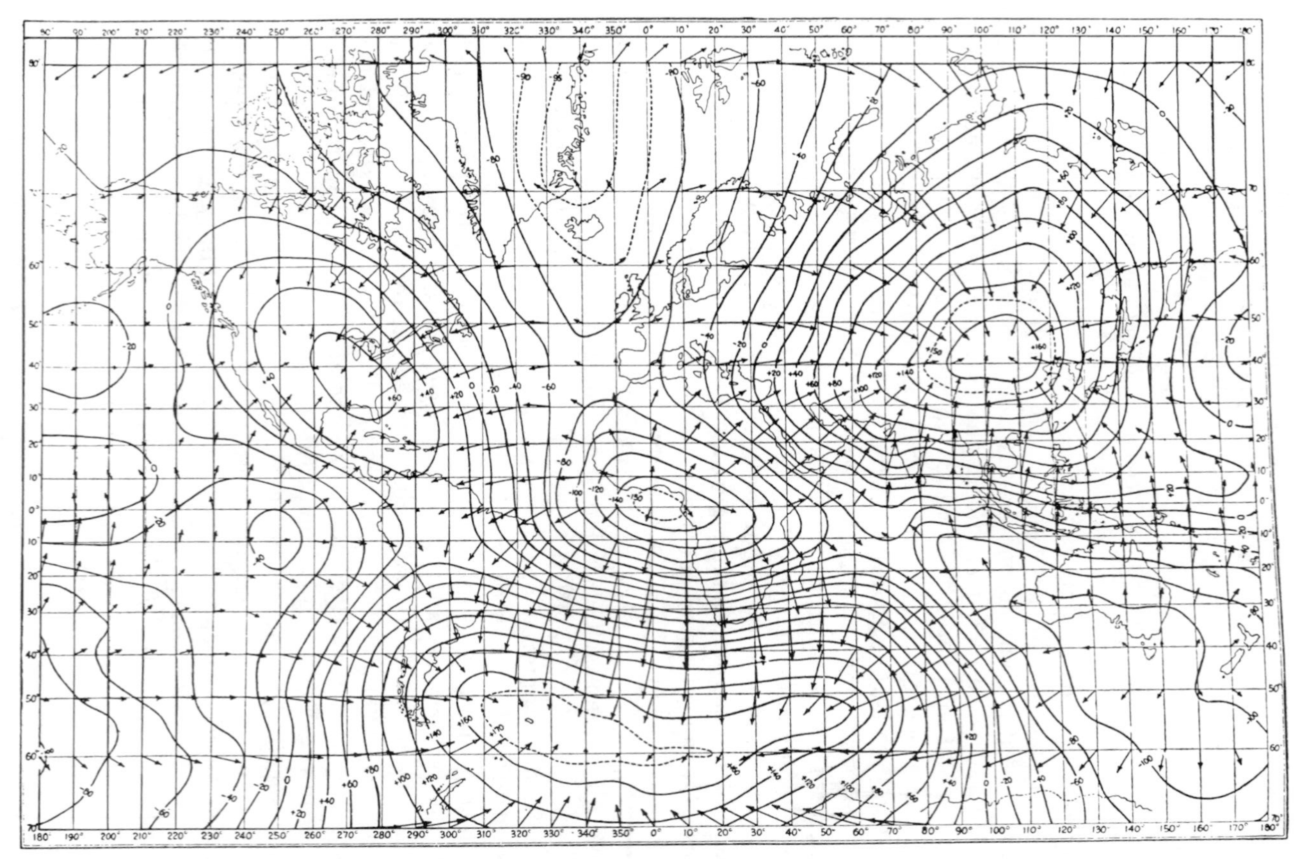

**Figure 8.1.** The nondipole field for 1945. Contours give the vertical component in intervals of 0.02 G ($2 \times 10^{-6}$ T) and arrows represent the horizontal component. Reproduced, by permission, from Bullard et al. (1950).

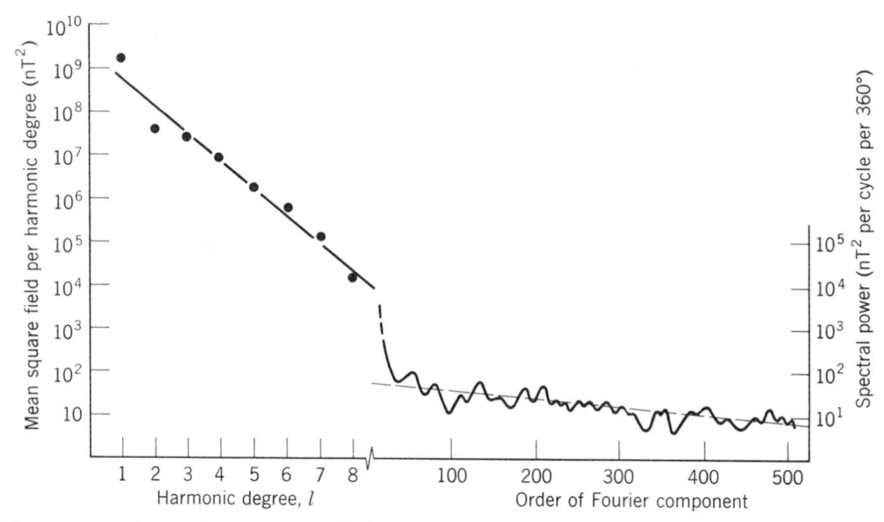


**Figure 8.2.** Spatial spectrum of the main geomagnetic field. The left part of the figure gives the mean square amplitudes of the field represented by each harmonic degree, for low degrees, according to Lowes (1974) and the right-hand part gives the higher-order terms of a Fourier analysis by Bullard (1967), corrected by Lowes (1974), of a world-encircling line profile by Alldredge et al. (1963).

where the factor $(l + 1)$ normalizes the field coefficients $g, h$ in the same sense as the potential coefficients are normalized when used with polynomials defined by Equation C.14 (Appendix C) (see Lowes, 1966). $R_l$ is similar to the square of the harmonic degree amplitudes of the gravitational potential which have been used to infer the depths of the mass anomalies responsible for the features of the geoid (Fig. 4.2). Lowes found that the plot of the harmonic degree powers $R_l$ (i.e., squares of amplitudes) could be represented well by a simple power law in $l$:

$$R_l = 4.0 \times 10^9 (4.5)^{-l} (\mathrm{nT})^2 \tag{8.17}$$

as shown by the left part of Figure 8.2. This is equivalent to a variation of the harmonic degree amplitudes, $A_l$:

$$A_l = (R_l)^{1/2} = 6.3 \times 10^4 (2.1_2)^{-l} \,\mathrm{nT} \tag{8.18}$$

Reference to Equation 8.15 shows that the potential due to terms of harmonic degree $l$ varies with radial distance $r$ as $r^{-(l+1)}$ and therefore the field strength varies as $r^{-(l+2)}$. Thus by extrapolation downward the harmonic degree spectrum represented by Equation 8.18 becomes progressively less dependent on $l$ and would appear "white," with all harmonic degrees equally represented, at a radius $a/2.1_2 = 0.47$ Earth radii (in the outer part of the core). While extrapolation this far should not be interpreted too literally, it clearly indicates a source of the main part of the geomagnetic field, as represented by the low degree spherical harmonics, in the outer part of the core, where its spatial spectrum is apparently "white," or nearly so, over the range that we are able to investigate. The dominance of the dipole field, as seen at the surface, would not be apparent at core level.

The spectrum of the short "wavelength" features of the field is shown in the right hand part of Figure 8.2, which represents the field as seen by an aircraft at 3 km altitude. Using $l$ for the order of the Fourier terms, corresponding to harmonic degrees, Lowes found the power spectrum for $l \gtrsim 25$ to be well represented by

$$A_l^2 = 50 \exp(-0.004l) \tag{8.19}$$

corresponding to an amplitude spectrum

$$A_l = 7.1 \exp(-0.002l) = 7.1(1.002)^{-l} \tag{8.20}$$

Applying the same argument as to the low-order harmonics, we infer that the spectrum would appear "white" at a radius smaller than that of the aircraft flight path by the factor 1/1.002, which is 13 km below the aircraft level and 10 km below the surface. Since rocks are magnetic to a depth of about 20 km (below which temperatures exceed the Curie points of the magnetic minerals) the source of the higher harmonics of the field is identified as the magnetism of crustal rocks. Thus the internal magnetic field has two components with quite different spatial spectra, one due to a strong source in the core, which is apparent at the surface as large-scale features or low-order harmonics, the second being much weaker and due to the relatively superficial effect of magnetism in crustal rocks.

The Lowes (1974) analysis of the field carries no implied assumptions as to the nature of the origin of the field but merely that, assuming its source to be in the outer core, its spatial spectrum or harmonic content at the level of the source is white. The observations cannot put a limit to the fineness of the scale of the features at source level. Elsasser (1941) argued the other way by supposing that within the Earth (radius $a$) there is a spherical volume of radius $r(<a)$ enclosing a large collection of small dipoles, distributed randomly and either oriented randomly or all radial. The statistical pattern of the amplitudes of the harmonic components of the observed field (at radius $a$), which is what Lowes plotted in the left half of Figure 8.2, was then used to constrain $r/a$, which Elsasser found to be about 0.5, slightly less than the radius of the core and precisely what Lowes deduced from the later, improved data. This is an interesting way of presenting the problem because the

concept of a magnetic field represented by a very limited number of radial dipoles buried in the outer core follows naturally from the appearance of the core field at the surface of the Earth (Fig. 8.1), whereas the field pattern at core level may be much finer than such a radial dipole model implies.

The separation of the main field into spherical harmonic components is a mathematically convenient way of representing quantitatively the broad scale pattern of the field, but individual harmonic terms are not identifiable with features of the field, except perhaps for the dipole terms. A radial dipole model, suggested by Lowes and Runcorn (1951), was presented by Alldredge and Hurwitz (1964) (see also Alldredge and Stearns, 1969), who found that the field was well represented by a central dipole plus eight radial dipoles distributed about the core at 0.25 Earth radii as in Table 8.1. It appears that this analysis may be meaningful in spite of the objection that at core level the greater emphasis of higher harmonic components would make this a poor representation. The fact that the centred dipole in this model is twice as strong as the spherical harmonic dipole, throws additional doubt on the significance of the dipole field as an independent entity (neither is it strikingly dominant in Fig. 8.2). However, paleomagnetism (see Section 9.3) leaves no doubt that an axial dipole is a statistically significant entity.

Table 8.2 compares the magnetic fields of the Sun, Moon, and

**Table 8.1** Magnetic moments $m$ and positions of radial dipoles at 0.25 Earth radii which, with the central dipole, give the best fit to the field, as represented by the Admiralty chart for 1955[a]

| | $ma^{-3}$ | Colatitude (deg) | East Longitude (deg) |
|---|---|---|---|
| Central dipole $m_0$ | −0.69711 | 23.6 | 208.3 |
| Radial dipoles at 0.25$a$ | 0.10250 | 13.7 | 341.9 |
| | 0.11440 | 46.0 | 179.9 |
| | −0.02724 | 54.9 | 40.1 |
| | 0.07704 | 77.4 | 241.7 |
| | 0.02879 | 91.3 | 120.8 |
| | −0.09469 | 139.8 | 319.3 |
| | −0.11795 | 141.1 | 43.0 |
| | 0.04103 | 102.9 | 180.1 |

From Alldredge and Hurwitz (1964).

[a]Moments are expressed in units of $ma^{-3}$ G ($10^{-4}$ T), where $a$ is the Earth's radius. The value for the dipole from spherical harmonic analysis is $ma^{-3} = B_{h0} = 0.307\ G$ ($3.07 \times 10^{-5}$ T). The rms residual field, or departure of the observed field from that represented by these dipoles is 760 nT, about 1.5% of the average field strength.

**Table 8.2** Magnetic dipole moments of planets, Sun and Moon (estimates are given relative to the magnetic moment of the Earth, $m_E$)

| Body | Moment | Reference |
|---|---|---|
| Sun | $3 \times 10^6\, m_E$[a] | Babcock and Babcock (1955), Allen (1963) |
| Mercury | $6.6 \times 10^{-4}\, m_E$ | Ness et al. (1974b, 1975) |
| Venus | $< 10^{-4}\, m_E$ | Bridge et al. (1967), Ness et al. (1974a) |
| Moon | $< 2 \times 10^{-6}\, m_E$[b] | Colburn (1972) |
| Mars | $3 \times 10^{-4}\, m_E$ | Dolginov et al. (1973) |
| Jupiter | $1.9 \times 10^4\, m_E$[c] | Smith et al. (1974, 1975) |
| Saturn | $< 2.5 \times 10^3\, m_E$[d] | |

[a]Observed field not well represented by a dipole.
[b]Nondipolar field due to local rock magnetizations (Dyal et al., 1970).
[c]Very strong nondipole components (Acuna and Ness, 1975).
[d]Upper limit imposed by lack of evidence of synchrotron radio emissions from magnetosphere.

planets with that of the Earth. This shows that the Earth appears unique among the terrestrial planets in having a strong internally driven field, although in the case of Mercury there is a possibility of an appreciable intrinsic field (Ness et al., 1974b, 1975). In considering the nature and origin of the terrestrial field it is therefore instructive to consider how the Earth differs from the other solid planets. As noted in Section 1.5, the high value of the moment of inertia coefficient, $C/Ma^2 = 0.37$, for Mars shows that there can be only a relatively small dense core, whether fluid or not. The weakness of the Martian field is therefore consistent with our understanding that the much larger core of the Earth is responsible for the terrestrial field. The same explanation applies even more strongly to the Moon, but it appears that the Moon did have an internally driven field early in its history and that this magnetized the surface rocks that are responsible for the residual field which we now see. The absence of a field on Venus cannot be similarly accounted for because in size and density it is very similar to the Earth (Table A1, Appendix A). Coupled with the typically granitic composition revealed by remote surface sampling (Vinogradov et al., 1973) this suggests strongly that its internal composition and structure are also similar to those of the Earth and therefore that Venus has a comparable core. But Venus is rotating very slowly indeed and rotation is believed to be a necessary condition for the establishment of a planetary magnetic field, although we have no secure estimate of the required speed of rotation.

Mercury is also rotating very slowly, although its high density indicates the presence of a core with three quarters of the planet's radius. Although the magnetic moment of Mercury, as listed in Table 8.2, appears very small, relative to the size of the planet it is not as small as the moments of Mars, Venus, or the Moon and corresponds to a field exceeding 200 nT at the surface. If this is found to be an internally maintained field and not residual permanent magnetism or a consequence of electric currents induced by the solar wind (particle radiation from the Sun) it may necessitate some basic rethinking about geomagnetic dynamo mechanisms.

The strong and extensive field of Jupiter, now directly observed (Table 8.2) and previously inferred from radiation from electrons accelerated in its magnetosphere, is presumed to be driven by a convective dynamo in its electrically conducting interior (hydrogen being converted to a metallic, conducting form at very high pressures). This is believed to be essentially similar in principle to the terrestrial dynamo; the relatively stronger nondipole field in the case of Jupiter may be due to the fact that the source of the field is proportionately nearer to the surface. The existence of a magnetosphere around Saturn has often been suggested but has not been demonstrated.

It is of interest to enquire what is the magnetic moment of the smallest terrestrial-type dynamo. If we suppose that the smallest of the radial dipoles listed in Table 8.1 represents an approximate limit below which dynamo action is not self-sustaining in a body with fluid velocities similar to those of the Earth's core, then this limit is about $0.08\, m_{\text{Earth}}$, and the moments of the other terrestrial planets are certainly smaller than this. This would make the Earth unique among the terrestrial planets in having a maintained magnetic field at the present time, although it appears that the Moon may have had an internal field early in its history (Dyal et al., 1974; Fuller, 1974) and this becomes a reasonable assumption therefore for the other planets. On the other hand if Mercury's field is internally driven then the supposed limit is quite invalid.

## 8.2 SECULAR VARIATION AND THE WESTWARD DRIFT

It has been known for over 400 years that the main geomagnetic field is not steady but undergoes a secular variation, which is coherent over large areas of the Earth. The original discovery is attributed to H. Gellibrand, who in 1635, recognized a steady progressive change in magnetic declination, or angle between magnetic north and geographic north, at London (Chapman and Bartels, 1940, p. 910). Vector plots

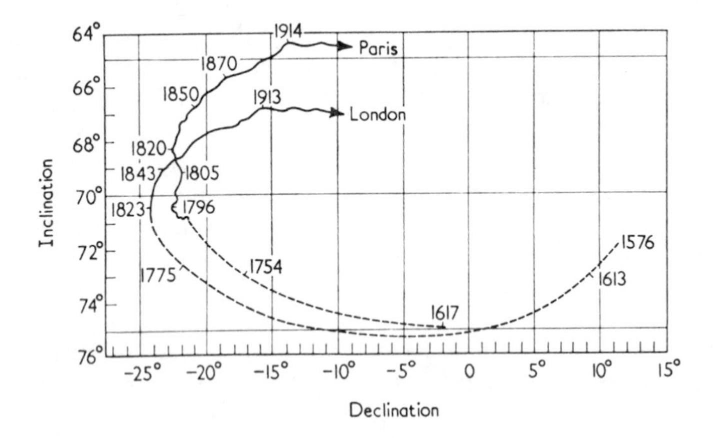


**Figure 8.3.** Vector plots of the inclinations and declinations of the geomagnetic field at London and Paris over several centuries. This plot is due originally to C. Gaibar-Puertas and is reproduced from Runcorn (1962). Another version appears as Figure 9.6.

showing the variations of inclination and declination over long periods of time indicate apparent cyclic variations at many observatories. This type of curve was first plotted for London by L. A. Bauer and has subsequently been used by a number of other authors. Curves for London and Paris are shown in Figure 8.3. Detail is poor for the early years but the general parallelism of the curves is clear; the scale of the secular variation is therefore large compared with the distance between London and Paris. It is important to note that we are considering variations in field which are much stronger than the transient changes due to disturbances in the magnetosphere or ionosphere. Such disturbance fields are discussed here in only one connection—they give information about the electrical conductivity of the upper mantle (Section 8.3).

The worldwide pattern of secular variation has been plotted in various ways by Vestine et al. (1947a), one of whose figures is reproduced as Figure 8.4. The general form is very similar to the nondipole field (Fig. 8.1) and can also be represented by a pattern of deep-seated sources. Comparison of Figures 8.1 and 8.4 suggests that strong centers of secular change tend to be situated west of the strong centers of the nondipole field, implying that the nondipole field is shifting westward. As early as 1692 E. Halley published his conclusion that the secular variation was due to a steady westward drift of the geomagnetic field (Bullard, 1956). Halley supposed that part of the field was due to magnetic poles situated in an inner core and he thus came very close to

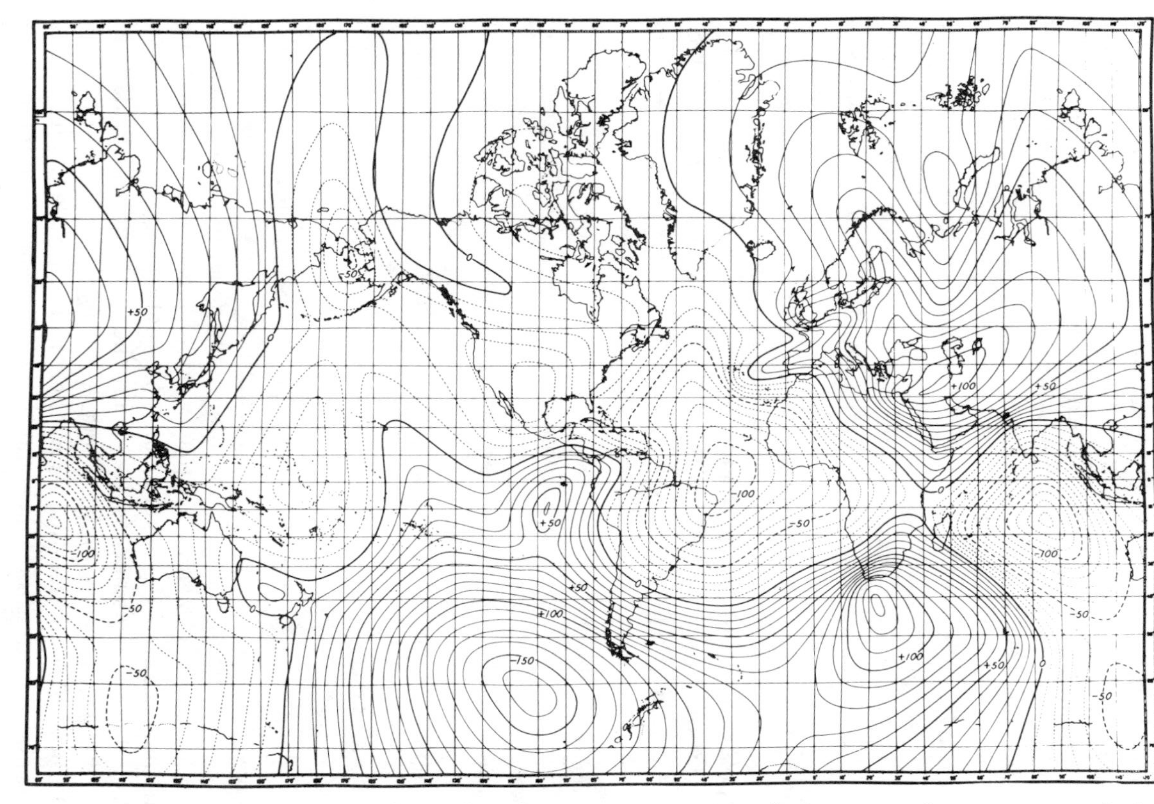


**Figure 8.4.** Contours of rate of secular change in intensity of the vertical component of the geomagnetic field at epoch 1942.5. Values are in nano tesla per year. Figure reproduced, by permission, from Vestine et al. (1947a).

present-day ideas; he pointed out that if the core rotated more slowly than the outer shell of the Earth, its field would drift westward. A detailed analysis of the westward drift was made by Bullard et al. (1950), using Vestine's tabulation of the field over a 40-year period (1905–1945). They concluded that the nondipole field was drifting westward at an average rate of 0.18° $year^{-1}$ at all latitudes, and for all harmonic components, but with a wide scatter in calculated rates; they found that the dipole field, i.e., the equatorial component, was drifting very little if at all, but Vestine (1953) reported a westward drift of 0.30° $year^{-1}$ of the eccentric (off center) dipole. The substantial scatter in the drift at different latitudes, or between different harmonic components, appears as uncertainty in the analysis but was due to a genuine variability in the features of the field rather than to inadequacy of the data. The features, as shown in Figure 8.1, form, deform, and disappear rather like eddies in a stream of water.

As noted in Section 8.1, it is probably unwise to attach too strong a physical significance to the separation of the dipole and nondipole fields. The dipole field itself is not constant but contributes to the secular variation. From a study of the field as a whole, Nagata (1965) concluded that there were five distinguishable features of the present secular variation:

1. A decrease in the moment of the dipole field by 0.05% $year^{-1}$.
2. A westward precessional rotation of the dipole at 0.05° of longitude per year.
3. A rotation of the dipole toward the geographic axis at 0.02° of latitude per year.
4. The westward drift of the nondipole field at 0.2° of longitude per year.
5. Growth and decay of features of the nondipole field, giving changes which average about 10 nT $year^{-1}$.

There are, of course, alternative ways of presenting these conclusions. We may refer to the decreasing strength of the axial dipole, that is, the component of the dipole field parallel to the geographic axis and to a decrease in strength, at a slightly greater rate, of the westward-drifting equatorial component of the dipole field. All analyses agree in finding a westward drift of the nondipole field, but the rate appears to depend upon how the data are analyzed. Thus Alldredge and Hurwitz (1964) examined the motion of the dipoles in their radial dipole model of the field and found that in the interval 1945–1955 the centered dipole grew by 22% and drifted west at more than 1° $year^{-1}$. Excluding the central dipole and also a very small one which apparently moved rapidly, they obtained a mean westward drift of 0.25° $year^{-1}$ but this mean probably has no significance because the variability was very great—four of the nine dipoles drifted eastward. This may indicate that the westward drift itself is much less significant than has often been supposed.

Evidently the secular variation is an essential feature of the action of the geomagnetic dynamo. We have no evidence, either elsewhere in the solar system or from stars, including the Sun, that a stationary dynamo exists. Thus the secular variation is as important as the geometrical form of the field to its interpretation in dynamo theory. The irregularity of the secular variation, most strikingly apparent in the randomness of intervals between geomagnetic reversals (Section 9.4), invites the conclusion that a feeling for the form of the core motions is conveyed by the word *turbulent*.

Vestine (1953) first suggested that the secular variation was correlated with the fluctuations in the rate of rotation of the Earth (Section 3.4). He considered the westward drift of the eccentric dipole field in the periods 1905–1925 and 1925–1945 and concluded that acceleration of the rotation (of the mantle) was accompanied by accelerated westward drift. This conclusion is not entirely dependent on a physical significance of the average westward drift, but only on the fluctuations in its rate. It is virtually impossible to be sure of correlations, such as this, which result from the comparison of sets of data which are very "noisy" or include fluctuations unrelated to the effect that is sought. Munk and MacDonald (1960b) accepted the reality of the correlation, but Malin and Clark (1974) considered it to be doubtful and Yukutake (1973) claimed that the significant correlation was between the rotation rate and the strength of the dipole field. As mentioned in Section 3.4, the necessity to invoke core-mantle interaction to explain the rotational fluctuations is in doubt anyway because variations in zonal wind patterns indicate that angular momentum exchange between the solid Earth and the atmosphere may suffice (Lambeck and Cazenave, 1973, 1974). Nevertheless some angular momentum exchange with the core is expected in terms of our understanding of core motions and dynamo action.

The motion of particular features of the geomagnetic field—especially the westward drift—is widely interpreted as evidence of motion in the Earth's fluid core and is basic to the discussions of Elsasser (1950) and Bullard et al. (1950), which provide a convenient starting point for an assessment of the significance of the drift. The core is considered to be undergoing convective, that is, radial, motion, but the tendency to conservation of angular momentum in the rising and sinking columns of material ensures that there is also a differential rotation. Thus if we consider convection to start in a fluid core that is initially rotating coherently, the rising material carries with it relatively little angular momentum and its angular velocity is reduced, while the sinking material, also conserving its angular momentum, increases in angular velocity. The outer part of the core then rotates more slowly than the inner part.

Now consider the effect of differential rotation on a magnetic field in the core, which is supposed initially to be dipolar. The relative rotation of the "layers" of conducting core material draws out the field lines of the dipole field into an additional toroidal field, or alternatively, we may consider the motion of conductor through the dipole field to cause circulating currents that are responsible for the toroidal field. The interaction between the dipole field and these currents tends to reduce the differential rotation, that is, it couples the inner and outer parts of the core. But the currents generating the toroidal field are not entirely confined to the core but "leak" into the lower mantle, which is an electrical conductor (Section 8.3) although a poor one by comparison with the core. The interaction between the mantle component of the current and the dipole field establishes a relative torque between the core and mantle until a sufficient differential rotation is established between the mantle and outer part of the core. The angular velocity of the mantle is then intermediate between the inner and outer parts of the core. Alternatively Houben et al. (1975) suggested that tidal motion of the core-mantle boundary provided the driving force for a bodily westward drift of the core. Bullard et al. (1950) identified the features of the nondipole field (Fig. 8.1) with large-scale eddies in the outer part of the core, which is necessarily rotating more slowly than the mantle, that is, drifting westward with respect to it. They also inferred that the dipole field, which also drifts westward, but less slowly, had a deeper source. This model is undoubtedly too simple, but the complexity of the field at core level and the fact that the dipole field may not be distinguishable in origin from the other components do not invalidate the principle of the argument.

The necessity for a bodily movement of the outer part of the core with respect to the mantle has not been universally agreed. The westward drift has also been interpreted as a westward propagation of hydromagnetic waves around the core (Hide, 1966; Rikitake, 1966a); this is a suggestion that has received some support from the analysis by Yukutake and Tachinaka (1969) who explained the whole secular variation as a superposition of two fields, one stationary and the other drifting westward. However, much of the data that they used was so old as to be doubtfully adequate and the estimated drift rates for several harmonic components were so variable that it is difficult to assign physical significance to them. If a significance, in terms of eddies or current loops, is attributed to the radial dipoles in Table 8.1, then we may note that Alldredge and Hurwitz (1964) found four of them to be drifting eastward and four westward. This invited the conclusion that the westward drift is merely a consequence of the fact that westward-drifting components of the field happen to penetrate to the surface more readily at the present time and that at core level they are balanced by eastward-drifting components. This conclusion receives some support from archeomagnetic data (Fig. 9.6), which indicate that the cyclic variation of Figure 8.3, which is associated with a westward drift, is characteristic only of the most recent few centuries.

The relative weakness of the nondipole field over the Pacific region has attracted some attention. From paleomagnetic measurements on Hawaiian lavas, Doell and Cox (1971, 1972) inferred that the secular variation had been consistently low in the area for hundreds of thousands of years and suggested that it was electromagnetically screened from the more rapidly varying features of the field, including the westward-drifting nondipole field, by a highly conducting layer deep in the mantle under the Pacific Ocean. However there is disagreement as to whether the effect is real or an artifact of the paleomagnetic sampling (Tarling, 1967; Bingham and Stone, 1972; McElhinny and Merrill, 1975) and the supposition of slight secular variation is contradicted by the historical record from the Honolulu magnetic observatory.* Thus we may consider the geomagnetic secular variation to be a universal feature and the present weakness of the nondipole field over the Pacific to be fortuitous.

In anticipation of the discussion in Section 8.4, it is useful to refer to some of the conclusions of Chapter 9. Paleomagnetism extends the record of the geomagnetic field over an enormously greater time scale, although with much less precision. Over the past few thousand years the intensity of the field has changed by a factor of two or more. The extreme fluctuations of the field appear to be reversals of the dipole field, the most recent clearly established of which occurred 700,000 years ago. But, allowing for the ambiguity in polarity, the field appears to be that of an axial dipole when averaged over tens of thousands of years, that is on an extended time scale the average magnetic and geographic axes coincide.

## 8.3 ELECTRICAL CONDUCTION IN THE CORE AND MANTLE

There is no prospect that estimates of the electrical resistivity of the Earth's deep interior will ever be comparable in precision to the seismologically determined parameters, elasticity and density. Resistiv-

*During the period 1902–1938 the horizontal field component at Honolulu decreased by 800 nT and the vertical component by 1600 nT, an average change of 50 nT year$^{-1}$, which is by no means small.

ity of the crust is explored by magnetotelluric probing, in which the principal problem* is heterogeneity of the medium, which is affected very strongly by water content. Lateral inhomogeneities in electrical conductivity appear to extend 100 km or more into the mantle (Gough, 1973a,b), but observations are too imprecise to give more than a rough overall picture below that. However, the general increase in conductivity with depth is inferred from the effects of induced currents on geomagnetic variations. Lower mantle estimates are constrained by the penetration of the mantle by the secular variation of the main geomagnetic field, which is generated in the core. Mantle conductivities are characteristic of semiconductors (Tozer, 1959), in which conductivity is highly variable; in particular it depends very strongly on temperature, so that if the mineralogy and physics of the deep interior are believed to be sufficiently well understood, the conductivity estimates lead to an inferred temperature profile (Tozer, 1959, 1970).

In the core the situation is reversed. There is no effect that gives directly an estimate of conductivity, although geophysical constraints are imposed by the requirements that the core should be a sufficiently good conductor to allow dynamo action (Section 8.4) but not so good that the corresponding thermal conductivity gives an unacceptably large heat flux from the core (Section 7.3). More explicit estimates are obtained by extrapolating laboratory data on liquid iron and its alloys to the pressure and probable temperature of the core. The value of resistivity used here, $3 \times 10^{-6}\ \Omega$ m, is the one assumed by Bullard (1949) and deduced also by Gardiner and Stacey (1971) from an extrapolation of more recent laboratory data on iron alloys.

At laboratory pressures the effect of alloying elements on liquid iron is generally slight (Baum et al., 1967). This is because the increase in resistivity due to alloying results from the scattering of electrons by disorder of the atomic structure, but a liquid is already so disordered that impurities cannot effect much more reduction in the electron mean free path. However, at very high pressures, even in a liquid, the atoms are compressed to a state approximating a close-packed structure and so display greater order; thus pressure decreases the resistivities $\rho_e$ of pure metals by reducing the influence of thermal disorder (including melting). But impurity disorder is not reduced by pressure. Data by Bridgman (1957) indicate that for iron with alloying elements exceeding 13%, resistivity is increased by pressure instead of being decreased as in pure metals. This conclusion must be applied to the core, in which the alloying elements may amount to 15% (section 6.4). The assumption that

*For purposes of mineral exploration, this is a feature to be exploited rather than a problem.

the resistivity-pressure relationship for pure metals can simply be renormalized for application to alloys was the basis of the Gardiner-Stacey estimate. Jain and Evans (1972; see also Evans and Jain, 1972) suggested (1 to 2)$\times 10^{-6}\,\Omega$ m, but $2\times 10^{-6}\,\Omega$ m must be regarded as the absolute lower limit imposed by the consequent heat flux from the core (section 7.3). The higher value favored here and by Gardiner and Stacey (1971) is supported by the data of Keeler (1971) who made transient (shock wave) measurements on iron and silicon iron and estimated a resistivity of $3.3\times 10^{-6}\,\Omega$ m for 20% Si—80% Fe in the liquid state at 3000 K and $1.4\times 10^{11}$ Pa (1.4 Mbar—corresponding to the pressure just inside the outer core). A possibility that addition of sulfur to liquid iron causes a stronger increase in resistivity than do other alloying elements is implied by the data of Vostryakov et al. (1964), who found a pronounced peak in resistivity at 36 wt%S, corresponding to the composition FeS, at which the conduction electrons of iron are largely immobilized as valence electrons in bonding to sulfur. However, metallic conduction in the liquid persisted throughout the composition range and the bonding effect is presumed not to be a major one at 20%S or less, as in the core.

Thus it appears that $3\times 10^{-6}\,\Omega$ m is still about as good an estimate of core resistivity as we can make at present. The value may be as low as $2\times 10^{-6}\,\Omega$ m without implying an implausible core heat flux but values exceeding $4\times 10^{-6}\,\Omega$ m appear improbable in the light of Keeler's (1971) data.

The two methods of estimating mantle conductivity are complementary. Both depend on the electromagnetic skin effect. Geomagnetic disturbances of magnetospheric origin are modified by induced electric currents and so give a reasonable indication of upper mantle conductivity, but are increasingly difficult to apply to greater depths. Conversely, the spectrum of the secular variation of the main field is used to infer attenuation of the higher frequency components by conduction in the lower mantle, but is not significantly affected by the less conducting upper mantle. Electromagnetic core-mantle coupling of rotational fluctuations provides no constraint if atmospheric effects are admitted as a possible cause of length of day variations (Section 3.4). Except perhaps for the outer layers, the uncertainty at any depth is still of order a factor 10*.

A fluctuating electromagnetic field induces currents in a conductor, which oppose changes in the field (Lenz's law of electromagnetic

*Electrical resistivities of solids at ordinary temperatures encompass a range exeeding $10^{23}:1$, so that an uncertainty of a factor 10 still implies a reasonable understanding of the electrical state of a material.

induction). It follows that penetration of the conductor by the field is opposed by the electrical conduction and therefore that the fluctuations do not penetrate a conductor as effectively as they would penetrate an insulator, in which they are not attenuated. This is the *skin effect*. For derivation of the basic equations reference should be made to a text on electromagnetism (e.g., Harnwell, 1949) and for a comprehensive review of applications to geophysics see Rikitake (1966b). The principle is conveniently described in terms of the penetration of a plane electromagnetic wave of angular frequency $\omega$, into a semi-infinite conductor of conductivity $\sigma$, bounded by the plane $z = 0$, and extending indefinitely in the positive $z$ direction. Then if the magnetic field strength at the surface is $B_0 \sin \omega t$, the strength at depth $z$ is

$$B = B_0 e^{-\alpha z} \sin(\omega t - \alpha z) \tag{8.21}$$

where

$$\alpha = \frac{1}{z_0} = [(\mu_0/4\pi) \cdot 2\pi\sigma\omega]^{1/2} \tag{8.22}$$

and $z_0$ is known as the skin depth of the material (at frequency $\omega/2\pi$), the depth at which the amplitude of the field oscillation is reduced to $1/e$ of its surface value. In the Earth we may take the permeability to be the free space value ($\mu_0 = 4\pi \times 10^{-7}$ H m$^{-1}$), the influence of crustal magnetic materials being trivial and the temperature at greater depths being higher than the Curie point of any mineral. It is to be noted that not only is the field at depth attenuated but that it has a phase lag relative to that at the surface. It follows that the induced currents modify the field observed outside the conductor, which can be represented as a sum of two fields, one of external origin and the other internally generated by induction, and that there is a phase difference between these two components. The phase difference is sensitive to any variation of conductivity with depth.

The plane wave approximation is not directly applicable to the Earth as a whole, for which we must consider a spherical conductor, in a simple case with uniform conductivity, but in a better approximation nonuniform but spherically symmetrical. Induction in a conducting sphere within the Earth by geomagnetic variations was considered in detail by S. Chapman and co-workers, the calculation of greatest interest being that of a sphere with radially varying conductivity (Lahiri and Price, 1939). The spherical symmetry of the problem allows the components of a disturbance field that are of external and internal origins to be separated by spherical harmonic analysis. Referring to Equation 8.12 (or Eq. C.11 in Appendix C), the coefficients $C_l^m$ and $S_l^m$ describe the spatial variations of a field whose origin is internal to the Earth; $C_l'^m$ and $S_l'^m$

represent a field of external origin. In the analysis of the main geomagnetic field the primed coefficients are found to be zero, but in representing disturbance fields, that is, diurnal variations, magnetic storms, and similar effects, the primed coefficients are larger because the disturbance fields are generated outside the solid Earth, in the magnetosphere or magnetic envelope to which the geomagnetic field is confined by its interaction with the interplanetary plasma. However, the induced currents within the Earth cause fields which, being of internal origin, are represented by the unprimed coefficients in Equation 8.12. Thus both primed and unprimed coefficients are required to represent disturbance fields and the ratios $C_l^m/C_l'^m$, $S_l^m/S_l'^m$ give the ratio of internal (induced) field to external (inducing) field for any harmonic component. These ratios are determined by spherical harmonic analyses of disturbance fields, as are also the phase differences between the components of the field represented by primed coefficients and the corresponding unprimed coefficients.

The most reliable data and physically the simplest situation to analyze is the diurnal variation, which has a 24-hour period so that the phase difference between the induced and inducing fields appears as a phase difference in longitude as the diurnal magnetic wave travels around the Earth. The diurnal variation was given greatest weight in the analysis of Lahiri and Price (1939), who established the principles by which the electrical conductivity of the mantle is estimated from geomagnetic variations. They first produced a range of mathematically simple conductivity models that were consistent with the relative amplitudes and phases of the external and internal diurnal fields and then eliminated those that were not also consistent with magnetic storm data.

Subsequent investigators have confirmed the conclusion of Lahiri and Price that there is a strong increase in conductivity at a depth of a few hundred kilometers in the mantle. This is shown in Figure 8.5, which is a digest of several approaches to the conductivity problem. Upper mantle values are based on Banks (1969, 1972), but with an indication of the relatively thin conducting skin, which is probably due mainly to the presence of water in surface layers (especially sea water, which has a conductivity of $4\ \Omega^{-1}\ m^{-1}$ but is very unevenly distributed).

The upper mantle data do not extrapolate directly to the value of conductivity near to the base of the mantle that is required by consideration of the attenuation of secular variation as it penetrates to the surface from the core. This is also an electromagnetic skin effect problem, but in this case the source field and conductivity distribution are both unknown. Estimates of mantle conductivity from the observed secular variation spectrum depend on an assumed form of that spectrum at the

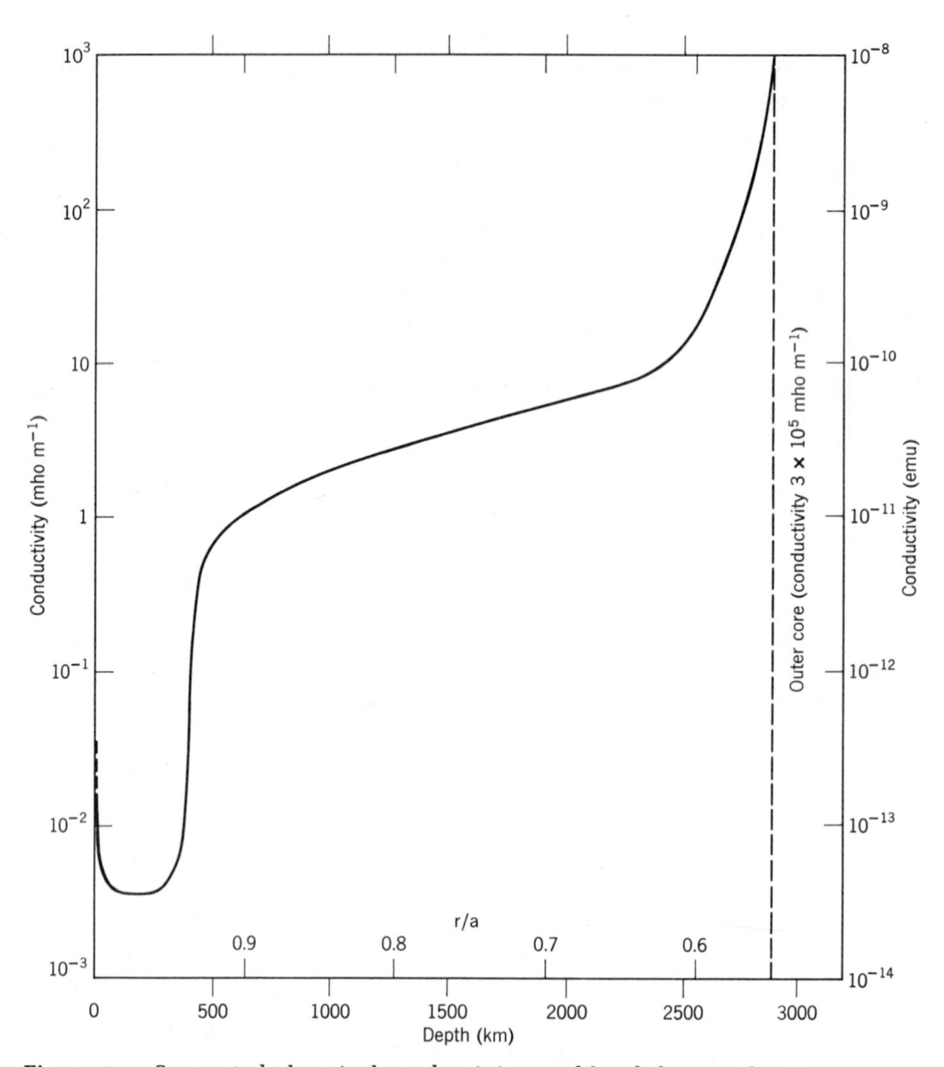


**Figure 8.5.** Suggested electrical conductivity profile of the mantle. Upper-mantle data are based on Banks (1969, 1972—see also Parker, 1971), with extrapolation to lower mantle values of McDonald (1957) and Currie (1968) and allowance for a conducting "skin."

surface of the core. The simplest approach, taken by Runcorn (1955), is to assume a "white" spectrum (all frequencies equally represented) and similar geometrical scales for all frequency components. Then the observed diminution of higher frequencies may be explained as attenuation by a layer of thickness $L$ and uniform conductivity $\sigma$, so that by

Equation 8.22 the amplitude spectrum at the surface has the form

$$B = B_0 \exp\{-[(\mu_0/4\pi)2\pi\sigma\omega]^{1/2}L\} \tag{8.23}$$

By fitting the constants of this equation to the secular variation spectrum, Runcorn (1955) obtained the result $\sigma L^2 = 1.1 \times 10^{16}\ \Omega^{-1}\ \mathrm{m}$, so that if the layer is considered to have a thickness of $10^6$ m its conductivity is $\sigma = 1.1 \times 10^4\ \Omega^{-1}\ \mathrm{m}^{-1}$.

This is an extreme upper bound on the lower mantle conductivity because the sources of secular variation are localized and are attenuated with distance purely geometrically as well as electromagnetically. Although we do not know the actual secular variation spectrum at the core, we can be sure that the higher-frequency components are relatively more localized and therefore that even without electromagnetic screening the higher frequencies will appear more attenuated at the surface. The preceding estimate of conductivity is therefore certainly too high. These difficulties were overcome in a more sophisticated calculation by McDonald (1957), who assumed a spatially random distribution of secular variation sources at the core surface and allowed for geometrical spreading as well as electromagnetic screening. McDonald assumed a power law variation of conductivity with radius in the lower mantle similar to that assumed by Lahiri and Price (1939) for the upper mantle, but the details of the assumed distribution are less important than the deduced values. McDonald's estimate for the base of the mantle was $200\ \Omega^{-1}\ \mathrm{m}^{-1}$, decreasing to about $60\ \Omega^{-1}\ \mathrm{m}^{-1}$ 1000 km above the core. A reconsideration of the secular variation spectrum led Currie (1968) to suggest an average value of $200\ \Omega^{-1}\ \mathrm{m}^{-1}$ for the whole lower mantle of 2000 km thickness. These estimates can be matched to the upper mantle curve only by postulating a steep rise in conductivity near to the core, as suggested in Figure 8.5. This is in accord with the temperature profile indicated in Figure 7.4.

Much higher estimates of lower mantle conductivity have been given by Kolomiytseva (1972) and Braginskiy and Nikolaychik (1973) who deduce from the screening of secular variation, values similar to those of the core. This appears to be implausible in terms of mantle properties and may result from a screening effect in the outer layer of the core itself.

The range of conductivities represented in Figure 8.5 is clearly indicative of semiconduction processes, whose relevance to the mantle has been reviewed by Tozer (1959) and Shankland (1967, 1975). The basic physics of semiconduction is discussed in texts on solid-state physics (e.g., Kittel, 1971), and it will suffice here to quote the significant results.

Semiconductors are characterized by the existence of an energy gap between the highest filled (valence) electron states and the next available states (the conduction band). An electron in a completely full band cannot give electrical conduction because to conduct it must respond to an applied electric field, that is, it must change from one state to another with greater momentum in the direction of the field; if there are no vacant states accessible to it, it cannot make this change and a material in which this is so for all electrons is an insulator. However, in a semiconductor the energy gap $E_g$ between the filled valence states and empty conduction states is sufficiently small, relative to the available thermal energy $kT$ at temperature $T$ that a few electrons are excited into the conduction band. The conductivity is proportional to the number of electrons thus excited and is a strong function of temperature. A material in which this occurs is an intrinsic semiconductor. Normally, however, impurities in a simple semiconductor give rise to additional electron energy levels within the gap. These impurity levels may be occupied by electrons at low temperatures, in which case they are known as donor levels, because they give electrons into the conduction band more readily than does the valence band; alternatively, the impurity levels may be vacant at low temperatures and are then known as acceptors, because electrons may be excited into them from the valence band. Conduction may occur either by virtue of electrons in the conduction band or "holes" in the valence band, or both. The situation is represented diagrammatically in Figure 8.6.

Both the intrinsic and extrinsic mechanisms are processes of electronic conduction. Also possible at high temperatures is ionic conduc-

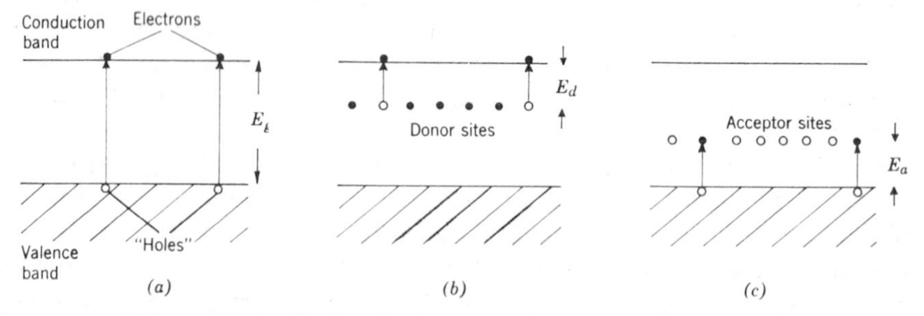


**Figure 8.6.** Illustration of electron energy levels in intrinsic and extrinsic semiconductors. Since the energy gaps, $E_d$ and $E_a$ for the excitation of electrons to or from impurity levels, are smaller than the energy gap between the valence and conduction bands, extrinsic (impurity) conduction occurs more readily than intrinsic conduction at low or moderate temperatures. However, at high temperatures impurity conduction saturates by virtue of the finite number of available states and intrinsic conduction takes over. (*a*) Intrinsic semiconductor. (*b*) *n*-type extrinsic (or impurity) semiconductor. (*c*) *p*-type extrinsic (or impurity) semiconductor.

tion, in which ions move bodily under the influence of an electric field. All three processes are described by very similar equations. Conductivity is proportional to the number of charge carriers, which increases with temperature according the Boltzmann distribution. Thus the number of intrinsic charge carriers is

$$n_i = A \exp\left(-\frac{E_g}{2kT}\right) \tag{8.24}$$

where $A$ is a factor that depends on temperature (as $T^{3/2}$ for simple band forms) and is of order $10^{26}\ \mathrm{m}^{-3}$. Allowing for the saturation of extrinsic conduction due to the finite number $N$ of impurity sites, the number $n_e$ of extrinsic charge carriers is given by

$$n_e = (N - n_e)^{1/2} A' \exp\left(-\frac{E}{2kT}\right) \tag{8.25}$$

where $E$ is $E_d$ or $E_a$ from Figure 8.6 and $A'$ is a constant equal to $A^{1/2}$ in the absence of any differences in effective masses of the charge carriers. We may assume mobilities of electrons or holes to be determined primarily by phonon scattering, as in metallic conduction, in which case they vary approximately as $T^{-1}$; the temperature dependence of conductivity in both intrinsic and extrinsic cases is thus dominated by the exponential terms. The temperature dependence of ionic conductivity $\sigma_3$ is similarly represented by a Boltzmann term with an activation energy, in this case the energy of ionic diffusion $Q$. The general form of conductivity as a function of temperature is therefore a sum of three exponential terms, representing intrinsic, extrinsic, and ionic conduction, respectively:

$$\sigma = \sigma_i + \sigma_e + \sigma_3 = \sigma_{i_0} \exp\left(-\frac{E_g}{2kT}\right) + \sigma_{e_0} \exp\left(-\frac{E}{2kT}\right) + \sigma_{3_0} \exp\left(-\frac{Q}{kT}\right) \tag{8.26}$$

In any limited temperature and pressure range one of the terms in Equation 8.26 is dominant; to determine which is principally a matter of estimating the activation energies. At laboratory pressures the intrinsic band gap, $E_g$, in olivines, which are representative of upper mantle composition, is about 8 eV (Shankland, 1967, 1968). This is too high a value to allow intrinsic conduction to be significant at any plausible temperature, so that upper mantle conduction must be explained in terms of extrinsic or ionic conductivities. Ionic conduction is found to be significant in measurements at laboratory pressures on polycrystalline oxides and silicates, but much less so in good single crystals, implying that ionic conduction is facilitated by grain boundaries (Shankland, 1967). It is probably the most important conduction mechanism in the crust and may remain significant throughout the range in which the mantle material is very close to melting, but pressure strongly inhibits movements of ions (i.e., it increases $Q$), which can have no relevance to conduction in the lower mantle (Tozer, 1959).

The conditions of the lower mantle are not sufficiently well understood definitely to exclude intrinsic conduction at those levels, but we have no evidence that the intrinsic band gap $E_g$ is greatly reduced from the upper mantle value of 8 eV. We can estimate what $E_g$ must be if intrinsic conduction is dominant in the lower mantle by extrapolating from observations on silicon and germanium to estimate $\sigma_{i_0} = 7 \times 10^5\ \Omega^{-1}\ \mathrm{m}^{-1}$ ($7 \times 10^{-6}$ emu). Then if the temperature at the base of the mantle is 3300 K (as in Fig. 7.4) and the conductivity is $300\ \Omega^{-1}\ \mathrm{m}^{-1}$ (see Fig. 8.5) we require $E_g = 9.3 \times 10^{-19}\ \mathrm{J} = 4.5\ \mathrm{eV}$. Observations by Mao and Bell (1972) of electrical conductivity and optical absorption in $Fe_2SiO_4$ in both fayalite (olivine or low pressure) and spinel (higher pressure) crystal forms showed a strong increase in conductivity and an extension of optical absorption through the visible range to infra-red wavelengths as pressure was increased to $2.7 \times 10^{10}$ Pa (270 kbar). These effects are unconnected with phase transitions and are too violent to be attributed to pressure narrowing of an intrinsic band gap; the conductivity at 200 kbar would be consistent with an extrinsic energy gap less than 0.3 eV. Shankland's (1967) experiments on very lightly iron-doped forsterite ($Mg_2\ SiO_4$), the Mg-rich end member of the olivine [$(Mg_x\ Fe_{1-x})_2\ SiO_4$] solid-solution series, point to onset of a process of charge-transfer either between iron ions (which may be either $Fe^{2+}$ or $Fe^{3+}$) or between $Fe^{2+}$ and $O^{2-}$. Charge transfer is likely to be the dominant process of electronic conduction in the upper mantle, as well as being responsible for its opacity (as considered in Section 7.2). Conventional extrinsic electronic conduction may take over in the lower mantle.

## 8.4 GENERATION OF THE MAIN FIELD

Study of the secular variation reveals that the features of the main field move both with respect to the surface of the Earth and with respect to one another. If the average westward drift is taken to be representative of the rate of motion, then the velocities observed at the surface are of order 20 km year$^{-1}$. This is more than $10^5$ times faster than prolonged large-scale motion evident from geology and virtually precludes the participation of the solid part of the Earth. Since the large-scale features of the field indicate a deep internal origin and seismological evidence (Chapter 6) favors a fluid core, we can reasonably suppose the secular variation to be a consequence of core motions. Furthermore, geochemical and density considerations (Sections 1.5 and 6.4) are consistent with a core composition largely of iron, a good electrical conductor. A moving fluid conductor and a magnetic field exercise a mutual control on one another. The study of this problem is known as magnetohydrodynamics or hydromagnetics, and is important not only in connection with the magnetic fields of the Earth and other astronomical bodies, but also in laboratory plasma physics, in which the conductor is a tenuous, highly ionized gas. The concept is the same, although the problem of the Earth's core is in one respect simpler—it is virtually incompressible. Roberts (1967) has given a readable and comprehensive review of the subject.

Magnetohydrodynamic theories of the Earth's field are developments of an idea, due originally to J. Larmor, who pointed out that suitable internal motion of a large fluid conductor could cause it to act as a selfexciting dynamo. But there are still many unknowns and too few observational constraints to allow the supposition that we are close to an unambiguous theory. In spite of elimination of some early hypotheses, the number of possible alternatives has grown in recent years. The mathematical and physical complexity of the problem is forbidding but significant progress is apparent and there is no promising alternative to magnetohydrodynamic action as the origin of the Earth's main field. Elsasser (1950, 1956), Hide and Roberts (1961), Rikitake (1966b), and Gubbins (1974) have reviewed the subject.

The necessity for a mechanism of regeneration of the field is seen by making an order-of-magnitude calculation of the time constant for free decay of electric currents in the core. We can consider the simple case of a core-sized conductor in the form of a toroid (anchor ring), to which standard formulas for inductance $L$ and loop resistance $R$ may be applied (see, for example, Harnwell, 1949, p. 330). Assuming that the mean radius of the loop is $r_1$ and the radius of its cross section is $r_2$, as in Figure 8.7, and that the loop just fits within the boundaries of the outer, fluid core, whose inner and outer radii are 1300 km and 3500 km, we have $r_1 = 2.4 \times 10^6$ m, $r_2 = 1.1 \times 10^6$ m. Then, approximately,

$$L = 4\pi \left(\frac{\mu_0}{4\pi}\right) r_1 \left[\ln\left(8\frac{r_1}{r_2}\right) - \frac{7}{4}\right] = 3.6 \text{ H} \tag{8.27}$$

$$R = 2\rho \frac{r_1}{r_2^2} = 1.2 \times 10^{-11}\,\Omega \tag{8.28}$$

where resistivity $\rho \approx 3 \times 10^{-6}\,\Omega$m in the core. The time constant for decay of current in the loop is then

$$\tau = \frac{L}{R} = \left(\frac{\mu_0}{4\pi}\right)\frac{2\pi}{\rho} r_2^2 \left[\ln\left(8\frac{r_1}{r_2}\right) - \frac{7}{4}\right] \approx 3 \times 10^{11} \text{ sec} = 10^4 \text{ years} \tag{8.29}$$

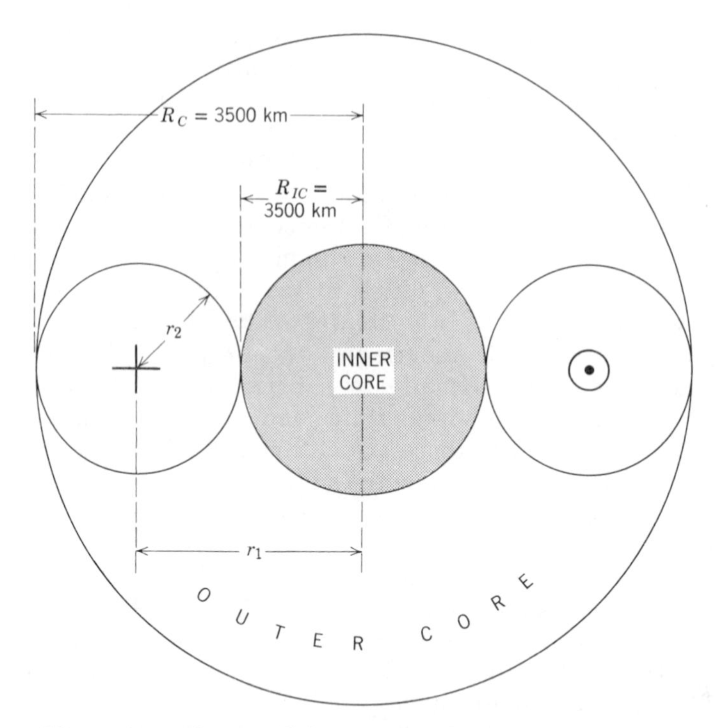


**Figure 8.7.** Dimensions of a toroidal current in the outer core that is the basis of order-of-magnitude estimates of spontaneous decay time and energy dissipation for core currents.

This is a rough estimate, based on a simplified model of core currents; in particular since it considers the largest possible loop and since the characteristic decay time depends upon the square of loop size, as in Equation 8.29, $10^4$ years should be regarded as an upper limit. Furthermore, if we suppose that features of the nondipole field are due to geometrically smaller current loops, then their decay times are correspondingly smaller. These conclusions are compatible with the present-day observations of secular variation, which show that the dipole moment of the Earth is changing by 0.05% $\text{year}^{-1}$, a rate which could be represented by a time constant of 2000 years. The geomagnetic field cannot be a vestige of the Earth's early history, several thousand million years ago, but it must be continuously regenerated.

The rate of decay of the field has been calculated assuming ohmic dissipation by core currents. This is almost certainly the dominant dissipation mechanism because viscous drag must be negligible for such a large, slow-moving body, assuming any reasonable value of viscosity, $\eta$. If we take the rate of core motion to be 10 km/year in cells of radius

1000 km the rate of shear is $\dot{\epsilon} \approx 10^{-2}$ year$^{-1} = 3 \times 10^{-10}$ sec$^{-1}$ and the stress required to overcome viscous drag is $\eta\dot{\epsilon}$, giving a total viscous power dissipation for the volume $V$ of the core

$$\left(-\frac{dE}{dt}\right)_v = \eta\dot{\epsilon}^2 V \approx 16\eta \text{ W} \tag{8.30}$$

An upper bound, $\eta < 10^3$ decapoise (kg m$^{-1}$ sec$^{-1}$) is imposed on core viscosity by lack of evidence for its effect on the obliquity of the ecliptic (Rochester, 1976). Extrapolating laboratory observations of material properties, Bukowinski and Knopoff (1976) gave a lower bound of 1 decapoise, although Gans (1972) suggested $10^{-2}$ decapoise. But whichever value is favored viscous dissipation in the core by Equation 8.30 is insignificant. A lower bound to the ohmic dissipation in the core is obtained by considering the current loop of Figure 8.7. If a total current $I$ flows in the loop then its magnetic moment is approximately*

$$m = \pi r_1^2 I \tag{8.31}$$

which may be equated to the observed dipole moment, $m = 7.94 \times 10^{22}$ A m$^2$, to obtain the total toroidal current, $I = 4.4 \times 10^9$ A (corresponding to a current density of $1.2 \times 10^{-3}$ A m$^{-2}$). Using the approximate effective resistance of the loop given by Equation 8.28, the power dissipation is

$$-\frac{dE}{dt} = I^2 R = \frac{2\rho \text{m}^2}{\pi^2 r_1^3 r_2^2} = 2.3 \times 10^8 \text{ W}\dagger \tag{8.32}$$

Equation 8.32 is a convenient starting point for order-of-magnitude calculations of the power dissipated by core currents, but as it stands it is an unrealistically low estimate. Applying equations with the form of Equation 8.32 to the radial dipoles of Alldredge and Hurwitz (Table 8.1), a further dissipation at least equal to that of the dipole field is obtained (and the central dipole in this model has a dissipation four times as great). Furthermore, the *poloidal* (dipole and observed nondipole) features of the field are necessarily associated with toroidal fields‡ which

*If the current density is assumed to be uniform through the toroid then the more exact relationship is

$$m = \pi(r_1^2 + r_2^2/4)I$$

†Parker (1972) showed that the minimum dissipation by the optimum current distribution in a sphere with the radius $r_c$ and magnetic moment $m$ of the core is $\left(-\frac{dE}{dt}\right)_{\text{Min}} = \frac{15\text{ m}^2\rho}{2\pi r_c^5} = 0.88 \times 10^8$ W (see Problem 8.5 in Appendix J).

‡A toroidal field may be visualized as the field produced by a current pattern having the geometrical form of wire wound uniformly onto the anchor ring. If the wire is would along the ring and then back again to the starting point, the field is confined to the anchor ring.

form closed loops within the core (perhaps extending marginally into the conducting lower mantle). Until recently it was considered that the toroidal field was very much stronger than the dipole field but this is now doubted (Busse, 1974) and we suppose instead that the strengths (and therefore dissipations) are comparable. These considerations raise the estimated minimum dissipation to more than $10^9$ W and make $10^{10}$ W appear more likely. In Section 7.4 (see especially Table 7.3) the power available from core convection is assigned an approximate value of $2.75 \times 10^{11}$ W, which very adequately covers what is required if we accept approximate equality of the toroidal and poloidal field strengths.

Alternative power sources for the dynamo that have been suggested are precessional torques between the core and mantle (Malkus, 1963, 1968), latent heat of a progressively solidifying inner core (Verhoogen, 1961) and, most recently, tides (Houben et al., 1975). Rochester (1974), Rochester et al. (1975), and Loper (1975)—see also Section 3.2—find that the precessional dissipation does not exceed $10^8$ W, at least for stable precessional flows, which is clearly inadequate even without considering the problem of maintenance of the fluidity of the core against the conducted heat loss (Section 7.3), so that although this once seemed a promising mechanism it should now be discounted (with the proviso that precession driven turbulence has not been rigorously eliminated). The latent heat suggestion conflicts with the argument that the thermal profile of the Earth is maintained essentially constant by a convective balance (Section 7.2), but cannot be firmly discounted in terms of our insecure numerical data on relevant core parameters. The latent heat of solidification itself is probably less important than either the direct thermal capacity of the whole core or the gravitational energy release by shrinking of the Earth if the denser solid inner core is growing. The total energy for a minimal core power of $3 \times 10^{12}$ W* over $4.5 \times 10^9$ years is $4 \times 10^{29}$ J and with a core heat capacity of $10^{27}$ J deg$^{-1}$ implies a total core temperature drop of only 400°, without appealing to either latent heat or gravitational energy. Since the only argument against a slow, progressive fall in temperature of this magnitude is the convective balance hypothesis used in Section 7.2, it is safest to admit the cooling core suggestion as a possible dynamo mechanism.

From the point of view of the dynamo mechanism, the cooling core and radioactive core hypotheses hardly differ. Both imply core convection (and the thermodynamic inefficiency of this process—Section 7.4). But even this restriction on the ultimate driving mechanism leaves wide open the discussion of still-basic questions such as the scale of core motions: is the important motion steady, turbulent or oscillatory, small

*Most of this is conducted heat, as in Section 7.3.

scale or corewide? A review of current thinking, with some emphasis on mathematical manipulations and difficulties is given by Gubbins (1974), who expresses the general concensus that "any sufficiently complicated and vigorous motion will operate as a homogeneous dynamo," echoing the quotation from Elsasser (1963) at the head of this chapter. Nevertheless the magnetohydrodynamic principles are clear and may be summarized in simple terms.

Formal theories of magnetohydrodynamics (see, for example, Elsasser, 1956; Roberts, 1967) express in a more general way the conclusion, represented by Equation 8.29, that a magnetic field can diffuse out of an electrical conductor with a time constant proportional to its conductivity (or reciprocal of resistivity $\rho$) and to the square of its dimensions $l$. The characteristic time constant is

$$\tau = \frac{l^2}{\eta} \tag{8.33}$$

where $\eta$ is termed the magnetic diffusivity and is given by

$$\eta = \left(\frac{4\pi}{\mu_0}\right)\frac{\rho}{4\pi} = 2.4 \text{ m}^2 \text{ sec}^{-1} \tag{8.34}$$

For the whole core $\tau \approx 10^4$ years* as previously deduced. The time constant represented by Equation 8.33 can be compared with the characteristic time $\tau'$ for the internal motion of a fluid conductor of dimension $l$ with velocity differences $v$:

$$\tau' = \frac{l}{v} \tag{8.35}$$

Then if $\tau' \ll \tau$, that is, if

$$lv \gg \eta \tag{8.36}$$

the field does not have time to diffuse out of the fluid and is carried along (and deformed) by the fluid motion. The field is then said to be "frozen in" to the conductor. This effect is basic to all magnetohydrodynamic theories of the geomagnetic dynamo. That it can lead to an intensification of a magnetic field is shown in Figure 8.8. The applicability of the inequality (8.36) to the Earth's core is demonstrated by considering the features of the nondipole field to originate in core volumes of order 1000 km in radius with internal velocities of about 10 km yr$^{-1}$ ($3 \times 10^{-4}$ m sec$^{-1}$), corresponding to the extrapolation of the westward drift down to the core, so that

$$lv \approx 300 \text{ m}^2 \text{ sec}^{-1} \approx 140\ \eta \tag{8.37}$$

*Taking $l$ equal to the half thickness of the outer core.

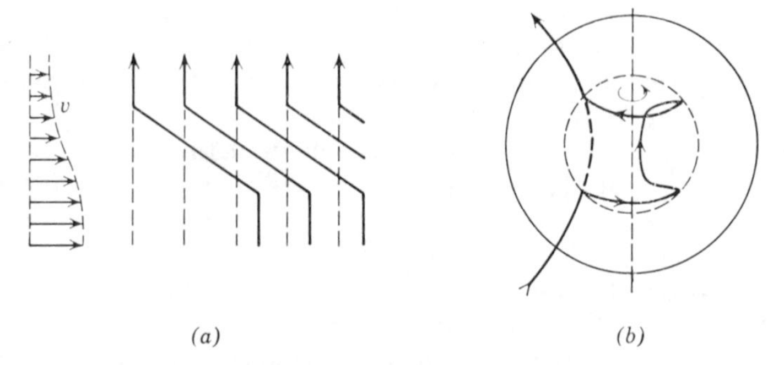


**Figure 8.8.** (*a*) The deformation of a magnetic field by a velocity shear intensifies the field (the strength of the field being represented by closeness of the lines of force). (*b*) Differential rotation of the inner and outer parts of the core, which is a consequence of convection, draws out the field lines of an initial poloidal field to produce a toroidal field. Both figures after Elsasser (1950).

The inequality represented by Equation 8.36 is frequently represented in terms of a "magnetic Reynolds' number" $R_m$:

$$R_m = \frac{lv}{\eta} \tag{8.38}$$

Until about five years ago it was generally supposed that dynamo action was only possible for $R_m \gg 1$, but this is now questioned. This condition does apply to large-scale, steady dynamos of the kind considered by Bullard and Gellman (1954), but not to the more probable alternative of small-scale local motions (turbulence?) that may develop the field cooperatively. Roberts (1972) has shown the possibility of dynamo action in which the motion is restricted to small cells of scale such that $R_m < 1$, but the pattern of motion is a repeating one, so that if we take the dimension $l$ in Equation 8.38 to be that of the whole core, while $v$ is the small scale velocity, the condition $R_m > 1$ probably still applies. In terms of Equation 8.37 this constitutes a justification for the assumption that the rate of the westward drift gives an estimate at least of the order of magnitude of the speed of core motions.

Several theoretical models have been devised to show that self-exciting dynamo action is possible in spherical conducting fluids with suitable internal motions, but of particular interest is a laboratory model by Lowes and Wilkinson (1963, 1967, 1968), which demonstrates the principle, using metal cylinders rotating inside a metal casting, in the manner of Figure 8.9*a*. Electrical connection of the cylinders to the casting was accomplished by filling the spaces with mercury. To scale their dynamo to laboratory size with attainable speeds of rotation, Lowes and Wilkinson used a ferromagnetic metal of high permeability,

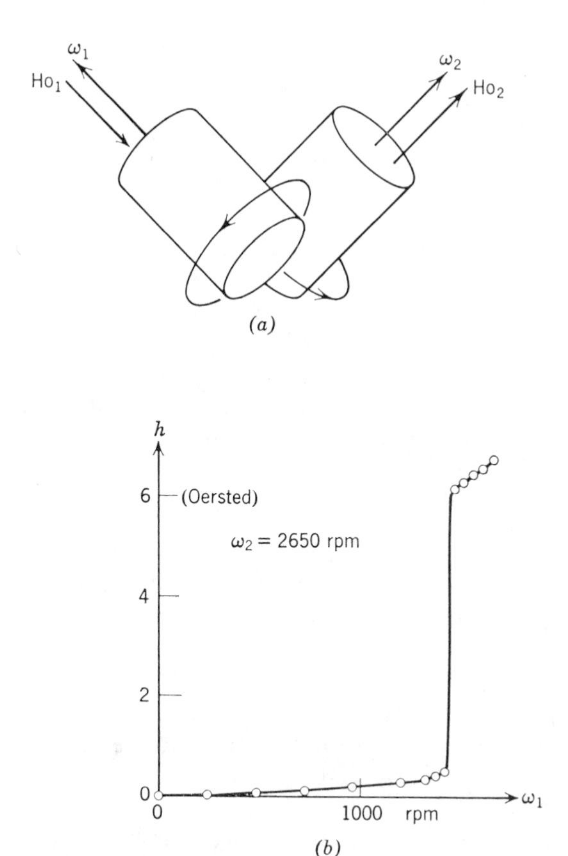


**Figure 8.9.** (*a*) Rotating cylinders in the Lowes-Wilkinson laboratory model of geomagnetic dynamo action. (*b*) Externally observed field (in oersteds or $10^{-4}$ T) as a function of rotational speed of cylinder 1, with cylinder 2 maintained at a fixed speed. Reproduced, by permission, from Lowes and Wilkinson (1963).

but this is merely a necessary experimental technique and was not responsible for the sudden onset of a self-excited field at a critical speed of rotation, as shown in Figure 8.9*b*.

The essential feature of the Lowes-Wilkinson model is the feedback between the two magnetically connected cylinders. An initial axial field in one causes it to act as a homopolar disk, generating an emf between its axis and periphery. The consequent circulating current produces a magnetic field of toroidal form which links with cylinder 2, from which a further current flow is initiated. This produces a second toroidal field linking with cylinder 1 and reinforcing the initial field if the rotations are in the correct sense, as shown in Figure 8.9*a* or both reversed. For both of the correct combinations of cylinder rotations, a field of either

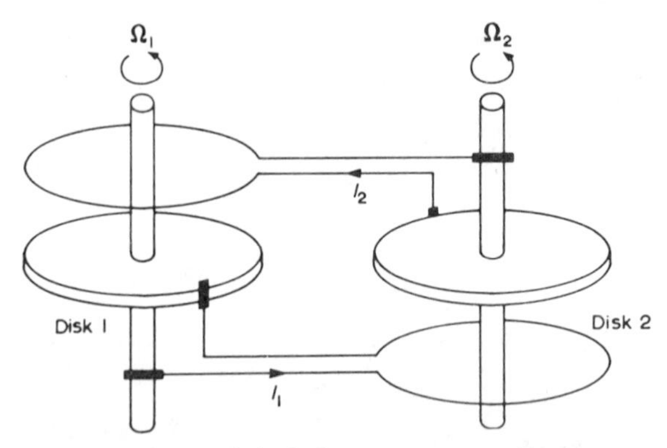


**Figure 8.10.** Two interconnected disk dynamos. At sufficient speeds of rotation these dynamos are self-excited but calculations by Allan (1958) and Rikitake (1958) show that with different experimental parameters the current generation may have instabilities leading to oscillation and even field reversal. Figure after Rikitake (1958).

sign may be self-maintained, the sign being determined by an initial, stray excitation, but the state of zero field is unstable and self-excitation of a dynamo is inevitable if the motion is suitable. This is relevant to the polarity reversals of the Earth's field which have occurred many times (Section 9.4). Opposite polarities of the field are equally likely to result from a particular pattern of motion and the paleomagnetic record shows that over geological time the Earth's field has been "normal" or "reversed" equally often.

The problem of stability of the geomagnetic dynamo action has been reviewed by Rikitake (1966b), using particularly calculations by Allan (1958) on the stability of two interconnected disk dynamos (Fig. 8.10). Repeated numerical integration of the equations of electromagnetic induction for this model, which is similar in principle to that of Lowes and Wilkinson (1963), show oscillations about the mean field and less frequent complete reversals, not unlike the apparent behavior of the earth's field. In these models the motion and the field are constrained by the symmetry of the assumed or imposed geometry, so that the field may have either of two opposite polarities, but not some random orientation. In the case of the Earth's field the orientation does wander off axial symmetry, as is apparent in the present misalignment of magnetic and rotational axes, but this must be regarded as a transient departure. Paleomagnetic data considered in Section 9.3 show that when averaged over $10^4$ years or more the axes coincide. This is important to dynamo theories, all of which invoke the coriolis force associated with rotation as a feature of dynamo action, although a steady, axially symmetric dynamo could not be self-maintained.

# 9

# Paleomagnetism:

# Prehistory of the Earth's Magnetic Field

"If the facts are correctly observed there must be some means of explaining and coordinating them . . ."

BULLARD, 1965, p. 323

## 9.1 MAGNETISM IN ROCKS

The direct record of the secular variation of the geomagnetic field extends over about four centuries to the present time, with comprehensive cover of the Earth's surface only very recently achieved. While the present observations allow us to infer a great deal about processes occurring within the Earth, we add another dimension to the subject by expanding the time scale of observation back by a factor of $10^6$ or so to the remote geological past. This is paleomagnetism or "old magnetism," the study of the ancient magnetic field by measurement of the magnetizations of rocks. It laid the groundwork for and is still basic to our understanding of global tectonics (Section 10.1). Several books give comprehensive discussions of the principles of paleomagnetism and the important results (Irving, 1964; Strangway, 1970; Tarling, 1971; McElhinny, 1973).

Most rocks contain a small percentage (0.1 to 10%) of iron oxide and sulfide minerals which have ferromagnetic or, more correctly, ferrimagnetic properties. These minerals occur as small grains dispersed through the magnetically inert, that is, paramagnetic or diamagnetic, matrix provided by the more common silicate minerals that make up the bulk of the rocks. The physical study of rock magnetism is thus concerned with the properties of individual grains of strongly magnetic material that are sufficiently well dispersed to be magnetically independent of one another. They are normally so diluted by the nonmagnetic minerals that the magnetism of rocks is slight, but it is readily measurable in most cases. Since the magnetic properties of rocks are due to ferromagnetic components, rocks have all of the normal properties of ferromagnetics: coercivity, remanence, magnetostriction, and so on. It is the property of remanence that is of interest in paleomagnetism. The physical principles of rock magnetism are discussed comprehensively by Stacey and Banerjee (1974) and a more elementary and still valuable introduction is the first edition of Nagata's *Rock Magnetism* (1953). Details of experimental techniques appear in a volume edited by Collinson et al. (1967).

The mineralogy of rock magnetism has been reviewed by Nicholls (1955). It is complicated by a multiplicity of phases and solid solutions of iron oxides, particularly with titanium dioxide; most magnetic minerals are within the ternary system $FeO - Fe_2O_3 - TiO_2$. For many purposes it is sufficient to distinguish two types of mineral: (1) the strongly magnetic cubic oxides magnetite ($Fe_3O_4$), maghemite ($\gamma - Fe_2O_3$), and the solid solutions of magnetite with ulvospinel ($Fe_2TiO_4$), which are known as titanomagnetites; and (2) the more weakly magnetic, rhombohedral minerals based on hematite ($\alpha - Fe_2O_3$) and its solid solutions with ilmenite ($FeTiO_3$). Pyrrhotite ($FeS_x$, $1 < x < 2.14$) is magnetically similar to hematite.

Magnetite is representative of the cubic minerals with spontaneous magnetizations comparable to the familiar ferromagnetic metals (Fe, Co, Ni). The $Fe^{2+}$ and $Fe^{3+}$ ions are arranged interstitially in a face-centered cubic oxygen lattice on two types of lattice site, *A* in fourfold coordination with the oxygen ions and *B* in sixfold coordination, such that antiferromagnetic coupling between the *A* and *B* ions causes antiparallel alignment of their magnetic moments. But the *B* ions are twice as numerous as the *A* ions, so that the lattice has a strong spontaneous magnetization ($4.8 \times 10^5$ Am$^{-1}$ or 480 emu cm$^{-3}$), as in commercially developed ferrites. Hematite is representative of the more weakly magnetic, uniaxial minerals, in which the oppositely magnetized sublattices of interacting $Fe^{3+}$ ions are equally balanced, that is, antiferromagnetic, but canted at a small angle to give a slight spontaneous magnetization

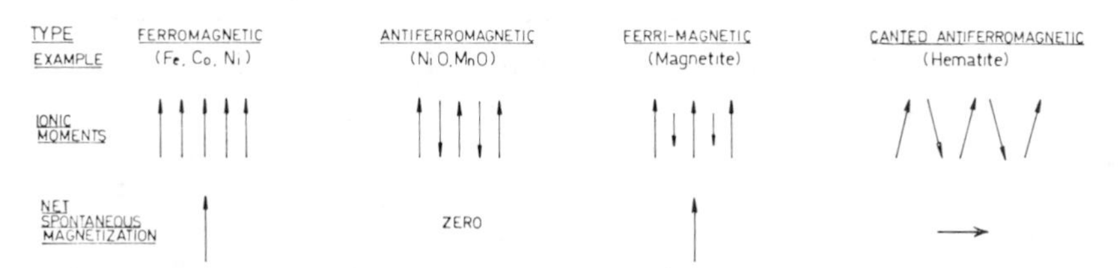


**Figure 9.1.** Four basic patterns of alignment of atomic magnetic moments by mutual interaction. In paramagnetics the interactions are too weak to cause mutual alignment, and in diamagnetics there are no atomic moments in the absence of a magnetic field.

perpendicular to the ion moments, as in Figure 9.1. Its spontaneous magnetization (2200 $Am^{-1}$ or 2.2 emu $cm^{-3}$) is confined to the basal plane except at temperatures below $-25°C$.

The strong spontaneous magnetization of magnetite gives rise to magnetostatic forces which cause the magnetic structure to be divided into domains, locally magnetized to saturation but, except in the case of extremely fine grains, arranged to minimize the magnetic moment of the whole body of magnetite. The principles are well understood in terms of ferromagnetic domain theory, as reviewed by Kittel (1949) and applied to rocks by Stacey and Banerjee (1974), although the grain size is normally so small that ferromagnetic domains in rocks have rarely been subject to direct experimental observation. The multiplicity of domains depends on grain size in the manner of Figure 9.2. For equidimensional magnetite grains the critical size for the transition between states (*a*) and (*b*) is about 0.03 $\mu$m but larger grains may be single domained if they are elongated.

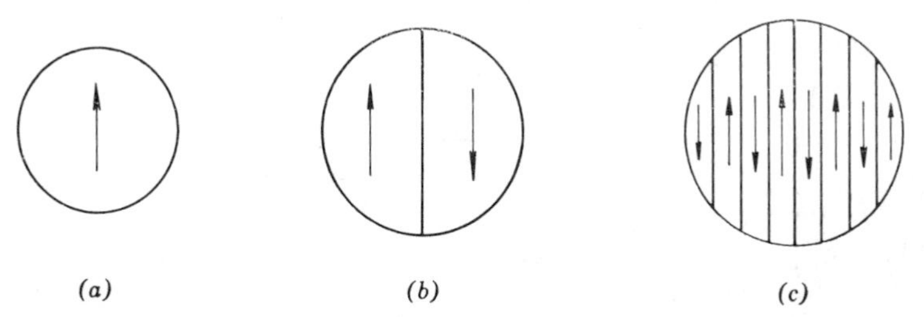


**Figure 9.2.** Diagrammatic representation of the domain structures of magnetic grains. (*a*) Single domain. (*b*) Two-domain grain. (*c*) Multidomain grain. Hematite commonly behaves as single-domained material and magnetite grains are generally multidomained but there is special interest in small multidomains [such as (*b*)] that have some single-domain properties, including high stability of remanence.

In paleomagnetism there is a particular interest in small, multidomained grains (diameters less than about 15 $\mu$m) which behave in some important respects as single domains and are referred to as pseudo-single domains. The basic reason for this is illustrated in Figure 9.3, which shows a two-domained grain in which the single domain wall interacts with crystal defects (or perhaps with an asymmetry in grain shape) so that none of its possible stable positions coincides with the position which gives zero total moment for the grain. Although the moment of such a grain can be reversed, it cannot be demagnetized. Stacey and Banerjee (1974) suggested that the surfaces of multidomained magnetic grains have a very fine domain structure* with irregularities which give rise to such pseudo-single domain moments and that this effect is noticeable in grains up to 15 $\mu$m in diameter. The important result is that such grains have high magnetic stabilities, characteristic of single domains. Since it is the highly stable component of remanence in rocks which is of interest in paleomagnetism, it is appropriate to discuss the

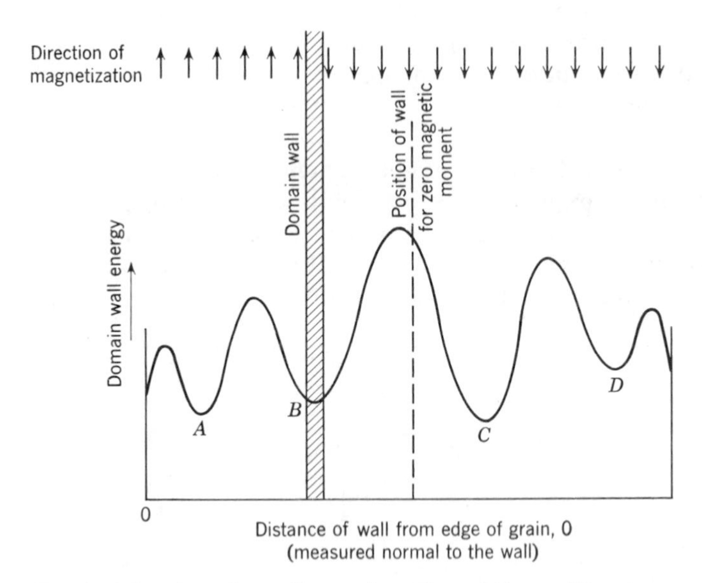


**Figure 9.3.** Energy of a domain wall as a function of its position in a small grain due to its interaction with crystal defects. The stable positions are at the potential energy minima *A*, *B*, *C*, *D*. The displacement of the wall from the position of zero total magnetic moment is responsible for the pseudo-single domain (psd) moment of the grain.

*Small *closure* domains are normally formed on the surface of a magnetic material to provide paths of magnetic flux closure to the larger domains inside it.

process by which rocks acquire remanent magnetism in terms of the original single-domain theory (Néel, 1955).

As a magnetic grain is cooled, so it acquires a spontaneous magnetization at a critical temperature, the Curie point (580°C for magnetite). If it is a single domain it then has a magnetic moment, which progressively increases in strength with further cooling*. But thermal agitation suffices to reorient the moment repeatedly until the *blocking temperature*, $T_B$, is reached, generally some tens of degrees below the Curie point. Now consider initially the simple case of an assembly of identical grains whose magnetic moments may be oriented either one way or in the opposite direction but not between. For true single domains this means that we assume a grain anisotropy parallel for all grains; in the case of small multidomains we assume their domain structures are all oriented the same way. At the blocking temperature thermal agitation just suffices to cause spontaneous reversal of the moments $\mu_B$ and the average alignment of moments in a field $B$† (parallel to the magnetic axis) is given by the Boltzmann probabilities $P_+$ and $P_-$ for finding a particular grain parallel or antiparallel to $B$:

$$P_+ = \frac{\exp(\mu_B B/kT_B)}{\exp(\mu_B B/kT_B) + \exp(-\mu_B B/kT_B)} \tag{9.1}$$

$$P_- = \frac{\exp(-\mu_B B/kT_B)}{\exp(\mu_B B/kT_B) + \exp(-\mu_B B/kT_B)} \tag{9.2}$$

If there are $N$ such grains per unit volume of rock, then its magnetization at the blocking temperature is

$$M_B = N\mu_B(P_+ - P_-) = M_{SB} \tanh\left(\frac{\mu_B B}{kT_B}\right) \tag{9.3}$$

where $M_{SB} = N\mu_B$ is the saturation magnetization at the blocking temperature. On cooling to room temperature all of the magnetic moments remain oriented as at the blocking temperature and the magnetization increases by the factor $M_{SO}/M_{SB}$' where $M_{SO}$ is the room temperature saturation magnetization, so that the *thermoremanence* or *TRM* (remanent magnetization acquired by cooling in a field) is

$$M_{\text{TRM}} = M_{SO} \tanh\left(\frac{\mu_B B}{kT_B}\right) \tag{9.4}$$

Equation 9.4 is the basic equation in the theory of TRM. Generally, but not invariably, $(\mu_B B)$ is sufficiently small in the field range of

*A multidomained grain has a domain structure, such that its moment is zero in the absence of an externally applied field.

†In electromagnetic units use the field $H$ in these equations.

interest (0 − 1 oe or < $10^{-4}$ T) for the linear approximation to be valid:

$$M_{\mathrm{TRM}} = M_{SO} \frac{\mu_B B}{k T_B} \tag{9.5}$$

The more general case of an assembly of randomly oriented grains gives

$$\begin{aligned} \frac{M_{\mathrm{TRM}}}{M_{SO}} &= \int_0^{\pi/2} \tanh\left(\frac{\mu_B B \cos\theta}{k T_B}\right) \cos\theta \sin\theta \, d\theta \\ &= \int_0^1 x \tanh(ax)\, dx \end{aligned} \tag{9.6}$$

where $a = \mu_B B / k T_B$. This function does not have an analytical form, but is tabulated by Stacey and Banerjee (1974) and in low fields approximates to

$$\frac{M_{\mathrm{TRM}}}{M_{SO}} \approx \frac{\mu_B B}{3 k T_B} \tag{9.7}$$

A further generalization which allows also for a uniform distribution of the magnitudes of the grain moments, up to some limit $\mu_{B\,\mathrm{max}}$ (at $T_B$), gives

$$\frac{M_{\mathrm{TRM}}}{M_{SO}} = \int_0^1 \int_0^1 xy \tanh(axy)\, dx\, dy \tag{9.8}$$

where now $a = \mu_{B\,\mathrm{max}} B / k T_B$. This is also tabulated by Stacey and Banerjee (1974) and in low fields approximates to

$$\frac{M_{\mathrm{TRM}}}{M_{SO}} \approx \frac{\mu_{B\,\mathrm{max}} B}{9 k T_B} \tag{9.9}$$

The parameters $\mu_B$ and $T_B$ can be expressed in terms of the temperature dependence of spontaneous magnetization and the energies of the barriers opposing reversals of the grain moments. However, the important feature is that below the blocking temperature the relaxation time for spontaneous decay of TRM (when the inducing field is removed) increases very rapidly with decreasing temperature. If we represent the energy barrier by $E$, the probability of reversal of a grain moment in time $dt$ is

$$dP = C e^{-E/kT}\, dt \tag{9.10}$$

where $C$ is a frequency factor* of order $10^8$ sec$^{-1}$. The relaxation time is then

$$\tau = \frac{1}{C} e^{E/kT} \tag{9.11}$$

*This is the frequency of a *spin wave*, or thermal wave in the magnetic alignment, having a half wavelength equal to the dimension of a single domain or of a domain wall.

which is comparable to the time scale of laboratory cooling, say 1000 sec, if $E/kT \approx 25.3$. This is the situation at the blocking temperature. Then if we consider grains with blocking temperatures at about 800 K (50° below the Curie point of magnetite) cooling to atmospheric temperature, about 300 K, the relaxation time increases to

$$\tau_0 = 1000 \exp\left[25.3\left(\frac{800}{300} - 1\right)\right] \text{sec} \approx 7 \times 10^{13} \text{ years} \qquad (9.12)$$

even without allowing for the intrinsic increase in barrier energy at the lower temperature. Thus the appeal of paleomagnetism is to the occurrence of grains with high blocking temperatures, which are magnetically stable even on the geological time scale of $10^9$ years or more.

The magnetization of sediments requires another explanation because sediments have not normally been heated sufficiently to acquire thermoremanence. Those sediments that are accumulations of debris from the weathering of igneous rocks frequently contain grains of magnetite whose magnetic moments are partially aligned by the Earth's field during or immediately after deposition and thus impart a detrital remanent magnetization (DRM) to the deposit, as is observed in varved clays. However, the processes of compaction and solidification of a sediment are frequently accompanied by chemical changes in which at least some magnetite is oxidized to hematite. The hematite is formed chemically at low temperature in the Earth's field and if the grain size becomes large enough it may acquire chemical remanent magnetization (CRM) in the process. As a single-domain grain of hematite grows, the energy barrier opposing spontaneous (thermal) reversal of its magnetic moment increases, so that it becomes magnetically stable at a critical size, at which the barrier energy is of the order 25 kT. It has a "blocking size" at constant temperature, just as it has a blocking temperature for a particular size and the CRM acquired by an assembly of developing grains is determined by the Boltzmann distribution of grain moments at the blocking size. The theory of CRM is thus essentially the same as the theory of TRM. Chemical remanence has been demonstrated in magnetite by laboratory reduction of hematite and its stability is similar to that of thermoremanence. Chemical remanence in the hematite of sedimentary rocks is presumed to occur similarly during the oxidation of magnetite.

The intensities of natural remanent magnetizations (NRMs) of rocks that are of thermal or chemical origin, cover a very wide range, with no definite limits, but most are between 10 and $10^{-4}$ $\text{Am}^{-1}$ ($10^{-2}$ to $10^{-7}$ emu). By comparison, the remanent magnetization induced isothermally by exposure to a field comparable to that of the Earth is insignificant. However, rocks normally acquire secondary magnetizations, superimposed on the original or primary magnetizations that are of paleomagnetic interest. Secondary magnetizations have a number of causes; examples of the effects of currents due to lightning strikes have been found but are probably not very common. Mild reheating may introduce a low temperature TRM, or isothermal exposure to a field over geological time may cause viscous magnetization of less-stable magnetic constituents. But since such secondary magnetizations are due to the alignment of domains that are softer that is, they respond more easily to a field than the components of primary interest, they may often be removed by application of an alternating field, which is slowly decreased to zero (with the Earth's field canceled by additional coils) or by moderate heating and cooling in a field-free space, leaving a substantial part of the primary remanence. A very good example of the effect of partial demagnetization in reducing the scatter of directions of magnetization is shown in Figure 9.4. In this case the secondary magnetizations of the samples were probably due to lightning strikes; when they were removed the consistent direction of primary magnetization was revealed.

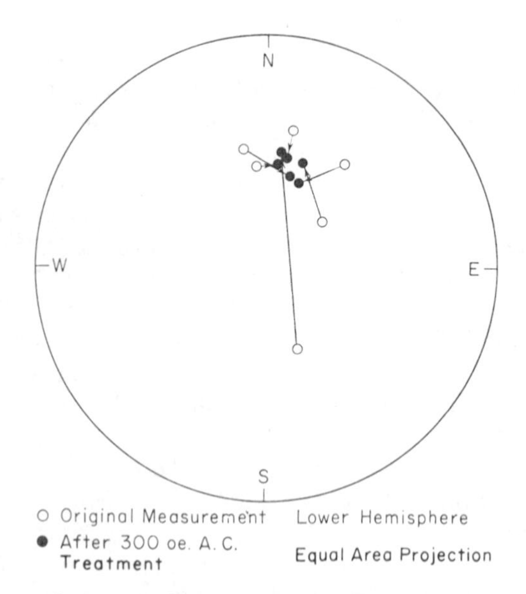


**Figure 9.4.** Effects of partial demagnetization on directions of natural remanence in six specimens from a single lava flow. Open circles represent directions before treatment and solid circles after partial demagnetization in an alternating field that was decreased slowly from 300 Oe (0.03 T). Figure from Cox and Doell (1960).

## 9.2 ARCHEOMAGNETISM AND THE SECULAR VARIATION

Proceeding backward in time past the beginning of the direct record of secular variation (1540), we come to the period from which dated archeological remains can be studied. Pottery and, more usefully, bricks from pottery kilns, whose last dates of firing can be estimated from the carbon-14 contents of ashes, have thermoremanent magnetizations that can be measured by the usual techniques of paleomagnetism, subject to the conditions that the samples may have awkward shapes and in some cases only nondestructive tests are allowable. Measurements on such materials constitute a special branch of paleomagnetism known as *archeomagnetism*, pioneered by G. Folgerhaiter in France and now a subject of detailed study also in Britain, the United States, the Soviet Union, Czechoslovakia, and Japan.

The measurements are concerned semi-independently with two aspects of the past geomagnetic field: its direction and magnitude. The directions in England, as measured by M. J. Aitken and co-workers, are plotted in Figures 9.5 and 9.6. In Figure 9.5 the data on inclination and declination obtained from measurements on bricks from reliably dated kilns are plotted independently to show the magnitude of the uncertainty in the measurements. In Figure 9.6 the data are combined to give an extension of Bauer's representation of the secular variation at London (see Fig. 8.3). Although the time span represented is not continuous, the data are sufficient to dismiss the idea of a simple cyclic secular variation. Doubt about this has also been aroused by paleomagnetic data from earlier periods (Denham, 1974). Furthermore, measurements of the intensity of the geomagnetic field in the past show that the secular variation has involved first-order changes in the dipole field as well as the nondipole field. The paleointensities can be obtained by comparing the NRMs of baked clays (pots, kiln bricks, etc.), with laboratory-induced TRM in the same samples, using techniques first developed by J. G. Koenigsberger and studied in detail by Thellier and Thellier (1959). Due to the superposition of dipole and nondipole fields, intensity measurements from a single area do not uniquely represent the strength of the dipole field, but intensity data from widely separated areas show a mutually consistent trend over the last 2000 years (Fig. 9.7). The intensity variation in Czechoslovakia over a more extended time scale is shown in Figure 9.8. Since the total change represented exceeds a factor 2, whereas the present nondipole field is only 20% of the total, it is almost certain that the changes of Figure 9.8 represent the dipole field.

The archeomagnetic evidence leaves us in doubt about the significance of the separation of the geomagnetic field into dipole and nondipole components. The secular variation involves both together and is

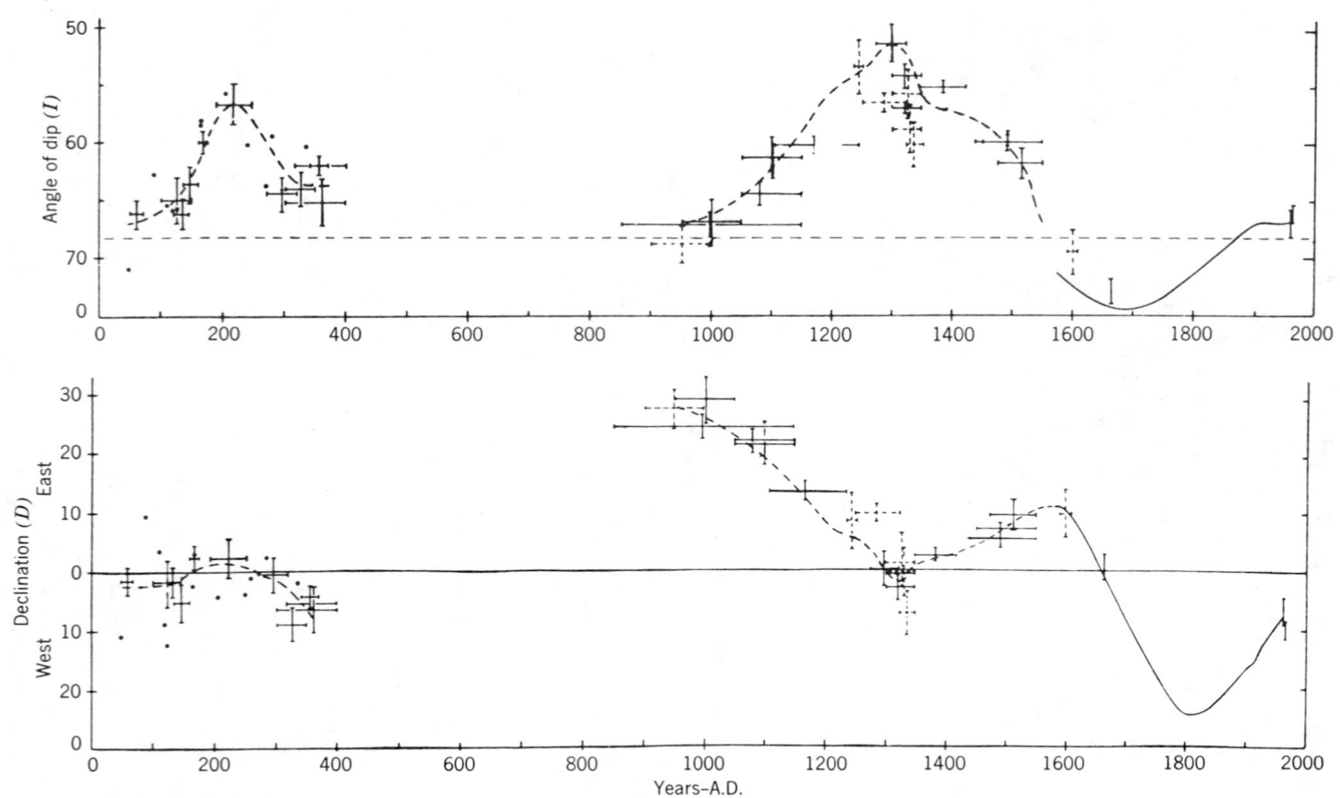


**Figure 9.5.** Secular variations in magnetic inclination and declination at London, deduced from the remanent magnetism of archeological specimens and compared with the direct observations since 1540. Uncertainties in measured values and in dates are indicated. Broken crosses and dots represent less reliable values. Figure from Aitken and Weaver (1965), by permission of the Society of Terrestrial Magnetism and Electricity, Japan, and the authors.

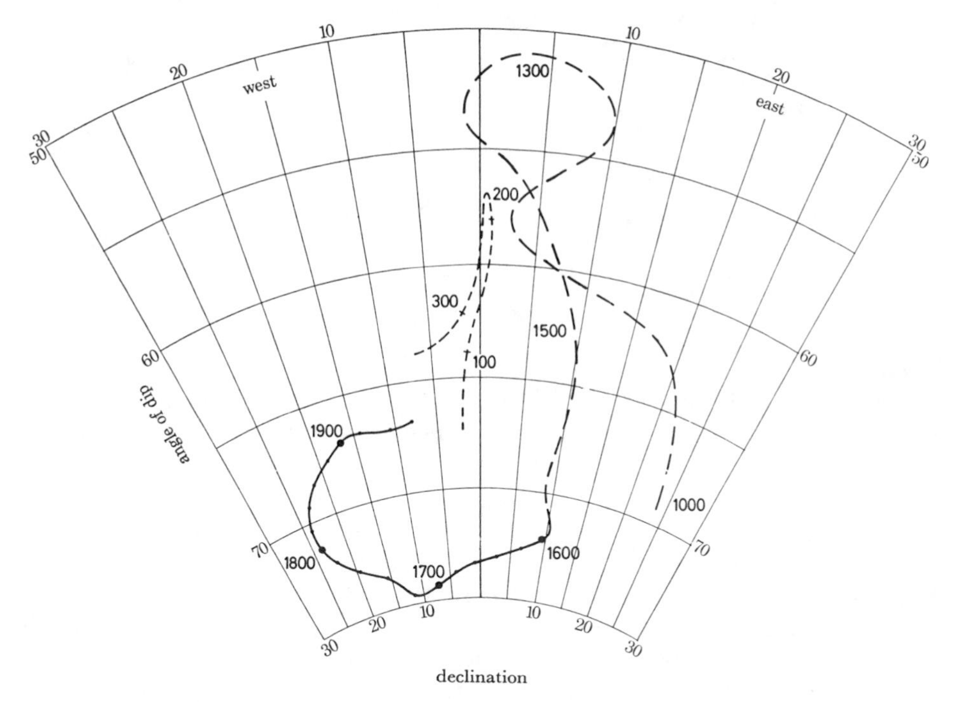


**Figure 9.6.** A combination of the data of Figure 9.5, plus some more recent data, showing that the cyclic secular variation of the past four centuries is not representative of the historical past. Reproduced by permission from Aitken (1970).

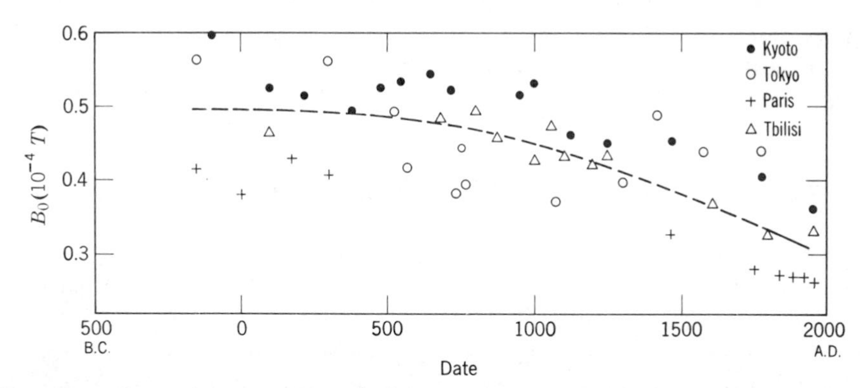


**Figure 9.7.** Intensity variation of the geomagnetic field over 2000 years from measurements by four archeomagnetic research groups. All values have been adjusted to give field intensities at the equator, assuming that the field was a simple dipole, a procedure that is expected to introduce up to 20% errors in individual observations but cannot affect the trend apparent in all of the data taken together. Figure from Sasajima (1965) by permission of the Society of Terrestrial Magnetism and Electricity, Japan, and the author.

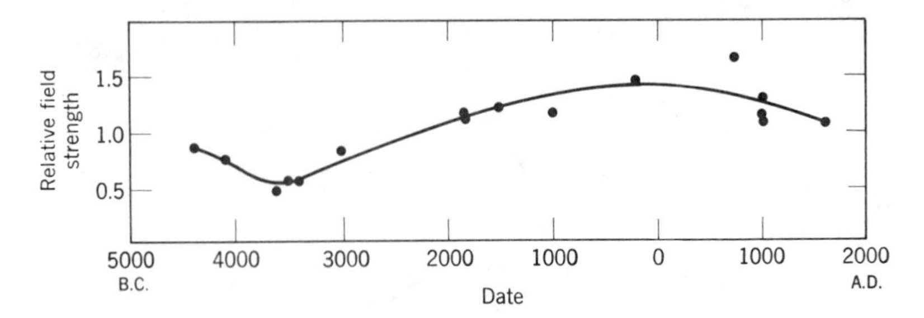


**Figure 9.8.** Secular variation in the intensity of the geomagnetic field in Czechoslovakia over a period of 45 centuries. Figure from Bucha (1965) by permission of the Society of Terrestrial Magnetism and Electricity, Japan, and the author.

not due simply to the nondipole field, although it is reasonable to suppose that fluctuations of the dipole field may be slower (see Section 8.2). This does not necessarily mean that the westward drift of the nondipole field is not real or even that it is not a significant indication of the nature of the core motions, but merely that the westward drift is not a feature of the field that must be accepted as a strong constraint on theories of the dynamo mechanism.

The time constant for major changes in the dipole field appears from Figure 9.8 to be a few thousand years. This is consistent with other evidence, in particular the reversals of the field, which are discussed in Section 9.4, and may be regarded as the extreme fluctuations of the dipole field. It is of the same order as the time constant for decay of an unmaintained electric current in the core, which, by the approximate calculation in Section 8.4, appears to be about 10 000 years.

## 9.3 PALEOMAGNETIC POLES AND THE AXIAL DIPOLE HYPOTHESIS

Paleomagnetic results are normally expressed in terms of pole positions. The pole position deduced from measurements on a particular rock means the coordinates of the axis of a hypothetical dipole field that would produce in the rock a local field parallel to the measured direction of its natural remanence. The pole must lie on a great circle defined by the ancient magnetic declination and is situated at an angular distance $\theta_m$, which is the ancient geomagnetic colatitude and is related to the measured paleomagnetic angle of dip $I$ by a result from Section 8.1:

$$\cot \theta_m = \tfrac{1}{2} \tan I \qquad (9.13)$$

Measurements on a single rock sample give a poor estimate of the orientation of the ancient dipole field because the local field has also a nondipole component for which there is no means of applying a correction. However, by taking an average pole position from a number of measurements on a geological formation, which is sufficiently extensive for its history of cooling or deposition to cover a time span of several thousand years, we expect to average out the effect of the variable nondipole field. The procedures of rock collection and statistical treatment of data necessary to ensure that a proper average is obtained are discussed by Irving (1964), Tarling (1971), and McElhinny (1973).

A satisfactory mean-pole position for the period represented by the British archeomagnetic collections is obtained by averaging the data of Figures 9.5 and 9.6. Irving (1964) gives its position as 82°N, 172°E, which is well removed from the present geomagnetic pole (78.5°N, 69°W) and is nearer to the geographic pole. A more significant average is obtained by using the archeomagnetic pole as one of several, covering a longer period, each pole being, of course, an average of numerous measurements. This has been done by numerous authors with variously selected data, but always with the same general conclusion that the paleomagnetic poles cluster about the geographic pole, with the present geomagnetic pole to one side of the distribution, as in Figure 9.9. This conclusion, first reached by Torreson et al. (1949) on the basis of measurements of magnetization in sedimentary rocks, led to a hypothesis on which the more important conclusions derived from paleomagnetism depend. This is the *geocentric axial dipole hypothesis*, according to which the geomagnetic field has always been predominantly dipolar and, when averaged over a sufficient time (10 000 years or more), the axis of the dipole coincided with the geographic or rotational axis of the Earth, and the dipole field was centered at the center of the Earth. A number of tests have been applied to data from remote as well as recent geological periods. Wilson (1972) found evidence of a slight persistent asymmetry (see Appendix J, Problem 9.1), which, if real, may be significant to the dynamo problem (Section 8.4), but is insufficient to invalidate use of the axial dipole hypothesis in paleomagnetism. The most important consequence is that paleomagnetic poles are ancient geographic or rotational poles and not merely geomagnetic poles.

Apart from its significance in paleomagnetism, the observation that the Earth's field averages to an axial dipole has an important bearing on the theory of the origin of the field. On the geological time scale the inclination of the dipole to the geographic axis appears merely as a transient excursion from the state of symmetry about the rotational axis. This means that we have no reason to seek a fundamental asymmetry in the Earth or in the geomagnetic dynamo mechanism as the cause of the present angle of the dipole field. Although the principles of symmetry

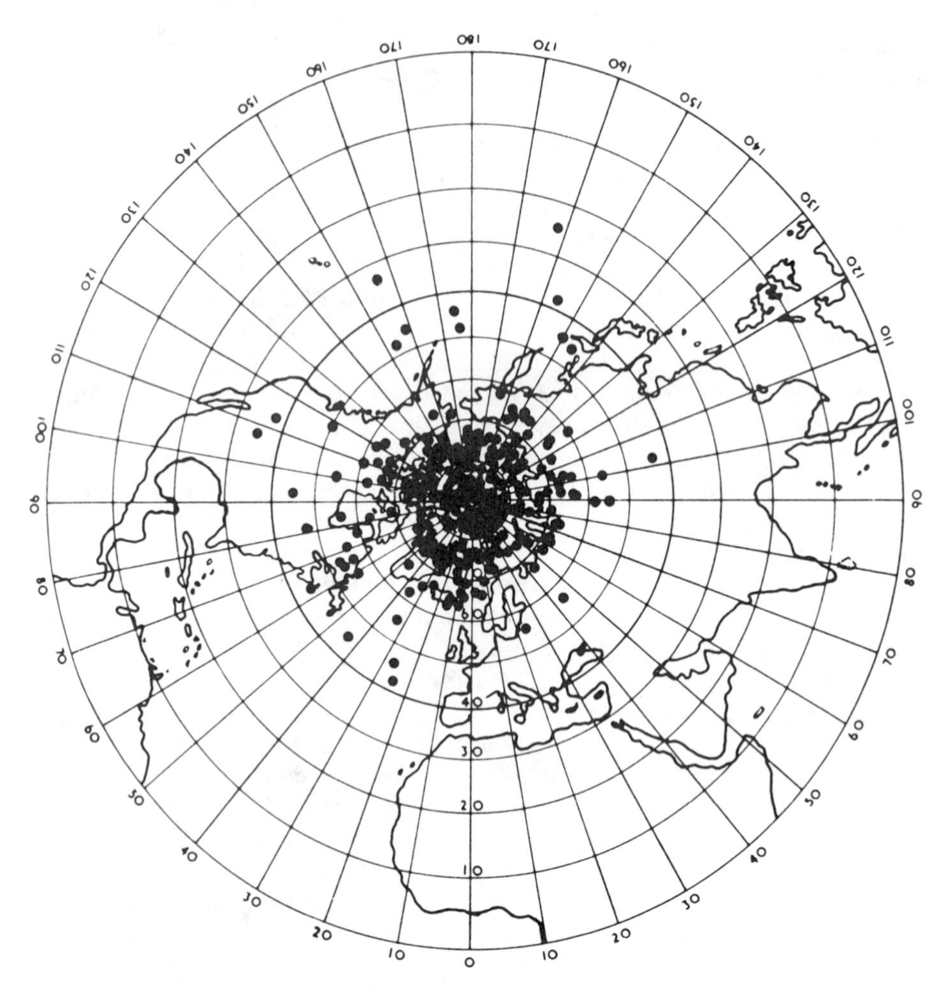


**Figure 9.9.** Pole positions of igneous rocks up to 20 million years old, showing that when averaged over a sufficient period the geomagnetic pole coincides with the rotational axis. The polarity of the field may have either sign, as considered in Section 9.4. Reproduced, by permission, from Tarling (1971).

are normally applied intuitively to physical cause and effect situations, it is worth considering them explicitly in the present connection. The principles were enunciated by Pierre Curie, whose statements have been summarized by Paterson and Weiss (1961) in the following terms:

"The symmetry of any physical system must include those symmetry elements that are common to all the independent factors (physical fields and physical properties of the medium) that contribute to the system, and it may include additional symmetry elements; however any symmetry elements absent in the system must be absent in at least one of the contributing factors."*

As far as we understand the relevant contributing causes of the geomagnetic field they are simply the rotation of the Earth, which has purely axial symmetry, and various motions, temperature gradients, and so on, in the core which, apart from the effect of rotation, has spherical symmetry. If the geomagnetic field were to deviate in a consistent way from symmetry with respect to the rotational axis, then we would have a field with lower symmetry than the combination of the causes we know about. The principles of symmetry would then compel us to seek an additional contributing cause of lower symmetry. If there is no consistent inclination or lateral or axial displacement of the dipole field then there is no necessity for any such additional basic cause. The principles of symmetry are relevant also to the consideration, in the following section, of the polarity of the geomagnetic field.

## 9.4 REVERSALS OF THE GEOMAGNETIC FIELD

Sequences of specimens from single geological formations, that is, a series of lava flows or sedimentary layers, through which there are progressive increases in geological age, are frequently found to have repeated alternations of magnetic polarity. The obvious explanation that the geomagnetic field underwent repeated reversals was recognized very early in the history of paleomagnetism (see Irving, 1964, p. 8), but doubt was raised by the discovery at Mt. Haruna, Japan, of a lava that acquired laboratory TRM in a sense opposite to the field in which it cooled (Nagata et al., 1952). It became important to know how common such self-reversals might be and an exhaustive theoretical search for possible mechanisms was conducted by Néel (1955). In reviewing subsequent studies, Stacey and Banerjee (1974) concluded that the only naturally occurring self-reversal mechanism that appeared important was the now well-established process of ionic reordering of ions in solid solutions of hematite ($Fe_2O_3$) and ilmenite ($FeTiO_3$). Ishikawa and Syono (1963) found that reversed thermoremance was a feature of a partially ordered state and that both high-temperature (completely disordered)

*Doubt about the validity of Curie's principle, as stated here, has been expressed by Elsasser (1966). However, this is a misinterpretation. Elsasser's concern is with unstable deformation of a body, which *locally* appears to have lower symmetry than the combination of its causes, although if viewed as an average of many individual events it does not. The problem is analogous to that of the geomagnetic field, which only complies with Curie's principle if viewed on a long time scale.

and equilibrium low-temperature (completely ordered) states had normal thermoremanence. Thus the strength of the reversed magnetization was a function of both composition and cooling rate. The reversal arises from a negative (antiferromagnetic) coupling between the ions and it is convenient to think in terms of a simple model of two interacting (*A* and *B*) atomic sublattices, which preserve the orientations of their spontaneous magnetizations during the ordering process, but the one which is initially weaker becomes more strongly magnetic during the process of ionic redistribution. This reverses the spontaneous magnetization of the lattice and consequently also the remanence.

Although known self-reversals are few, the possibility of other cases too sluggish to be observed in the laboratory, but important on the geological time scale, necessitates a careful appraisal of the evidence for reversals of geomagnetic field. There are four basic kinds:

1. There is a correlation between reversals observed on different continents and in different rock types, both igneous and sedimentary.
2. Sediments that were baked by contact with later igneous rocks have acquired TRM with the same polarity as the igneous rocks in almost all cases, regardless of the polarity of remanence in the unbaked sediments (Fig. 9.10 and Table 9.1). If the fact that approximately 50% of magnetically stable rocks are found to be reversed were to be explained in terms of self-reversals, then we would expect 50% of the baked contacts to disagree with the polarities of the intruding rocks.
3. The actual process of reversal has been traced both in rapid sequences of lavas (Fig. 9.11) and in deep-sea sediments (Fig. 9.12) in

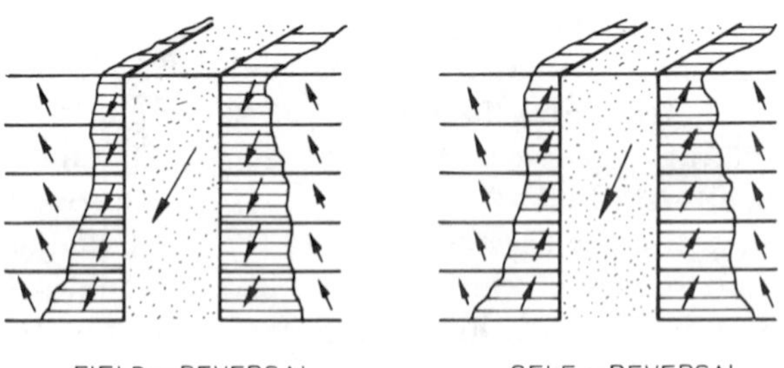


**Figure 9.10.** Coincidence of magnetic polarities of an intrusive rock and the baked (and remagnetized) intruded sediments is circumstantial evidence for a field reversal. Opposite polarities indicate a self-reversal. A statistical study of such baked contacts by Wilson (1962), as in Table 9.1, provided compelling evidence of field reversals.

**Table 9.1** Paleomagnetic polarities of igneous rocks and their baked contacts.[a]

| Igneous Polarity | Baked Contacts Polarity | No. of Cases |
|---|---|---|
| *N* | *N* | 47 |
| *R* | *R* | 104 |
| *I* | *I* | 3 |
| *N* | *R* | 3 |
| *R* | *N* | 0 |

[a]The original analysis by Wilson (1962) has been extended by later authors. The numbers given here are from McElhinny (1973). ($N$ = normal, $R$ = reversed, $I$ = intermediate).

which the field intensity is seen to diminish sharply for a short period during a reversal. This is usually taken to imply that the dipole field vanishes and reappears with opposite polarity but that the nondipole field persists through the transition. However, some authors have suggested that an equatorial dipole with a preferred orientation also persists.

4. The patterns of magnetic anomalies flanking ocean ridges (magnetic stripes) are consistent with the continuous spreading of ocean floors from the ridges (Section 10.1) with polarity of the ocean floor basalt alternating with age (which increases outward) in the manner determined both from dated igneous rocks and marine sediments.

The strength of these observations is such that one apparently significant piece of contrary evidence, the systematic differences in chemistry or oxidation state between reversed lavas and adjacent normal lavas, which have been reported by several authors (e.g., Ade-Hall and Wilson, 1963; Wilson and Watkins, 1967) but not by others (Larson and Strangway, 1966; Ade-Hall and Watkins, 1970), is admitted as paradoxical but not crippling to the field reversals hypothesis. The numbers in Table 9.1 allow us to make crude statistical estimates of the probability of self-reversal. The table indicates three cases of opposite polarity out of 157, that is, about 2%, or since each disagreement involves two rocks, the number of self-reversed rocks is about 1%*. This is reasonable. Also the number of intermediate directions is about 2% of the total and if we suppose that reversals occur on average every $10^5$ years, then the reversal process takes about 2000 years, which accords satisfactorily with evidence from sediment cores, such as Figure 9.12.

The detailed pattern of reversals has become a subject of consider-

*See Problem 9.2 (Appendix J).

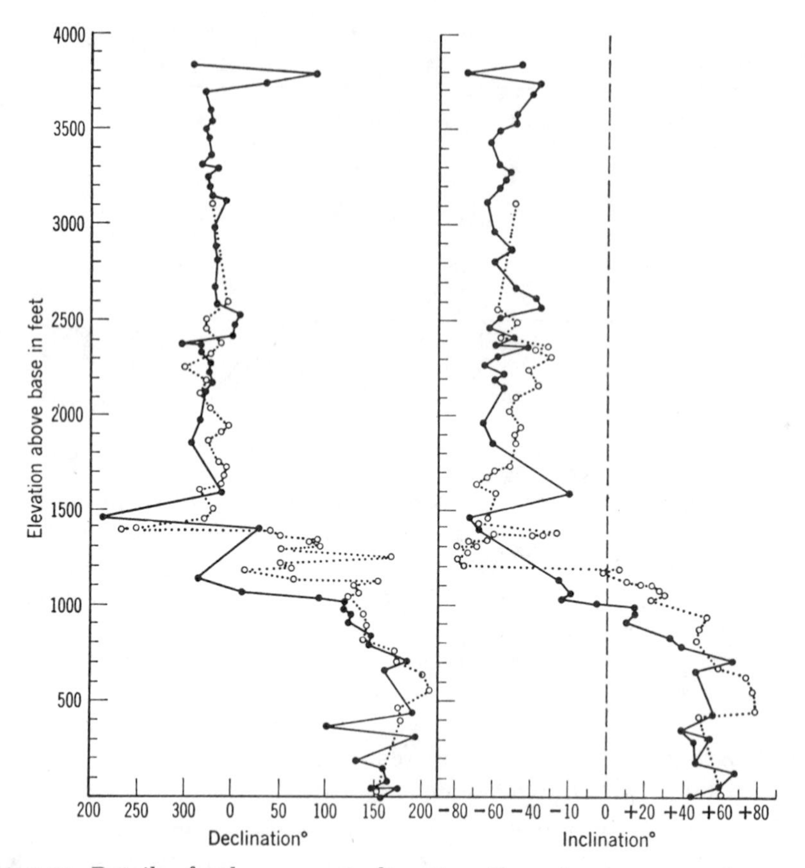


**Figure 9.11.** Details of paleomagnetic directions through a geomagnetic reversal, obtained by Van Zijl, Graham, and Hales (1962) and reproduced from the redrawn figure of Irving (1964). The measurements were on rocks from two localities, indicated by the continuous and dotted lines, from an extensive series of flat-lying lavas. The intensity decreased by a factor of 4 to 5 during the reversal.

able importance to stratigraphic correlation, especially the determination of relative dates of events recorded in marine sediments, and to the estimation of rates of sea floor spreading from the ocean rises. Potassium-argon dating of igneous rocks established the absolute time scale of reversals for the past 5 million years, which is naturally the period most intensively studied (Fig. 9.13). Dating precision does not suffice to identify particular events of greater ages but the polarity sequence has been extended by long sediment cores (Foster and Opdyke, 1970) and by inference from the pattern of marine magnetic anomalies (Fig. 9.14) as far back as 80 million years or more. McElhinny (1971) has

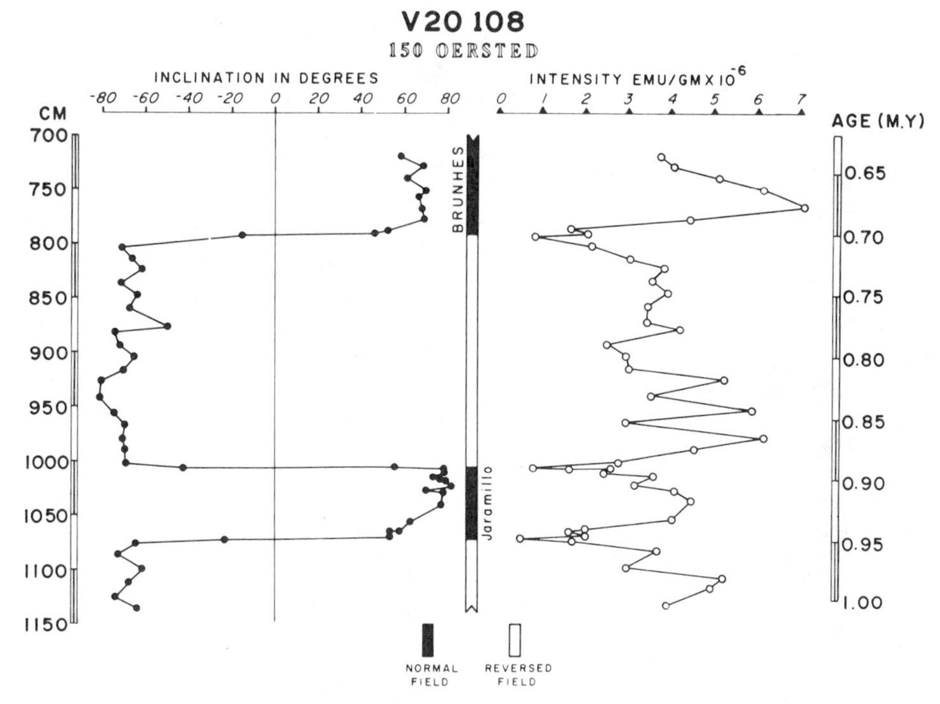


**Figure 9.12.** Plots of the inclination and intensity of magnetization in specimens from part of an ocean sediment core, after partial demagnetization in 150 Oe (0.015 T). This demonstrates clearly the diminution in field intensity during a reversal. Data from Ninkovitch et al. (1966), reproduced from the redrawn figure of Opdyke (1970).

studied the variation in frequency of reversals over more than 500 million years. There is no evident regularity.

Prolonged intervals of predominantly one polarity have been termed epochs, the most recent being named after pioneers of geomagnetism; the shorter periods of opposite polarity occurring within the epochs are termed events and bear the names of the localities where rocks that led to their discovery were obtained. The most prolonged epoch appears to have been at least 50 million years in duration, coinciding roughly with the Permian period, from which all rocks so far examined have been reversed. By contrast some polarity events are so brief that it is doubtful whether they should be recognized as such. It is even possible that an incomplete reversal followed immediately by a return to the original polarity may appear to have been a true polarity event in one area but nothing more than a weakening of the field in another, by virtue of quite

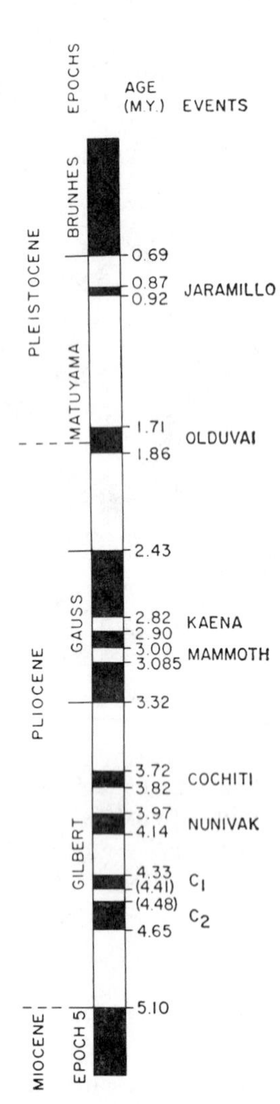


**Figure 9.13.** Polarity of the geomagnetic field for the past 5 million years. Figure reproduced, by permission from Opdyke (1972). See also an important review of the reversals time scale by Cox (1969). Note that this figure omits several probable very short "events." It is reasonable to class them with what are here termed "incipient reversals."

different contributions of the nondipole field in the two areas. Such events are referred to as paleomagnetic excursions, but the term incipient reversals appears appropriate. They are probably more common than has been recognized. If we regard reversals as disturbances of a bistable but delicately balanced system, then once disturbed, the system may settle for either polarity, but possibly has a preference for the original one. An example of a reversal involving apparent "uncertainty" or "hesitation" of the field in settling for a steady polarity is shown in Fig. 9.15.

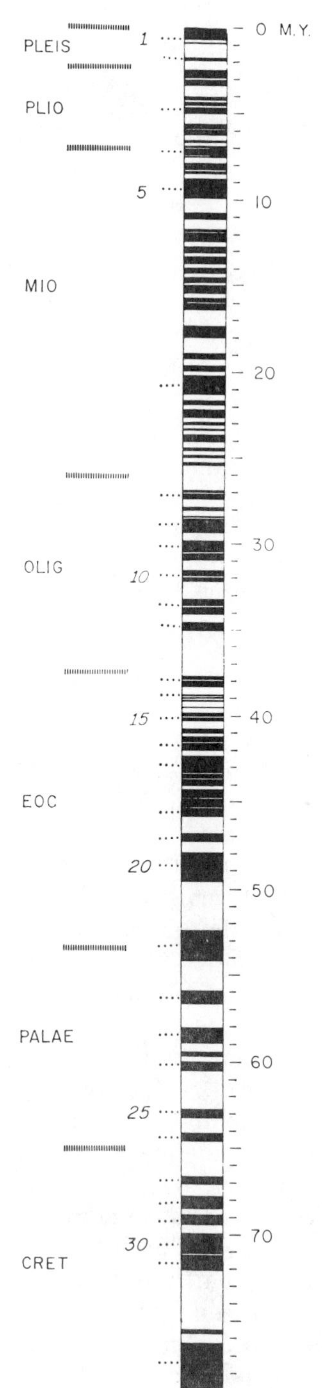


**Figure 9.14.** Polarity sequence of the geomagnetic field extended to 79 million years by inference from marine magnetic anomalies. Reproduced, by permission, from Heirtzler et al. (1968).

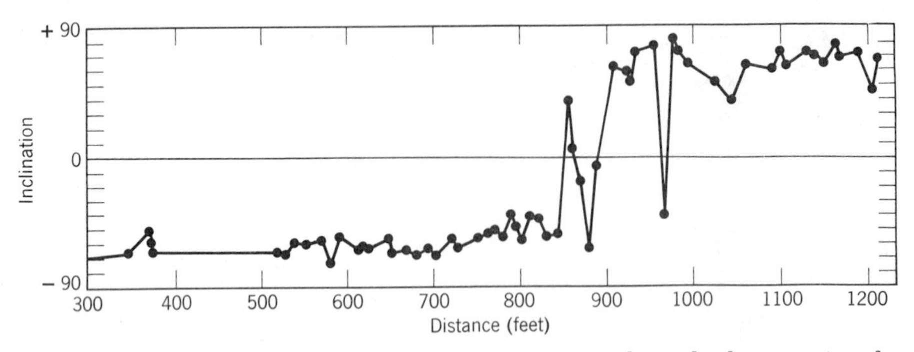


**Figure 9.15.** Inclination of remanent magnetization through the margin of a massive igneous intrusion. This is presumed to record the field direction at each point as the blocking temperature isotherm passed through it during cooling of the body. Similar vacillations of the field occurring either with or without true reversals but indicating numerous very brief or incipient reversals are reported by Kawai et al. (1973). Figure reproduced, by permission, from Dunn et al. (1971). Copyright 1971 by the American Association for the Advancement of Science.

Correlations which have been claimed between reversals and other geophysical or biological effects are difficult to pin down. Irving (1966) suggested that reversals occurred more frequently during periods of rapid polar wander and/or continental drift. This implies that the configuration of the mantle has a direct influence on the core motions that are responsible for the geomagnetic field. The dismissal of the precessional torque between the core and mantle as an effective agent in generating the magnetic field (sections 3.4 and 8.4), makes such an influence appear unlikely, but the existence of a correlation cannot be discounted by lack of an explanation.

A correlation between reversals and faunal extinctions has been claimed by several authors (Opdyke et al., 1966; Watkins and Goodell, 1967; Hays et al., 1969) on the basis of evolutionary changes in planktonic species found in marine sediments. This possibility has attracted considerable popular attention, but the correlation is not immediately obvious and McElhinny (1973) considered there to be no correspondence between long-term evolutionary trends and reversal frequency. If there is a relationship, the causal connection is obscure. Direct enhancement of cosmic ray bombardment due to a weakened field cannot be very significant at sea level and much less so within the sea (Waddington, 1967; Black, 1967; Harrison, 1968); active nuclides produced by bombardment of the upper atmosphere are probably also insufficient, although the production of carbon 14 is known to depend on the strength of the geomagnetic field and its decay to nitrogen in situ in biological materials appears to give it particular relevance. Coincidental climatic changes (Kawai, 1972), perhaps a sudden cooling triggered by massive injections of volcanic dust into the upper atmosphere (Kennett and Watkins, 1970), could have the observed effect.

An apparent coincidence between the ages of at least the most recent of the tektites (see Section 1.2) and reversals of the geomagnetic field was noticed as a result of the appearance in marine sediments of microtektites with ages similar to those of the tektites. They are found in limited ranges of depth at and immediately above the sedimentary levels where reversals are observed (Glass and Heezen, 1967; Gentner et al., 1970). The possibility of a causal connection must be allowed, the implication being that a violent meteorite impact, which produces tektites as fused splashes, disturbs the rotation of the Earth sufficiently to upset the geomagnetic dynamo, triggering a reversal. In all of these matters we can offer only intriguing speculations with the admitted possibility that we are trying to draw conclusions from chance coincidences.

## 9.5 POLAR WANDER AND CONTINENTAL DRIFT

When we look at paleomagnetic data on a time scale of many millions of years we see slow changes that are interpreted as evidence of polar wander and/or continental drift. On this time scale the relatively transient secular variation is averaged out (by averaging results from rocks whose ages may span many thousands or even millions of years) and even reversals are frequent, so that directions differing only in polarity are treated as equivalent. By the axial dipole hypothesis (Section 9.3) the averaged paleomagnetic poles are also geographic or rotational poles and we are considering slow movements of the pole with respect to sections of crust in which we find rocks suitable for paleomagnetic measurements. The paleomagnetic evidence for continental drift has revolutionised geophysical and geological thinking. While corroborative evidence is becoming quite extensive (Section 10.1) there is little doubt that it would not have been recognised without the development of paleomagnetism.

As demonstrated by Figure 9.9, pole positions determined from rocks younger than about 20 million years are clustered about the present geographic pole. Substantial departures from the present pole appear only in data from rocks that are more than about 30 million years old (lower Tertiary or earlier).

Since there are disparities between data from different continents we must now consider separately the paleomagnetic results from each of several large land masses. The term "continents" is comonly used rather loosely for these land masses, whose boundaries are not in fact everywhere coincident with continental divisions. The term "region" may be used where this is so. A plot of a succession of poles for a particular region that has remained coherent during the period in question gives the pole path for the region (e.g., Fig. 9.16). This is the *apparent* polar wander curve in the sense that it indicates a relative motion of the pole with respect to the continent without really indicating which has moved. The absolute orientation of the pole in space is fixed (except in so far as it is influenced by astronomical effects, considered in

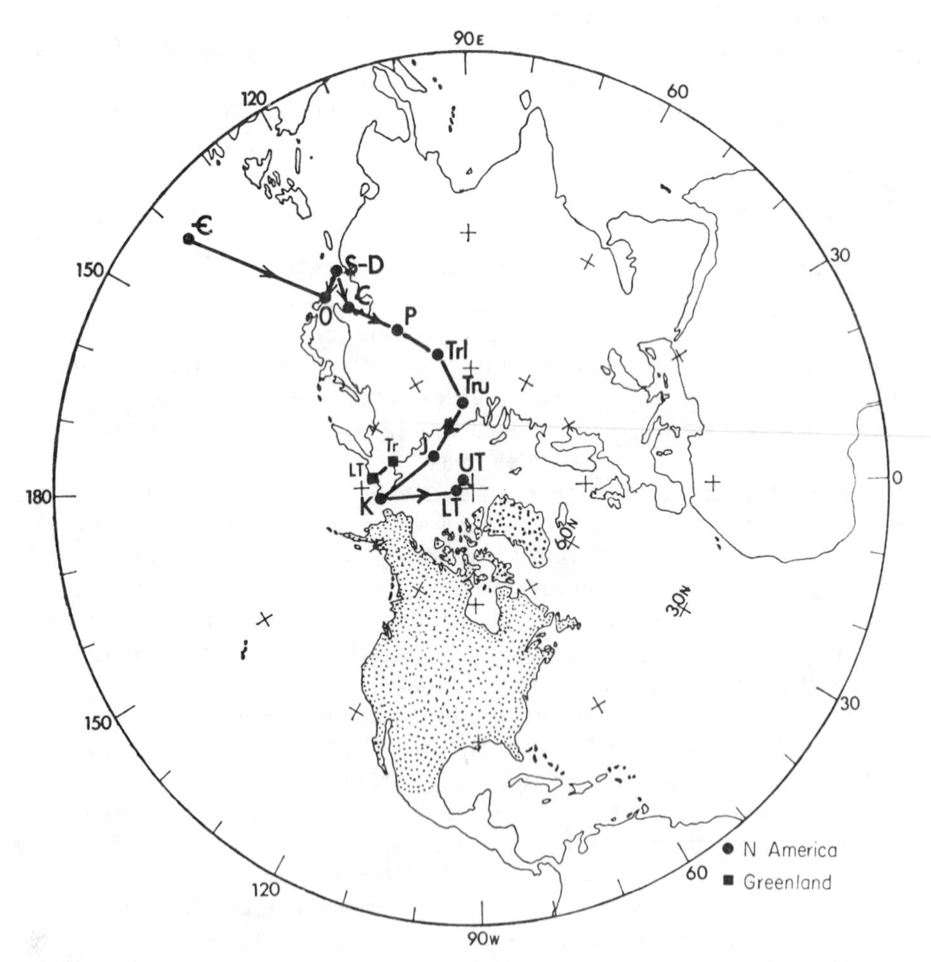


**Figure 9.16.** Phanerozoic (Cambrian to present time) pole path for North America, with two data for Greenland. Based on a plot by McElhinny (1973) with a corrected Ordovician pole by McElhinny and Opdyke (1973).

Chapter 3 but irrelevant here) and coherent motion of all terrestrial features with respect to it is mechanically possible (Goldreich and Toomre, 1969); this is true polar wander. Alternatively the sampled region may have moved (drifted) independently of the rest of the Earth, so that the apparent pole path is simply the inverse of the motion of that region. Wander of the pole and drift of the continent or any combination are not distinguished, although the occurrence of true polar wander should now be doubted (Jurdy and Van der Voo, 1975).

Only when we compare pole paths for different regions does the evidence for continental drift, as distinct from polar wander, emerge. The first continental blocks to be compared in this way were Western Europe and North America and an updated comparison of the apparent polar wander paths for these regions is reproduced in Figure 9.17. The two curves can be reconciled with a single dipole field only by supposing that the two regions were adjacent 200 million years ago and have drifted apart since that time. More dramatic is the relative drift of the southern continents, which coexisted as Gondwanaland until about 150 million years ago, before drifting apart. Figure 9.18 shows the coincidence of poles from the Cambrian to Mesozoic periods for a reconstruction of Gondwanaland. McElhinny (1973) has reviewed the paleomagnetic evidence for disposition of the continents during Phanerozoic time (since the beginning of the Cambrian period) and concluded that during Silurian time Gondwanaland merged with a previously independent block comprising what is now Europe plus North America to form Pangaea, which had an approximately common pole until the Mesozoic period when it broke up into the currently familiar fragments. The blocks comprising Siberia were not parts of Pangaea but collided with Europe at the time of the breakup of Pangaea. Pre-Cambrian drift is less well documented, but it appears that the separation and merging in different ways of continent-sized blocks has been a repeated process (McElhinny and Briden, 1971; Spall, 1972).

Unless we take the extreme and implausible position of asserting that the axial dipole hypothesis is invalid and that the geomagnetic field has been dipolar only for the past 20 million years with each region having an independent field in all previous time, continental drift is an unavoidable conclusion of paleomagnetism. For this purpose it is not even required that the field be axial, but simply dipolar. Evidence that it was in fact axial is obtained by comparing paleomagnetic and paleoclimatic data. Thus the clustering of the components of Gondwanaland around the south polar region, as indicated by high paleomagnetic latitudes of the southern continents, is consistent with the extensive glaciation of these regions in the Permian and Carboniferous periods (Fig. 9.19).

A general correspondence of paleoclimatic indicators with

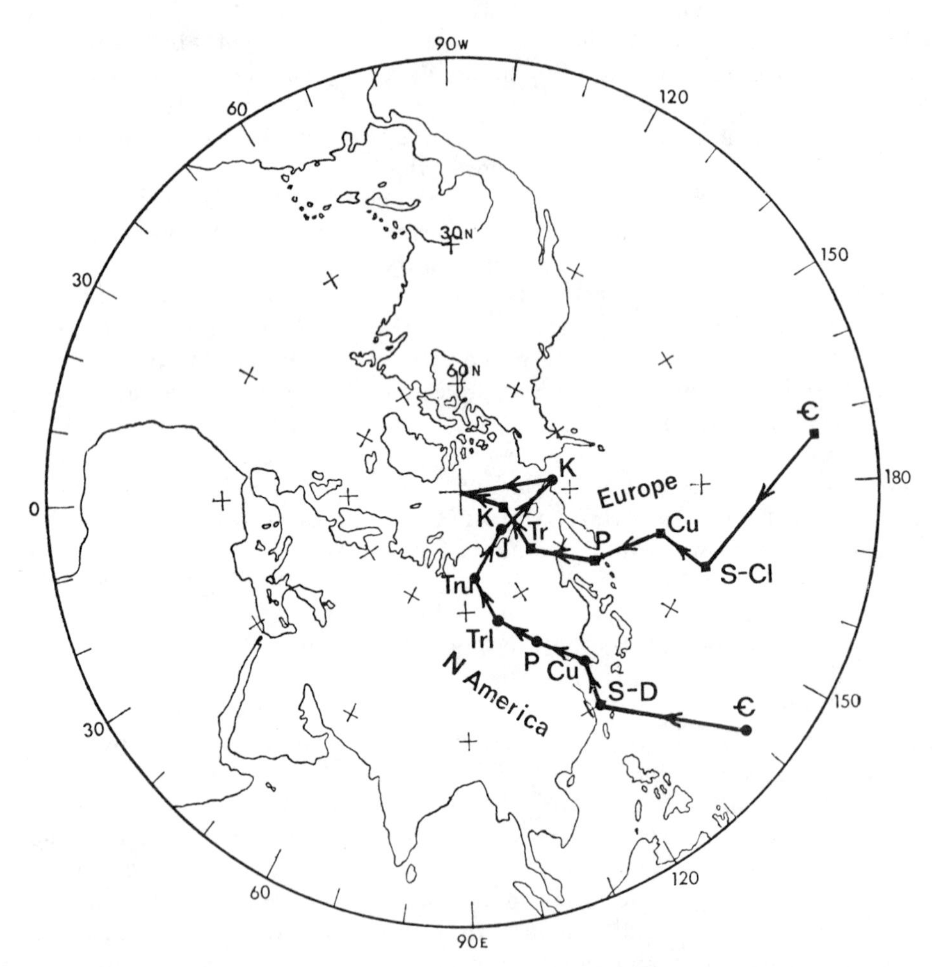


**Figure 9.17.** Comparison of the apparent polar wander paths for Europe (squares) and North America (circles). Reproduced, by permission from McElhinny (1973).

paleomagnetically determined latitudes is both expected and found (e.g., Briden and Irving, 1964) but it can hardly be perfect because quite rapid global climatic changes are known to have occurred. The most striking effect is the advance of glaciation from the polar caps (ice ages), which has now been correlated with other indicators, especially identification of dominant species of trees from fossil pollen and estimates of sea water temperature from oxygen isotope ratios in calcareous shells (e.g., Emiliani and Shackleton, 1974). The causes of these changes remain obscure, although several suggestions have strong advocates. Bryson (1974) has reviewed the field.

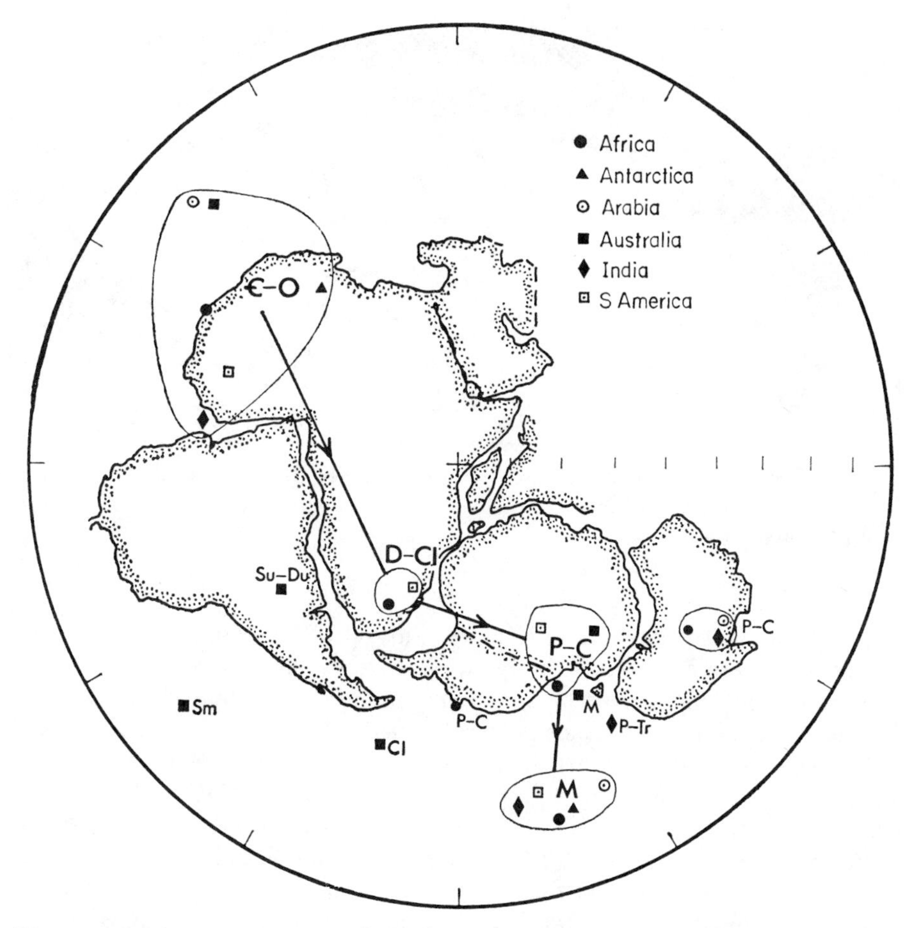


**Figure 9.18.** A reconstruction of Gondwanaland by Smith and Hallam (1970) with paleomagnetic poles plotted by McElhinny (1973). The coincidence of apparent pole paths for a 300-million-year period from Cambrian to Mesozoic times demonstrates the validity of this continental reconstruction. Reproduced, by permission, from McElhinny (1973).

A suggestion that global climatic changes arise from a superposition of several astronomical periodities in the range $10^4$ to $10^5$ years, due to the superposition of precession of the Earth on its elliptical orbit as well as planetary gravitational interactions, was originally advanced by M. Milakovitch. Although the annual average insolation or solar radiation energy falling on the Earth is constant, the supposition is that asymmetry of the land masses between the hemispheres unbalances the response of the Earth to the annual variation in insolation. Emiliani and Shackleton (1974) suggested that the cyclic temperature changes that

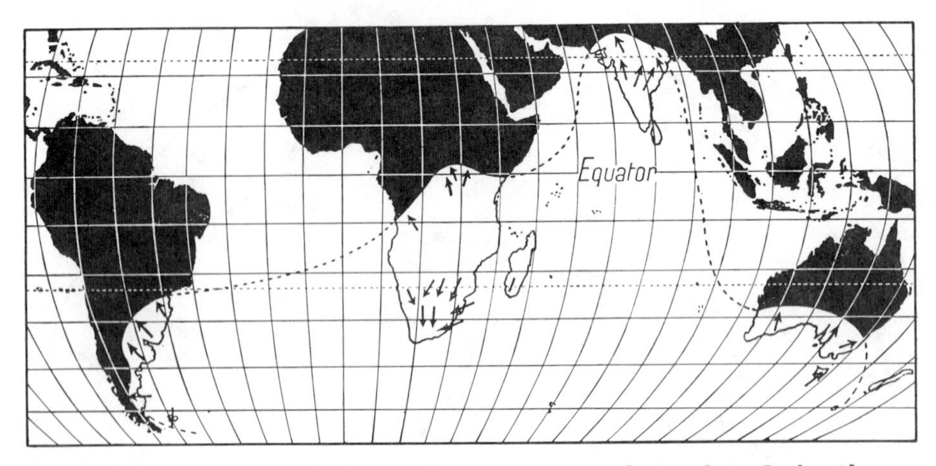


**Figure 9.19.** Glaciation of the southern continents during late Carboniferous times. Arrows indicate direction of ice movement. Reproduced, by permission, from Holmes (1965).

they found were consistent with Milankovitch cycles. But alternative suggestions are at least as persuasive. If precession were important we would expect the resulting ice ages to be out of phase in opposite hemispheres, whereas observations indicate that they are synchronised (Mercer, 1972). Öpik (1967) discounted the orbital variations as quite inadequate to account for the gross, longer-term variations in climate and even doubted their effectiveness in short-term glacial advances, favoring an intrinsic variability of the Sun, as could arise from variability of the interstellar dust that is swept into it in its motion through the galaxy. At least shorter term variations appear to be related to the injection of volcanic dust into the stratosphere (Lamb, 1970; Schofield, 1970) and consequent increased reflection of sunlight from the Earth. This appears particularly promising as an explanation of very sudden cooling. The evidence for a correlation between geomagnetic reversals and cooling is noted in Section 9.4. But all of these effects are rapid by comparison with continental drift and represent high-frequency noise in the comparison of pole paths with paleoclimates.

## 9.6 INTENSITY OF THE PALEOMAGNETIC FIELD

The variations in the strength of the geomagnetic field over the past several thousand years are discussed in Section 9.4. Essentially the same method of comparing natural remanence and laboratory thermoremanence has been used on numerous rocks, dated back to the early pre-Cambrian period, about 2700 million years ago. The results are summarized in Figure 9.20.

Since any plausible power source, such as the radioactivity of potassium (Section 7.4), would have been more vigorous in the early pre-Cambrian period, it is of interest to seek evidence of a decrease in average field strength over geological time. The thermal power produced $2.5 \times 10^9$ years ago by a fixed mass of potassium in the core would have been greater than at present by a factor 4 and, since ohmic dissipation by core currents varies as the square of the field strength, this corresponds to a field having only twice the present strength, which is not sufficiently out of line to establish an effect in view of the wide scatter of the data in Figure 9.20. But the conducted core heat would be the same now as in the remote past and this leaves a bigger factor for the change in convective heat. Furthermore, if convection was once more rapid, then

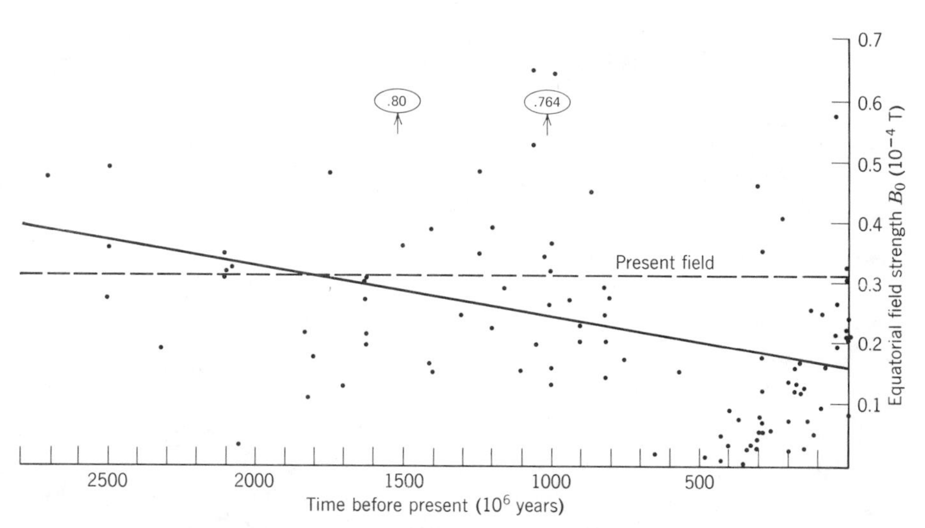


**Figure 9.20.** A plot of values of the intensity of the geomagnetic field through geological time, using data by several authors (Briden, 1966; P. J. Smith, 1967; Carmichael, 1967; Kobayashi, 1968; McElhinny and Evans, 1968; Schwarz and Symons, 1969). The values plotted are of $B_0$ as in Equation 8.8. Since field intensity varies quite rapidly, considerable scatter is expected on this wide time scale. Although long-term variations are also apparent from these data, the important conclusion is that the Earth had a magnetic field at least comparable to the present one early in its history. The value of the present field is indicated by a broken line and the solid line is an equal weight least squares fit to all of the data, which suggest that there has been a general decrease in average field strength.

we would expect the observed poloidal field to be stronger relative to the total (toroidal plus poloidal) field in the core. Thus we must look for a substantial decrease with time in the long-term average strength of the field. The straight line in Figure 9.20 is a least squares linear fit with all data equally weighted. We have no reason to expect a linear fit, but the scatter is such that more sophisticated treatment of the data would hardly be justified. However, we can conclude that, viewed on this time scale, the evidence is consistent with a general decrease in field strength. But the period of very weak field about 500 million years ago appears to have been very prolonged and may imply a special condition in the Earth's interior at that time. If this effect is removed, then the general decrease with time is very doubtful.

# 10

# Tectonics and Anelasticity

"... if I shall seem to advance any thing that looks like Extravagant or Romantick, the Reader is desir'd to suspend his Censure till he have considered the force and number of the many Arguments which concur to make good so new and so bold a Supposition"

EDMUND HALLEY, *MISCELLANEA CURIOSA*

## 10.1 PLATE TECTONICS

Global geology, as now understood, was developed only in the late 1960s. Nevertheless its general acceptance is essentially complete and comprehensive discussions in textbook form are available (Wyllie, 1971; LePichon et al., 1973). The salient features are summarized in this section. However, the basic underlying processes of plate tectonics are still the subject of hypotheses inadequately constrained by observations. In effect we claim to understand the kinematics but not the dynamics of global tectonics. Physical considerations basic to the dynamic problem are given in the following two sections and in Chapter 7.

Early advocates of continental drift were impressed by the evidence for relative horizontal displacements of sections of crust. Similarities both in shape and geology between continental margins now separated by wide oceans (e.g., Fig. 10.1) invited the conclusion that the continents have indeed drifted apart from a larger, common land mass and the continental reconstructions envisaged by Wegener (1929, reprinted 1966) and duToit (1937) have received remarkable general confirmation from paleomagnetism. The association of the seismic belts of the Earth with

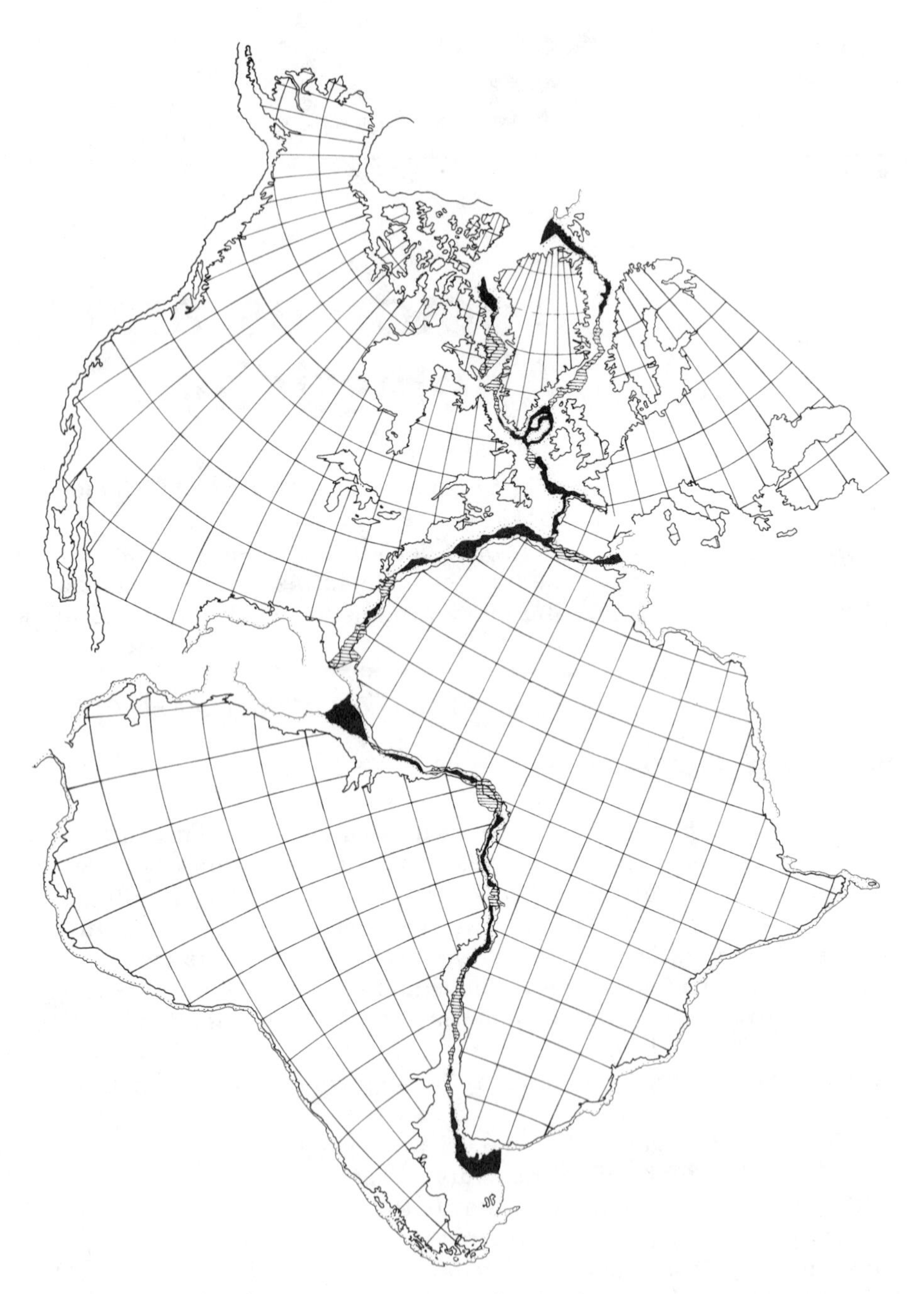

**Figure 10.1.** The fit of the continents around the Atlantic Ocean. Redrawn from Bullard et al. (1965). The best fit was adjusted to the 500-fathom (900 m) contour. Black areas represent overlap and shaded areas are gaps.

**Figure 10.2.** Mismatch of geological features across a branch of the San Andreas Fault near Indio, California. Photograph by Spence Air Photos.

horizontal displacements (Figs. 5.5 and 10.2) that could accumulate to hundreds or thousands of kilometers offered a suggested picture of the mechanism. But the apparent significance of horizontal motion along transcurrent faults or horizontal shear zones, as represented in Figures 5.5 and 10.2, and lack of evidence for associated vertical motion in the mantle obscured the true nature of the continental movements and delayed the general acceptance of continental drift by many years.

The interpretation of transcurrent faults as *transform* faults (Wilson, 1965a) is indicated in Figure 10.3. This shows how the rising and falling limbs of mantle convection appear as linear features on the Earth's surface. The rising limbs appear as ocean rises or ridges that are centers of spreading of the sea floor and the *subduction zones* where the surface layer plunges back into the mantle are marked by deep ocean trenches. Transform faults are connections across breaks in these linear features. The relegation of transcurrent faults to a secondary role in the tectonic pattern is important in terms of the source of driving power. As long as displacements were believed to be simply horizontal (perhaps with continents "floating" in the manner of rafts on the mantle, as is implied by the evidence of isostasy—Section 4.3) opponents of continental drift could point to the gross inadequacy of available forces or

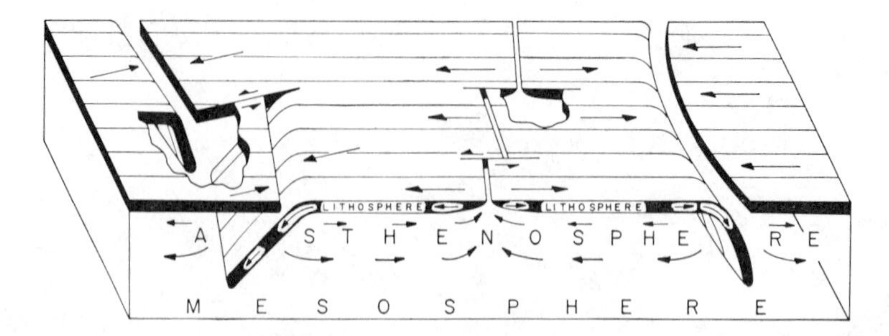


**Figure 10.3.** Schematic illustration of global tectonic movements. The motion is envisaged as being due to convection in the *asthenosphere* or "weak" layer of the upper mantle, with the harder *lithosphere* being carried along from line sources or spreading centers (ocean ridges) to line sinks ("subduction zones") which are marked by deep ocean trenches, and are associated with island arcs and similar structures. Transcurrent faults, where horizontal shearing occurs, appear as almost incidental to the motion, serving to connect separated sections of source or sink. Figure reproduced, by permission, from Isacks et al. (1968).

sources of energy. Recognition of large-scale vertical movements, whether strictly thermal convection or partly driven by progressive differentiation within the mantle, allows the appeal to gravitational energy arising from thermal (or chemical) differences. Section 10.3 gives a discussion of the energy balance and the convective stresses.

An interpretation of crustal features in terms of the convective pattern is given in Figure 10.4. This shows the emergence of basaltic ocean crust as a differentiation product at a rising convective limb along an ocean rise or ridge. The ocean rise is a source from which new sea floor spreads outward both ways. The lithosphere or rigid layer (plate), on which the crust itself is only a veneer (5 km), thickens as it cools in the spreading process and has a representative thickness of about 70 km. At the ocean trench, marking the disappearance of lithospheric material at a subduction zone, the rigid lithospheric slab plunges into the mantle, generally down an inclined plane (Benioff zone) marked by earthquakes with intermediate and deep foci (Benioff, 1949). It is progressively reheated and assimilated into the mantle and evidence of its existence is not seen below about 700 km depth. But the reworking, especially of the crustal layer, produces secondary differentiation products (the ocean floor basalts being a primary differentiate of the mantle). These emerge as *andesitic* lava, which may form a chain of islands (Japan, Aleutians) or build up the edge of a continent (Andes mountains of South America, from which andesite was named). A collision between two continental blocks is marked by the Himalaya mountains, where India has merged

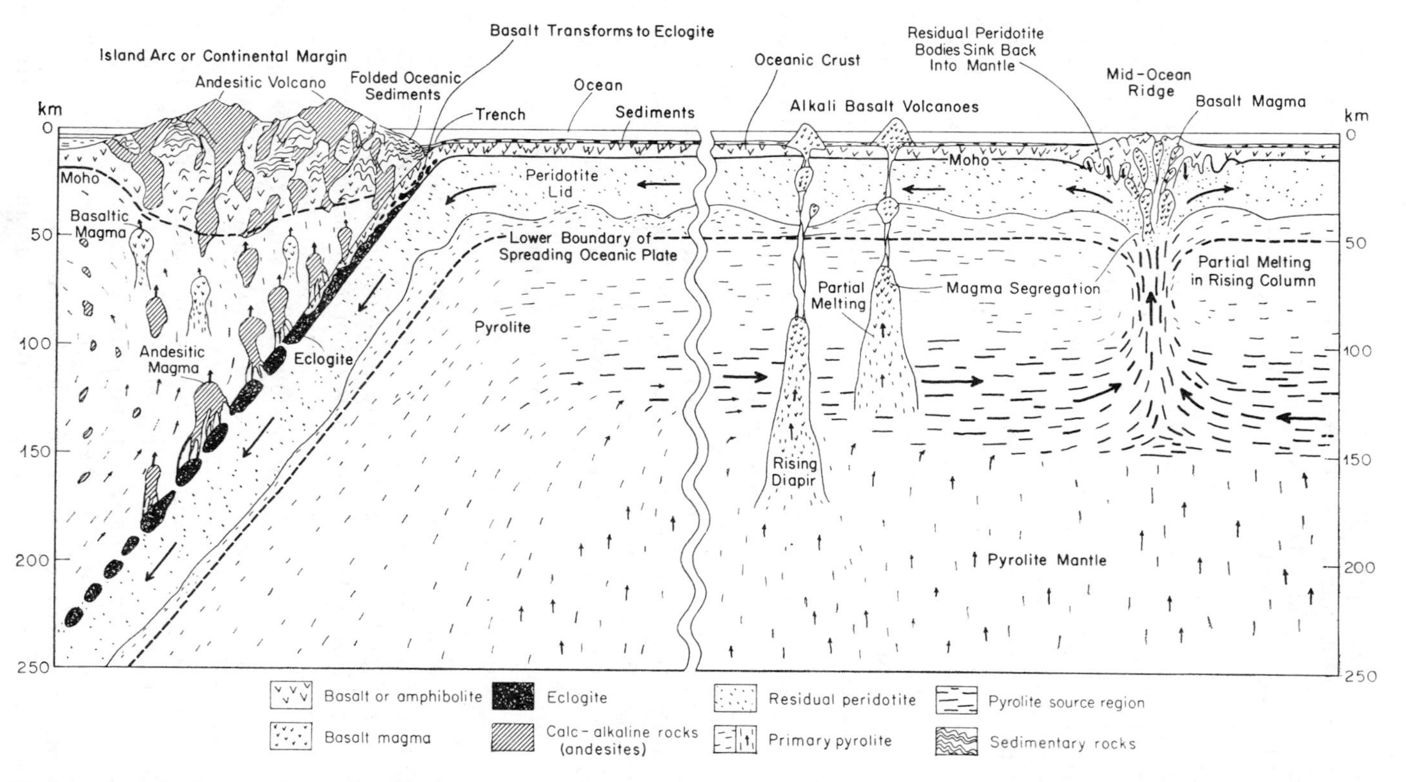


**Figure 10.4.** A picture of mantle convection, illustrating the formation of the two distinct types of crust, the basaltic ocean floor, and the more acidic continental type rocks that appear at island arcs as the second stage of a two-stage process of mantle differentiation. Reproduced, by permission, from Ringwood (1969).

with the main Asian plate (or plates). In this circumstance neither plate can plunge into the mantle complete because the density of the continental crust is too low (and its thickness too great). The result is a massive range of folded mountains. The Alps of Europe also appear to have been formed in this way.

The global pattern of sources and sinks, shown in Figure 10.5, is based on syntheses of the data by Isacks et al. (1968), Morgan (1972), and LePichon et al. (1973). The coincidence of plate boundaries with the belts of seismic activity is seen by comparing Figures 10.5 and 5.1; however, note that the majority of earthquakes appearing along the ocean ridges probably arise from shears across transform faults connecting sections of ridge (Sykes, 1967). Normal faulting at the ridges themselves does not appear to be a source of significant seismic activity. The rates of divergence at ocean ridges are estimated reasonably reliably from the scales of the magnetic stripes which flank them (Figs. 10.6, 10.7) in terms of the established time scale of geomagnetic reversals (Section 9.4). This is the hypothesis originally advanced by Vine and Matthews (1963) and independently by L. W. Morley and A. Larochelle (see Cox, 1973, pp. 224–226). A convincing confirmation of the validity of the magnetically determined spreading rates has come from the deep-sea drilling program, JOIDES. Maxwell et al. (1970) found the ages of the lowest sediments, immediately above the ocean floor basalt, to increase steadily with distance from the nearest ridge (Fig. 10.8) at precisely the rate deduced from the spacing of the magnetic anomalies. A best fit to all spreading and convergence rates has been calculated by Minster et al. (1974).

A semi-independent line of evidence has emerged in terms of the concept of lower mantle "hot spots," which establish active volcanic centers apparently unrelated to the upper mantle convection pattern. Wilson (1963, 1965b) first commented on the progressive increase in age along chains of (extinct) volcanic islands (e.g., Hawaii—McDougall, 1964) as evidence of plate motions relative to volcanic sources, and Morgan (1971, 1972) presented the explicit suggestion that island chains, such as Hawaii, mark narrow convective plumes originating in the lower mantle. Motion of a plate across such a lower mantle hot spot produces a chain of volcanic islands. Subsequently numerous hot spots have been recognized. They may occur in continental areas, notably at volcanoes Nyamlagira and Nyiragongo in central Africa, or in the oceanic areas, either coincident with ocean rises, as in Iceland which lies astride the mid-Atlantic ridge, or well removed from plate boundaries, as at Hawaii. Seismological evidence has been interpreted in terms of deep and narrow plume structures under such surface hot spots (Kanasewich et al., 1973; Kanasewich and Gutowski, 1975; Davies and Sheppard, 1972;

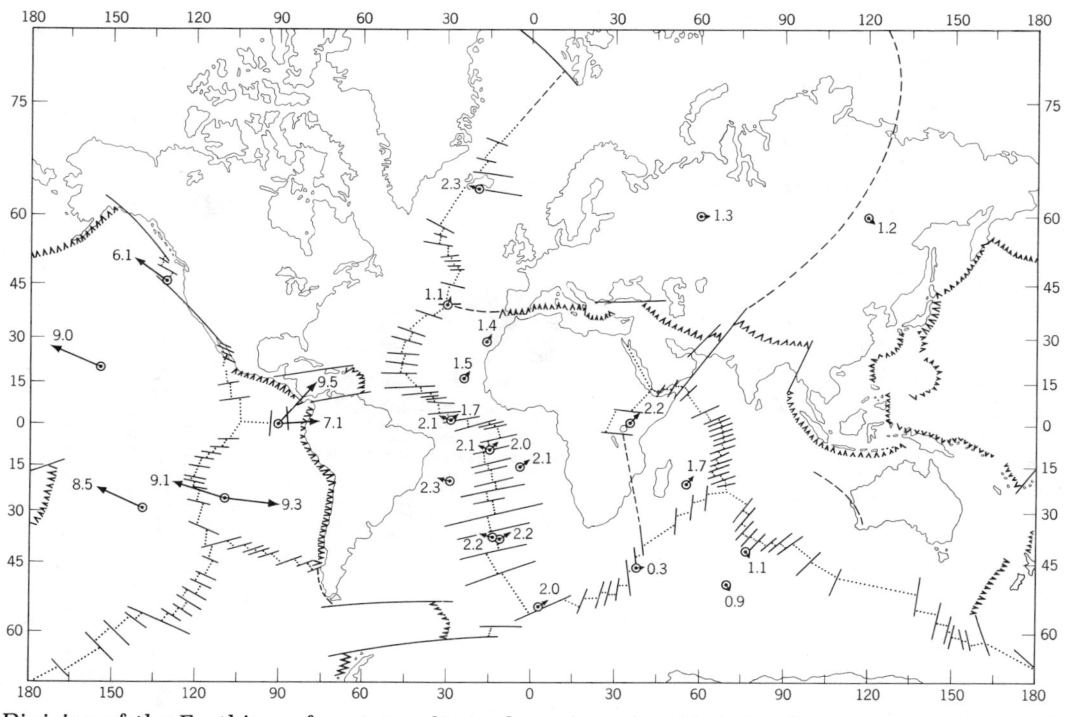


**Figure 10.5.** Division of the Earth's surface into a limited number of rigid plates that are in relative motion. Lines of divergence (ocean rises) are marked with lines of dots with extensions into the Red Sea and East Africa where new or incipient oceans are just beginning. The lines of convergence, marked by arrow heads showing directions of motion of downgoing slabs, are the zones of strongest seismicity (compare with Figure 5.1). By contrast the ocean rises proper probably produce few earthquakes, if any, and appear on seismicity maps by virtue of earthquakes on the lengths of transform fault separating sections of rise, as in Figure 10.3 (Sykes, 1967). Circled dots are the hot spots and arrows with numbers give motions of plates (in cm/year) relative to the hot spots, as estimated by Minster et al. (1974). The single lines mark transform faults or their extensions ("fracture zones") and broken lines indicate features that are unclear.

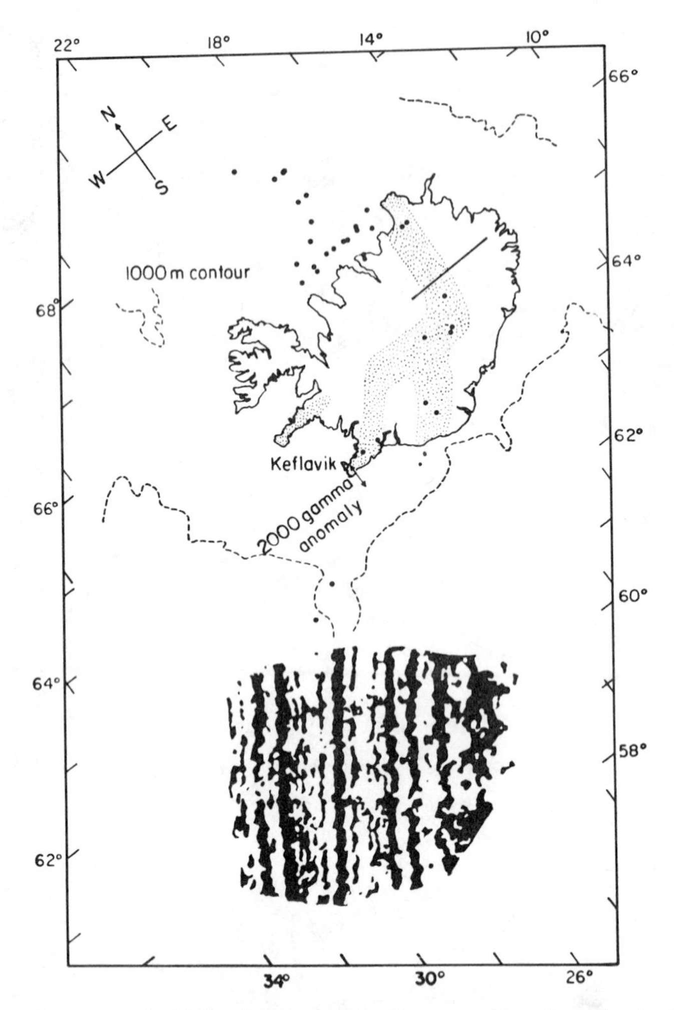


**Figure 10.6.** An example of the linear magnetic anomalies that flank ridges in all oceans. The anomaly pattern across the Reykjannes ridge in the north Atlantic, south of Iceland. Reproduced, by permission, from Heirtzler et al. (1966).

Iyer et al., 1974), although this is not a secure conclusion. Geochemical differences between the island basalts and ridge basalts (Hart et al., 1973; Schilling, 1975) imply distinct source regions for the island basalts, less depleted in alkali and alkali-earth elements and, notably, in radioactive species, which also tend to segregate upward in any differentiation process.

Grommé and Vine (1972) found that lava on Midway Atoll, an 18-million-year-old island of the Hawaii chain now at latitude 28°, had a

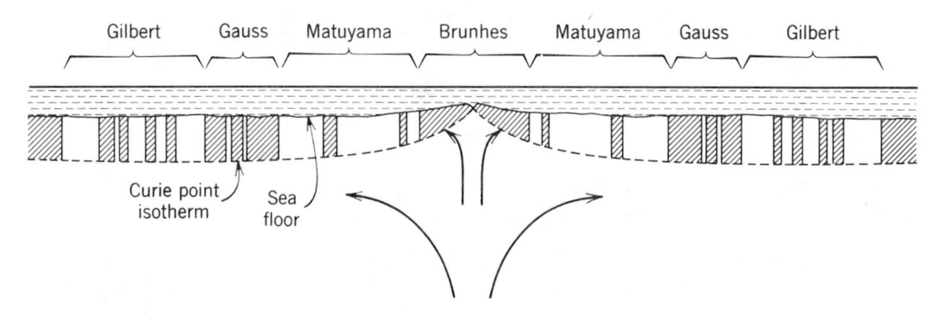


**Figure 10.7.** The ocean floor as a giant magnetic tape recorder. Knowing the time sequence of geomagnetic reversals the speed of the tape can be deduced. By assuming a constant spreading rate, the pattern of marine magnetic anomalies has been used to plot the sequence of geomagnetic reversals over a much longer time span than is available from dated continental rocks (Fig. 9.14). This procedure is justified by paleontological dating of the ocean floor (Fig. 10.8). Vacquier (1972) has reviewed the discovery and interpretation of marine magnetic anomalies.

paleolatitude of $15° \pm 4°$, coinciding within the uncertainty of measurement with the present latitude of the active island of Hawaii (19°). The suggestion is that the active point of this island chain is stationary or nearly so, being fixed by a feature of the rigid lower mantle. It receives strong support from comparison with plate motions relative to several other hot spots, which indicates that the hot spots remain in fixed relative positions (Minster et al., 1974). Thus we arrive at a summary of plate tectonics, as presented by Jordan (1974), who considered the lower mantle to provide an effectively rigid frame for the hot-spots whose points of volcanism serve as markers for the determination of absolute plate motions. Plates entirely of ocean floor are found to move more rapidly than those carrying continents, a conclusion in keeping with the somewhat hotter and therefore softer asthenosphere under the oceans (Fig. 7.4). While not all of the details are yet universally agreed, the picture is sufficiently complete and internally consistent to provide a basis for theoretical discussions of mechanisms (Sections 10.2 and 10.3).

## 10.2 STRESS-STRAIN BEHAVIOR OF THE MANTLE

Traditional discussions of the anelasticity of the Earth (e.g., Scheidegger, 1963; Jeffreys, 1970) have used a range of mathematical models that are attempts to adapt to solids the Newtonian viscous flow of liquids. The three common ones are represented in Figure 10.9. But the anelasticity of a crystalline solid is normally non-Newtonian (i.e.,

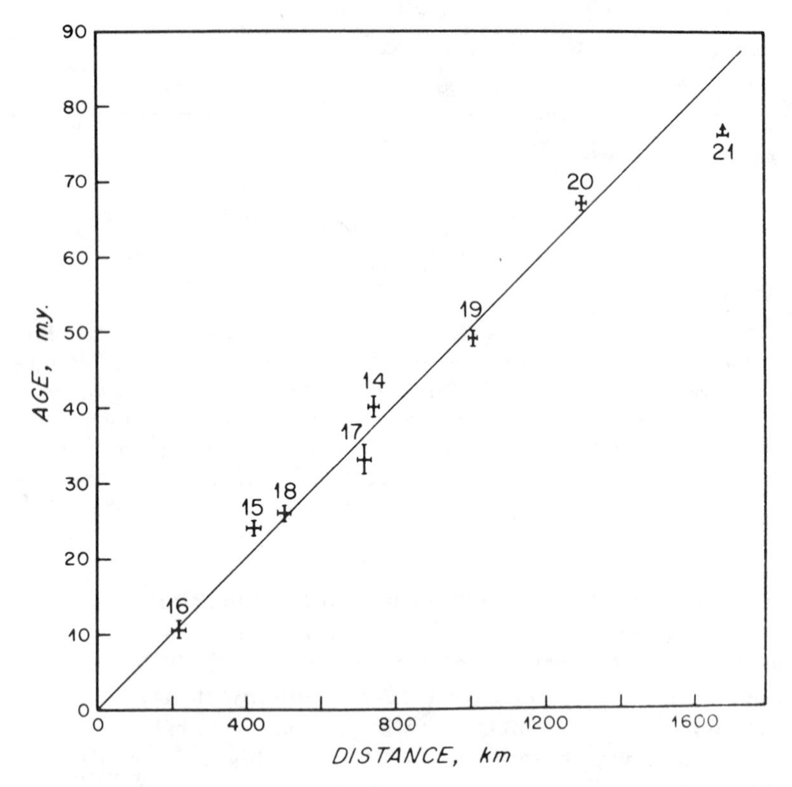


**Figure 10.8.** Paleontologically deduced age of the sediment immediately above the basalt basement as a function of distance from the mid-Atlantic ridge, from JOIDES drill cores from the south Atlantic. This shows the spreading rate (each way) from the ridge to be 2 cm/year. Reproduced, by permission, from Maxwell et al. (1970). Copyright 1970 by the American Association for the Advancement of Science.

nonlinear in stress) and we can expect to comprehend the rheological behavior of the Earth only by taking account of the fundamental atomic processes involved. Recent discussions by Weertman (1970), Boland et al. (1971), Hobbs et al. (1972—see also other papers in this volume), Kirby and Raleigh (1973), Stocker and Ashby (1973), and Kohlstedt and Goetze (1974) also recognize the importance of crystal dislocations in the deformation of the mantle and, in particular, the process referred to as dislocation climb, which has been studied by Weertman (1968).

Dislocations are extended linear crystal defects that occur naturally in all crystals (possibly except in very special cases, such as iron whiskers) and may be multiplied by deformation of a material (work

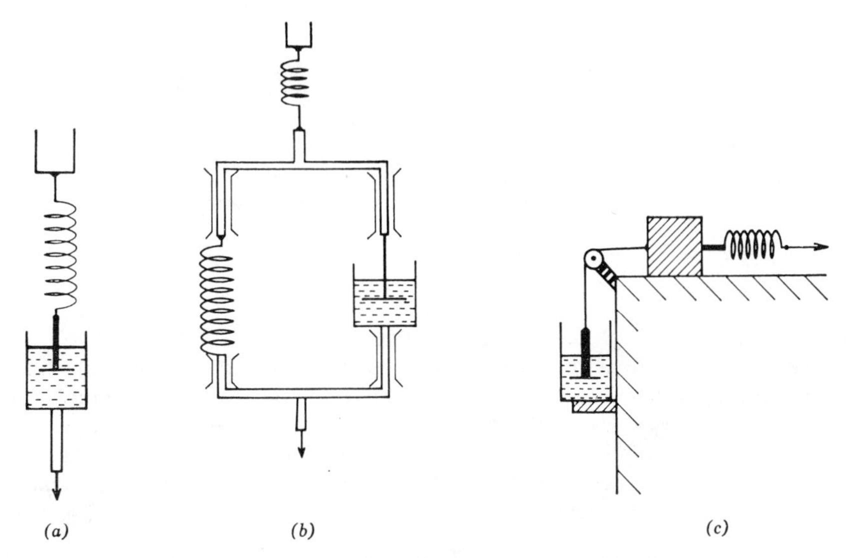


**Figure 10.9.** Mechanical models for ideal (*a*) Maxwell (elasticoviscous), (*b*) Kelvin-Voigt (firmoviscous), and (*c*) Bingham (plasticoviscous) solids. Firmoviscosity is often represented without the immediate elastic response of the top spring. Note that to move the block in (*c*) friction must be overcome, and this represents a finite yield point.

hardening) or removed by annealing or *recovery*. We are particularly interested here in high temperature, steady-state creep, or deformation in which spontaneous recovery keeps pace with dislocation multiplication and a material can deform indefinitely with a constant dislocation density and a constant creep rate for fixed stress. Crystal dislocations are discussed comprehensively by Cottrell (1953) and Friedel (1964) and a more elementary, textbook treatment is given by Kittel (1971). There are two basic types, represented in Figure 10.10. Both types can, in principle, be produced in a perfect single crystal by making a half cut through it and displacing atoms on opposite sides of the cut by a single lattice spacing. The cut defines the slip plane and the dislocation line (along the axis of the cylinder) is the boundary between the slipped and unslipped regions of the crystal. The displacement is specified by the slip or Burgers, vector, **b**. Since **b** is a lattice vector, the atoms on opposite sides of the slip plane are back in register. Thus the slip plane is not a discontinuity in the crystal. However, there is considerable distortion of the crystal around the dislocation line. The energy of a dislocation is proportional to $|\mathbf{b}|^2$ so that crystals tend to deform by slip with **b** the smallest lattice vector. Vector **b** is normal to the dislocation line in the

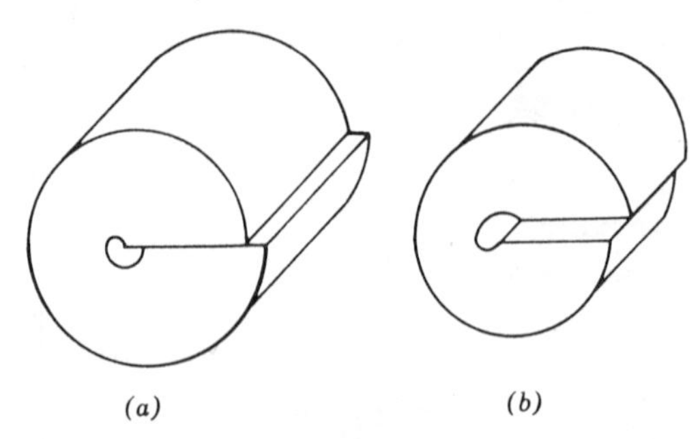


**Figure 10.10.** Forms of the atomic displacements in (*a*) edge and (*b*) screw dislocations. The dislocation lines or axes are the axes of the cylindrical crystals, which are drawn as hollow only for convenience in representation. Actual atomic displacements very close to the dislocation are non-Hookeian and are not quite as simple as represented here. Leftward movements of the dislocations in this figure cause increasing distortions of the originally cyclindrical bodies. The cumulative effect of movements of several dislocations is indicated in Figure 10.13.

case of an edge dislocation and parallel to the dislocation line of a screw dislocation. If the line has an arbitrary orientation with respect to **b**, the dislocation is termed "mixed."

Dislocations for which **b** is a lattice vector are known as total dislocations. However, dislocations for which **b** is not a lattice vector can occur in certain crystals. These are known as partial dislocations. In this case the movement of a partial dislocation leaves the crystal faulted—the atoms are not brought back into register on opposite sides of the slip plane as the dislocation moves through the crystal.

Dislocations may be observed directly in the electron microscope since the highly distorted region of the crystal around a dislocation causes the electron beam to be diffracted locally with a different intensity from that diffracted by the perfect crystal. A planar fault in the crystal gives rise to a fringe pattern whose characteristics depend on the phase shift suffered by the electron beam on crossing the fault. Transmission electron micrographs showing total dislocations in olivine and partial dislocations and faults in pyroxene are shown in Figures 10.11 and 10.12.

Motion of the dislocation line to the left in either of the dislocations in Figure 10.10 increases the deformation of the crystal in the manner indicated in Figure 10.13. To proceed from state (*a*) to state (*b*) in Figure 10.13 by displacing two planes of atoms coherently would require a stress of about $q/30 \approx 3 \times 10^9$ Pa ($3 \times 10^4$ bar), where $q$ is Young's modulus. This is much greater than the strength of all but a few specially

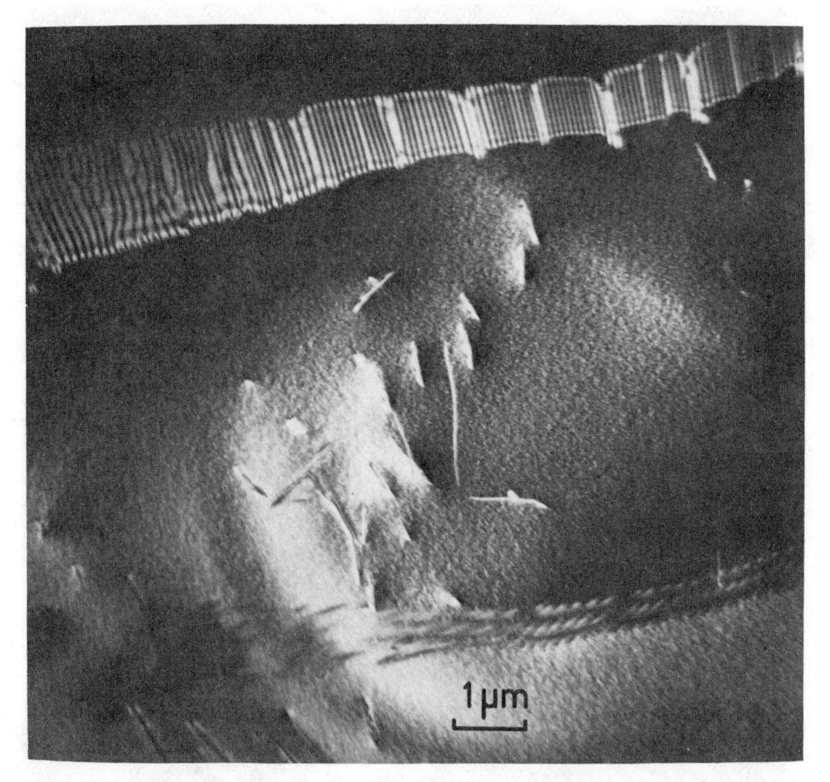


**Figure 10.11.** Transmission electron micrograph showing dislocations in olivine from Anita Bay, near the Alpine Fault, South Island of New Zealand. Most of the dislocations in this material are arranged in parallel arrays forming subgrain boundaries. Comparison with deformed metals indicates that this dislocation substructure is characteristic of deformation by very slow creep and is quite distinct from the dislocation arrangements observed in olivine deformed rapidly in the laboratory (at strain rates of $10^{-5}$ $sec^{-1}$). Reproduced, by permission, from Boland, McLaren, and Hobbs (1971).

prepared materials, such as iron whiskers, which are free of defects. However, the same displacement can occur relatively freely by the progressive movement of a dislocation in which the movement is a sum of individual atomic displacements. For this reason the creep behavior of most materials is determined by the dislocations. Dislocations are not completely free to move, however. They are impeded by interactions with one another and with other crystal defects (lattice vacancies and interstitial atoms, which are termed point defects) and the rate-controlling process, which determines how fast creep proceeds, is the overcoming of the obstacles. At the high temperatures and slow creep rates with

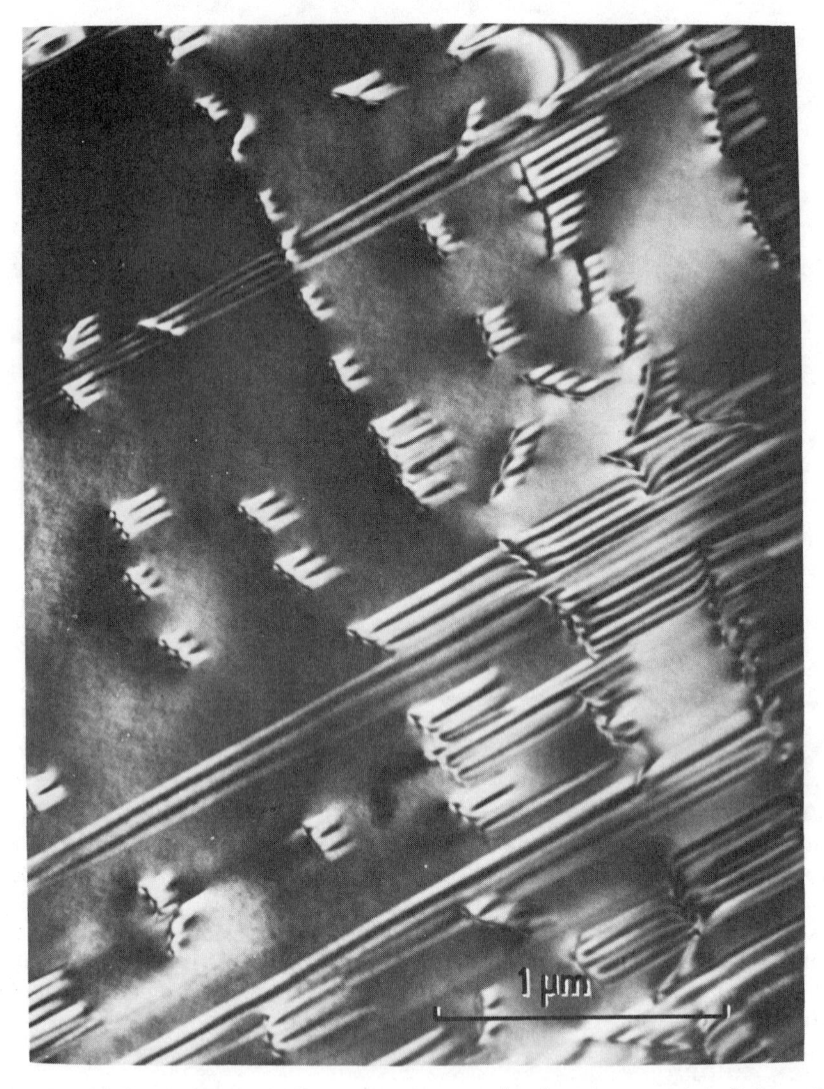


**Figure 10.12.** Transmission electron micrograph showing partial dislocations and associated faults in a naturally deformed orthopyroxene from the Giles Complex, Central Australia.
Analysis shows that the dislocations moved from right to left and that the faults left behind involved a transformation from ortho- to clinopyroxene. Photograph courtesy A. C. McLaren and M. A. Etheridge.

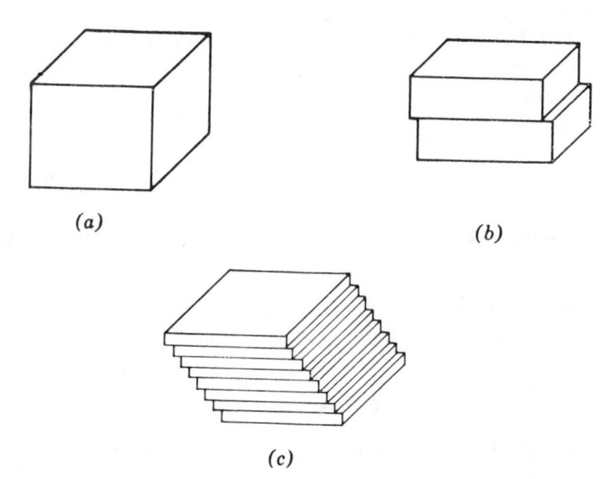


**Figure 10.13.** Cubical sample (*a*) before and after the movement through it of (*b*) one and (*c*) several dislocations. The deformation of single crystals may occur on preferred slip bands, on each of which there are many atomic displacements, so that a sheared specimen may have the macroscopic appearance of (*c*).

which we are concerned in the Earth, this occurs by atomic diffusion of point defects in the combined fields of the bulk stress on a material and the local stresses due to the dislocations, allowing the dislocations to move out of their slip planes. This is dislocation climb (Weertman, 1968) which, since it depends on a diffusion process, is controlled by the activation energy of self-diffusion, $Q$. The equation representing creep or shear strain rate, $\dot{\epsilon}$, controlled by dislocation climb in a material subjected to a shear stress $\sigma$ is

$$\dot{\epsilon} = Af\left(\frac{\sigma}{\mu}\right)\exp(-Q/kT) \tag{10.1}$$

where $\sigma$ has been normalised in terms of the rigidity modulus, $\mu$, and over a wide range of creep rates

$$f\left(\frac{\sigma}{\mu}\right) \approx \left(\frac{\sigma}{\mu}\right)^n \tag{10.2}$$

where $3 < n < 6$ (depending on the precise mechanism of diffusion within the stress field of a dislocation*) and $n \approx 4.5$ is a reasonable

*Dislocation lines are not perfectly straight but "jogged" or kinked in the sense that in a moving dislocation some sections are further "ahead" than others. Thus motion occurs at the "jogs" and not coherently along the length of the dislocation; what matters is the point defect diffusion near to the "jogs" and this is affected by their spacing. The parameter $n$ is also affected by composition.

approximation for the present purpose. The factor $A$ in Equation 10.1 is not strictly constant, but may be written

$$A = A_0 \left(\frac{\mu \Omega}{kT}\right) \tag{10.3}$$

where $\Omega$ is the atomic volume of the diffusing species. However, temperature dependence is dominated by the exponential term in Equation 10.1.

Dependence of creep on pressure, $P$, arises from the inhibition to diffusion, which may be represented in terms of an activation volume, $V^* \approx \Omega$, such that

$$Q = Q_0 + PV^* \tag{10.4}$$

However, it is more convenient to recognize the empirical relationship between diffusion, creep, and melting and write

$$Q = gkT_M \tag{10.5}$$

where $g \approx 20$* is an empirical constant and $T_M$ is the melting point (at any pressure), so that

$$\dot{\epsilon} = A_0 \left(\frac{\mu \Omega}{kT}\right)\left(\frac{\sigma}{\mu}\right)^n \exp\left(-g\frac{T_M}{T}\right) \tag{10.6}$$

Equation 10.6 is the creep law appropriate to material at temperatures above about $T_M/2$, subjected to stresses smaller than about $\mu/2000$, both of which conditions must be well satisfied in the Earth, except for a thin, cool skin. However, at temperatures very close to the melting point in a material which is very fine grained, creep by pure diffusion (Nabarro-Herring creep) would take over. This involves deformation by diffusion of atoms between grain boundaries and is characterized by a creep rate directly proportional to stress, that is, a Newtonian viscosity. While it appears unlikely that conditions within the Earth have been such as to produce a grain size small enough ($< \sim 0.05$ mm) to favor diffusion creep, our knowledge of the Earth's interior is not such that we can dismiss it with certainty.

Equation 10.6 is a creep law that can be applied at depth if the temperature profile and pressure dependence of melting point can be estimated. Uncertainties in appropriate values of the constants are magnified by uncertainty in composition, but creep strength is so strongly dependent on temperature that it is useful to invert the argument, using evidence from tectonics to estimate rates of motion and

*For metals Sherby and Simnad (1961) put $g = (K_0 + V)$ where $V$ is the valence and $K_0$ is 14 for body-centered lattices, 17 for face-centered and hexagonal close-packed lattices and 21 for diamond structures.

stress and thus calculate the temperatures corresponding to the strain rate and stress. It has been conventional to refer to the *viscosity*, $\eta$, of the mantle, but this is not a physically very meaningful concept for a material whose rheology is described by Equation 10.6, since this gives

$$\eta = \frac{\sigma}{\dot{\epsilon}} \propto \sigma^{-(n-1)} \propto \dot{\epsilon}^{-(1-1/n)} \tag{10.7}$$

Thus the specification of a viscosity at any depth in the Earth requires a statement of the strain rate to which it refers. However, this is not altogether inconvenient, because $n \approx 4.5$ and therefore we can very roughly think of $\eta$ as varying as $\dot{\epsilon}^{-1}$. The relationship, with numerical values, is shown in Figure 10.14. In the following section estimates of convective stresses are given in terms of the energy dissipation which, for unit volume, is

$$\frac{dE}{dt} = \sigma\dot{\epsilon} = \eta\dot{\epsilon}^2 \propto \dot{\epsilon}^{(1+1/n)} \tag{10.8}$$

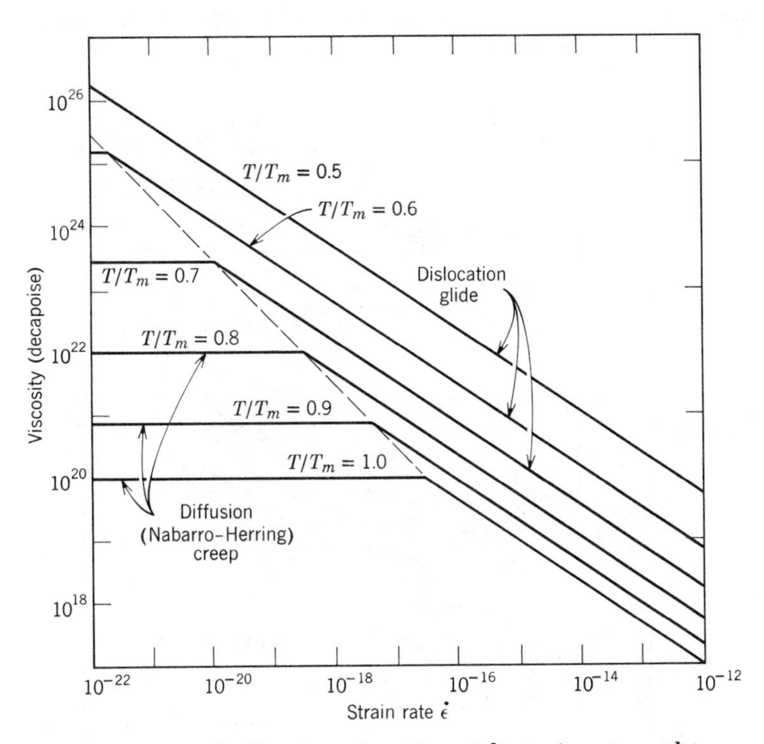


**Figure 10.14.** Variation of effective viscosity with strain rate and temperature (normalized in terms of melting point). After Weertman (1970) and Brennen (1974).

Since $n \simeq 4.5$ is assumed, a factor 100 change in $\dot{\epsilon}$ requires only a factor 3 change in $\sigma$ (at constant temperature) and the dissipation is almost linear in $\dot{\epsilon}$ instead of varying as $\dot{\epsilon}^2$, as in the case of a viscous fluid. But the essential point is that if both $\dot{\epsilon}$ and $\sigma$ can be estimated for any depth from a combination of tectonic observations and energy balance arguments, then $T/T_M$ can be inferred from Figure 10.14. The whole mantle appears to be quite close to its solidus temperature (Fig. 7.4).

The most directly quantitative indication of the rheology of the upper mantle is the delayed isostatic rebound of areas that were heavily glaciated and, as a consequence, depressed during the last ice age. The best documented is the Gulf of Bothnia and surrounding area (Fennoscandia—Figure 10.15) but also rebounding is the area of the Laurentide ice sheet, which covered eastern Canada, extending to the Great Lakes. Also it is hard to avoid the conclusion of McConnell (1968) that a more extensive residual depression of both polar regions contributes to the excess ellipticity of the Earth (Section 4.3). Data on Fennoscandian rebound that are suitable for calculations on mantle rheology are given in Figure 10.16. The uplift data, although piecemeal, is sufficiently extensive for the curve of rate of rise to be reasonably

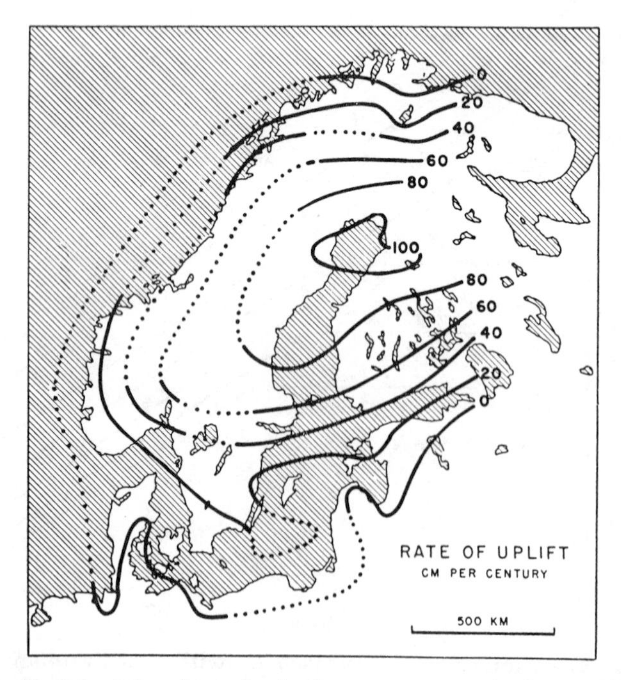


**Figure 10.15.** Postglacial rebound of the area around the Gulf of Bothnia (Fennoscandia). Reproduced, by permission, from Gutenberg (1958).

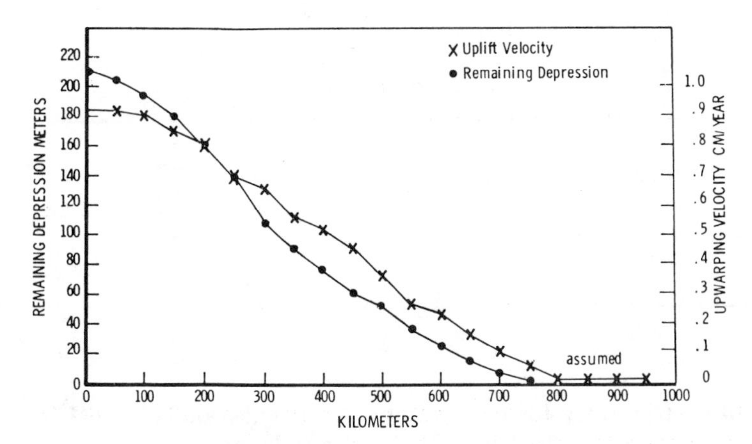


**Figure 10.16.** A comparison of the present rate of rise and the remaining depression of Fennoscandia as functions of distance from the center of depression. This figure, which is reproduced by permission from McConnell (1968), uses data by E. Kääriäinen and E. Niskanen. Note, however, that the assumed lack of movement outside the depressed area is unrealistic and a fall due to flow of material into the depression must be expected, although given sufficient time this too would be filled.

secure. The remaining depression is deduced from the isostatic anomaly and is therefore based on an interpretation of gravity data, which is less certain, but the general form of the depression is clear enough to justify quantitative calculations. McConnell (1968) considered both a homogeneous viscous half space and flow restricted to an asthenosphere of thickness 100 km or 200 km and deduced viscosities of $4 \times 10^{18}$ decapoise* to $6 \times 10^{20}$ decapoise. Post and Griggs (1973) and Brennen (1974) agree that the observations can be explained only in terms of non-Newtonian creep in the upper mantle, that is, an equation of the form 10.6 with $n > 1$.

A model for rebound based on a rheologically homogeneous asthenosphere of thickness $h$, much less than the lateral extent of the depressed area (Fig. 10.17), allows the effective viscosity of the asthenosphere to be estimated from very simple equations, using the data in Figure 10.16. The model has circular symmetry, with flow radially inward, causing upwarping of the lithosphere without elastic constraint and the lower boundary (mesosphere) is assumed to be rigid. If the rate of uplift at radius $r$ is $\dot{z}(r)$ then the volume rate of asthenospheric flow

*The cgs unit of viscosity, the poise ($\equiv 1\,\mathrm{g\,cm^{-1}\,sec^{-1}}$) is 0.1 decapoise (1 decapoise $\equiv 1\,\mathrm{kg\,m^{-1}\,sec^{-1}}$).

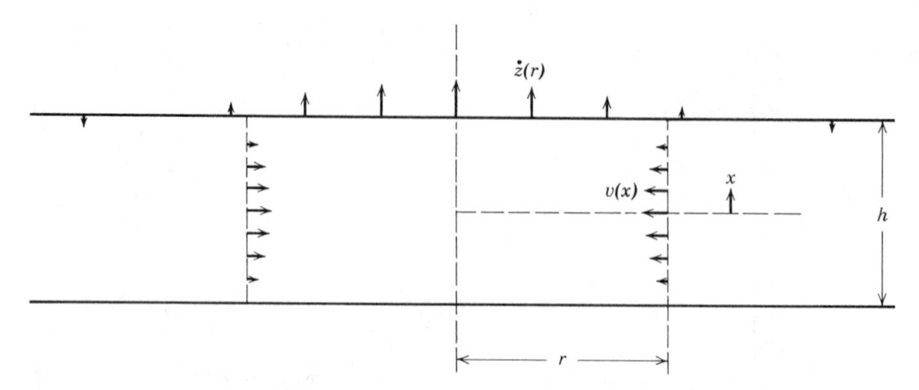


**Figure 10.17.** Geometry for interpretation of Fennoscandian uplift in terms of flow in a rheologically uniform asthenospheric layer.

through an annulus of radius $r$ is

$$F(r) = \int_0^r \dot{z}(r) \cdot 2\pi r \, dr \tag{10.9}$$

which can be tabulated directly from the observed values of $\dot{z}(r)$ in Figure 10.16. The flow is driven by the radial pressure gradient

$$\frac{dP}{dr} = \rho g \frac{dz}{dr} \tag{10.10}$$

where we may assume an upper mantle density $\rho = 3350 \text{ kg m}^{-3}$ and gravity $g = 9.9 \text{ m sec}^{-2}$. The "anomalous" surface elevation, $z(r)$, is the negative of the remaining depression, also plotted in Figure 10.16. The pressure gradient equals the vertical gradient of the shear stress, $\sigma$, across the layers of flowing asthenosphere, that is,

$$\frac{d\sigma}{dx} = \rho g \frac{dz}{dr} \tag{10.11}$$

where $x$ is distance measured vertically from the center of the asthenosphere and, since $dz/dr$ is constant at fixed $r$, Equation 10.11 can be integrated directly:

$$\sigma = \rho g \frac{dz}{dr} x \tag{10.12}$$

To relate this to the vertical variation of flow speed, $v$, we need to assume a creep law:

$$\frac{dv}{dx} = \dot{\epsilon} = A\sigma^n = A\left(\rho g \frac{dz}{dr} x\right)^n \tag{10.13}$$

where the constant $A$ is simply the reciprocal of viscosity, $\eta^{-1}$, in the simple Newtonian case, $n = 1$. Then, integrating with respect to $x$, using the boundary condition that $v = 0$ at $x = h/2$, we obtain the velocity profile

$$v = \frac{A}{(n+1)}\left(\rho g \frac{dz}{dr}\right)^n \left[\left(\frac{h}{2}\right)^{n+1} - x^{n+1}\right] \qquad (10.14)$$

and by a second integration the volume rate of flow

$$F(r) = 2\pi r \cdot 2\int_0^{h/2} v\,dx = \frac{4\pi r A}{(n+2)}\left(\rho g \frac{dz}{dr}\right)^n \left(\frac{h}{2}\right)^{n+2} \qquad (10.15)$$

Figure 10.18 is a plot of $[F(r)/2\pi r]$ against $(dz/dr)$ from the data in Figure 10.16. If the model were completely valid and the data perfect this would be a smooth, monotonically increasing curve through the origin. In the Newtonian case ($n = 1$) it would be a straight line of gradient

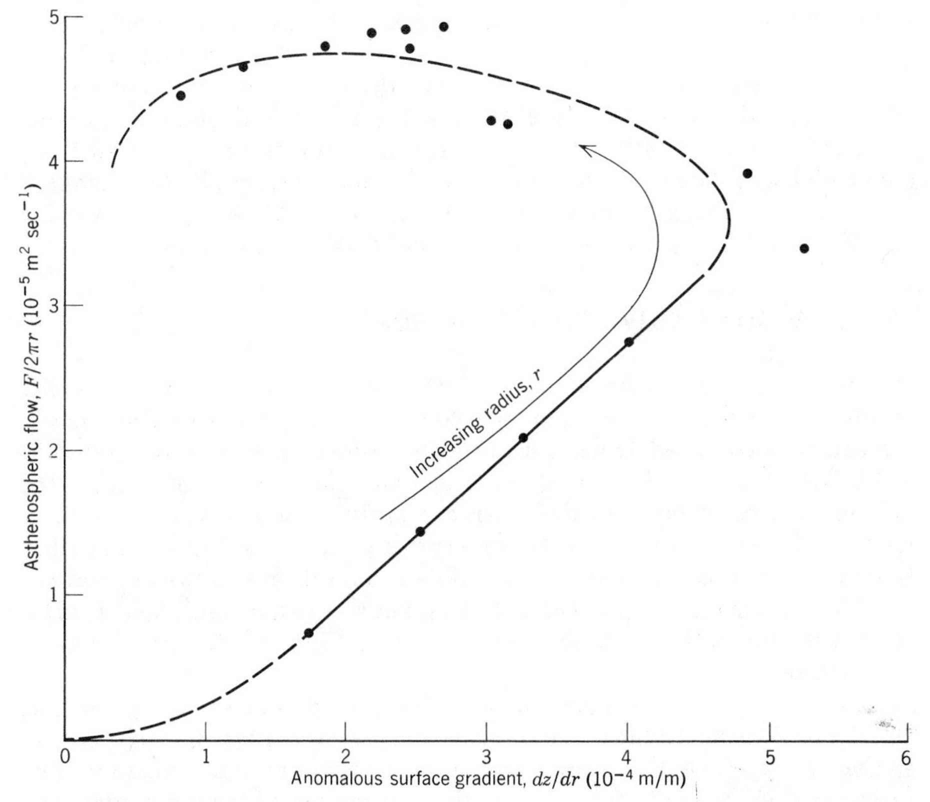


**Figure 10.18.** Relationship between asthenospheric flow and pressure gradient from the data of Figure 10.16.

$(\rho g h^3/12\eta)$. Defects in the data are more likely to appear in $(dz/dr)$, since this is an interpretation of gravity data rather than a direct observation. Thus the values are most reliable at small $r$ where the anomaly is greatest. Similarly the most obvious defect in the model is due to the neglect of flow at great depths, which is likely to affect least the data at small $r$. Thus for estimating viscosity the solid section of the curve may be used (the broken section at large $r$ being physically meaningless in terms of the model). Taking the gradient to be $0.08\ \mathrm{m^2\ sec^{-1}}$, $h = 100$ km, we can deduce an effective viscosity of $\eta = 3.5 \times 10^{19}$ decapoise ($3.5 \times 10^{20}$ P). This is an order of magnitude estimate and does not imply either linearity or nonlinearity of the creep mechanism. The same data can be used as a test for nonlinearity by plotting log $[F(r)/2\pi r]$ versus log $(dz/dr)$ which, according to Equation 10.15 gives a graph of gradient $n$. Replotted in this way, the data of Figure 10.18 give $n \approx 1.7$ for small $r$, clearly less than the values indicated by the dislocation climb theory of creep, but still nonlinear. However, neither the data nor the model are sufficiently reliable to make this a secure conclusion.

The increase in strength, or viscosity, in the deep mantle is also of interest. An extreme upper bound of about $10^{25}$ decapoise is imposed by the possible delay in the adjustment of the equatorial bulge to slowing rotation. Satisfying alternative explanations for the excess ellipticity (Goldreich and Toomre, 1969; McConnell, 1968; O'Connell, 1971) make it unlikely that the lower mantle viscosity exceeds $10^{24}$ decapoise. A value of $10^{22}$ to $10^{23}$ decapoise is as good a guess as we can make.

## 10.3 THE CONVECTIVE ENERGY BUDGET

Independently of knowledge of the creep law appropriate to the mantle, the stresses associated with convection and the effective viscosity can be estimated from the permitted power dissipations listed in Table 7.3. It should be noted that, of the numbers in this table, the efficiencies must be regarded as reasonably secure, being deductions from a very straightforward thermodynamic analysis, but the distribution of heat sources is simply a plausible assumption and the estimate of available mechanical power is only as good as the assumed heat source. However, this is likely to be as precise as any other estimate of mantle strength or viscosity.

Consider a slab of material of thickness $d$ and area $A$ which is slowly deforming under a shear stress $\sigma$ applied across the surfaces, $A$, so that the top face is moving at speed $v$ relative to the bottom face. The force causing this motion is $(\sigma A)$, so that the power which it dissipates is

$$\frac{dE}{dt} = \sigma A v = \sigma A \dot{\epsilon}\, d = \sigma \dot{\epsilon} V \tag{10.16}$$

where the creep rate is $\dot{\epsilon} = v/d$ and the volume is $V = Ad$. We may also refer to an effective viscosity of the material

$$\eta = \frac{\sigma}{\dot{\epsilon}} = \left(\frac{1}{V}\frac{dE}{dt}\right)\Big/\dot{\epsilon}^2 = \left(\frac{1}{V}\frac{dE}{dt}\right)\frac{d^2}{v^2} \tag{10.17}$$

These equations can be applied to any assumed pattern of deformation of the mantle, so that if the motion and the energy dissipation are specified, $\sigma$ or $\eta$ are determined.

For an application to the upper mantle we must doubt whether any of the numbers in Table 7.3 is entirely appropriate. The contribution of self-heating ($0.36 \times 10^{12}$ W) must be included, but in Section 7.1 it is suggested that core heat causes convective plumes that are largely independent of the upper mantle convective pattern, and this may also be true for heat originating in the lower mantle, so that for upper mantle convective power we may assume a rounded figure of $1.0 \times 10^{12}$ W. Referred to sections of crust, this is conveniently expressed as convective power per unit area

$$\frac{1}{A}\frac{dE}{dt} = 2 \times 10^{-3}\ \mathrm{W\,m^{-2}} \tag{10.18}$$

An order of magnitude estimate of the stress associated with this dissipation is obtained by considering a largely oceanic plate moving at 8 cm/year ($v = 2.5 \times 10^{-9}$ m/sec), which with Equations 10.16 and 10.18 gives $\sigma = 8 \times 10^5$ Pa (8 bar). The figure obtained by considering a representative continental plate, moving at 2 cm/year, is $3.2 \times 10^6$ Pa (32 bars). These estimates of stress, although crude, have the advantage that they are not very dependent on details of the convective model, such as the thickness of the zone in which the shear is concentrated. If we suppose that the lithosphere slides across the lower mantle with shearing movement concentrated in a relatively weak asthenospheric layer and the return path involves much slower motion at much greater depths (as envisaged, for example, by Andrews, 1975) then the stress estimates are uncertain only to the extent that the heat sources are unknown.

To convert estimated stresses to effective viscosities we need an assumption about the thickness of the zone of shearing. For direct comparison with the isostatic rebound estimate of Section 10.2 we may consider an asthenosphere 100 km thick, in which case $\dot{\epsilon} = v/d = 2.5 \times 10^{-14}\ \mathrm{sec^{-1}}$ (oceanic case) or $6 \times 10^{-15}\ \mathrm{sec^{-1}}$ (continental case)*, yielding estimates of effective viscosity (at these creep rates):

Oceanic asthenosphere: $\eta = \sigma/\dot{\epsilon} = 3 \times 10^{19}$ decapoise

Continental asthenosphere: $\eta = 5 \times 10^{20}$ decapoise

*These rates are comparable to those estimated for Fennoscandia.

These values must be taken as upper bounds in the sense that they suppose the only constraint to plate motion to be asthenospheric viscosity, whereas the constraint operating on the return path is likely to be at least as important. Thus the estimates of upper mantle viscosity from convection are similar to that obtained from isostatic rebound data in Section 10.2.

A similar argument can be applied to the lower mantle to examine the question as to whether it is convecting at all. It must be if the thermal estimates of Table 7.3 are even approximately correct because, even disallowing the core heat that may be carried upward by narrow plumes without affecting the bulk of the lower mantle, the heat originating in the lower mantle itself can only escape by convection. The evidence of an apparently rigid lower mantle frame for the "plumes" (Section 10.1) implies that the lower mantle convection rate is much slower than the upper mantle rate, but does not demand that it be zero. If we suppose that the lower mantle provides a very broad return path for the shallow flow that is apparent as sea floor spreading, then an average effective speed of 2 mm/year ($6 \times 10^{-11}$ m sec$^{-1}$) is the minimum that must be considered. Then, disposing of the $1.2 \times 10^{12}$ W of internally generated power over an effective area of about $2.4 \times 10^{14}$ m$^2$ in the lower mantle, we have $5 \times 10^{-3}$ W m$^{-2}$ and, by Equation 10.16, a *maximum* stress of $10^8$ Pa ($10^3$ bar) and a corresponding maximum viscosity of $3 \times 10^{26}$ decapoise. These are the maximum possible values consistent with lower mantle participation in convection. It is evident that they represent a serious overestimate because if lower mantle viscosity were so high the excess ellipticity of the Earth due to delayed response of the lower mantle to slowing rotation would be much greater than is observed. Accepting substantially lower values of stress and viscosity, we are compelled to conclude that the available power suffices to make the lower mantle convect and therefore that it must be convecting, if very slowly, in accord with the conclusion of Andrews (1975).

The foregoing argument strongly supports the hypothesis that global tectonics are driven by thermal convection of the mantle. This hypothesis originated many years ago (Pekeris, 1935) but suffered prolonged neglect before becoming widely accepted. The energy budget approach can be used to examine another possibility, which has also been neglected, that release of gravitational energy by progressive differentiation of the mantle to form the crust provides an additional power source for the tectonic engine. If we suppose that the crust, volume $V = 10^{19}$ m$^3$, having a density contrast $\Delta\rho$ of 500 kg m$^{-3}$ with respect to the mantle, has differentiated from the top $d = 1000$ km of the mantle in $\tau = 4.5 \times 10^9$ years, then the average energy release is

$$\frac{dE}{dt} = \frac{V\Delta\rho g d}{2\tau} = 1.7 \times 10^{11}\ \text{W} \tag{10.19}$$

This is clearly significant, being about 50% of the power available from convection due to self-heating of the upper mantle. The present net rate of production of new crust may be less than that allowed for here, but we are justified in supposing that the word *convection* applied to tectonics does not mean pure thermal convection, but partly a process of differentiation.

We may now use the stress estimates to determine the magnitude of probable density and temperature differences between upwelling and downgoing convective limbs. The boundaries of a vertical layer of thickness $d$, having a density difference $\Delta\rho$ with respect to its surroundings, perhaps due to a temperature difference $\Delta T$, are subjected to a stress

$$\sigma = \tfrac{1}{2}g\, d\,\Delta\rho = \tfrac{1}{2}g\, d\rho\alpha\,\Delta T \tag{10.20}$$

where $\alpha$ is volume expansion coefficient and $g$ is gravity. Stresses of 1 to $3 \times 10^6$ Pa (10 to 30 bars) are suggested by the energy balance argument and taking $\alpha$ for mantle material to be $1.5 \times 10^{-5}\,\text{deg}^{-1}$, we find for a slab of thickness 70 km $\Delta\rho = 0.3$ to $0.9\,\text{kg m}^{-3}$, $\Delta T = 60°$ to 180°C. Calculations of thermal diffusion into a downgoing lithospheric slab (e.g., McKenzie, 1970; Minear and Toksöz, 1970) suggest that these figures err on the low side and that stresses of order $10^7$ Pa (100 bars) must be contemplated. By the energy balance argument, this is permissible in fast moving (8 cm/year) material only over a limited part of the convective cycle, but it is reasonable that the cool plunging slab should be subject to stresses greater than the average for material in the convective cycle.

The estimation of temperatures and stresses in a downgoing slab is complicated by the phase changes that occur at shallower depth within the slab than in surrounding material by virtue of its lower temperature. The maximum permissible elevation of the boundary may be estimated from the stress which it causes. Accepting an average $10^7$ Pa (100 bars) stress over a slab of depth 700 km and effective thickness 50 km (reduced from its thickness at the surface by reheating), caused by an elevated phase boundary with a density contrast of $200\,\text{kg m}^{-3}$, we find the elevation to be about 140 km, in accord with the elevation suggested by Turcotte and Schubert (1971) and with the magnitude of the global scale gravity anomalies (Fig. 4.3). However, this is probably an upper limit as the permitted maximum stress is the result of the general thermal contraction of a slab as well as the density increment resulting from the phase change.

Verhoogen (1965) considered the possible effects of the mantle phase transitions on convection and concluded that whole mantle convection was not necessarily inhibited. From the point of view of the thermodynamic argument of Section 7.4 the problem is quite a simple one. If a phase transition is perfectly reversible, that is it occurs in both directions at the same pressure and temperature conditions, which we must expect to be the case at the high temperatures of the mantle, then although the form of the convection may be affected, the efficiency of the convective engine is still given simply in terms of adiabatic temperature differences. This is because the mechanical energy contributed or absorbed by the phase change is matched by the temperature change associated with an adiabatic transition and so correspondingly modifies the adiabatic temperature differences over the convective column.

The dynamics, heat transfer, and thermal contraction of the lithosphere as it spreads from an ocean ridge have been considered by numerous authors, including Sclater and Francheteau (1970), Parker and Oldenburg (1973) and Davis and Lister (1974). The essence of their conclusion is that the ridge crest is elevated by virtue of being hot and therefore thermally expanded and that as the lithosphere loses heat by diffusion to the surface (and so thickens—Leeds et al., 1974), it contracts, deepening the ocean away from the ridges. Apart from a complication close to the ridge itself, the ocean bottom profile conforms well to the simple relationship that follows intuitively from the appeal to thermal diffusion:

$$\Delta h \propto t^{1/2}$$

where $t$ is the age of the ocean floor (taken to be proportional to distance from the ridge) and $\Delta h$ is water depth relative to that over the ridge. However, for quantitative agreement with the thermal expansion argument these authors consider thermal expansion coefficients much higher than are characteristic of igneous rock, especially when it is under pressure. Thus we conclude that elevation of ocean ridges is only partly explained by lithospheric cooling and that it is also an indication of the convective driving force away from the ridges. This conclusion accords with the seismic evidence that the tectonic plates are under compression (Sykes and Sbar, 1973).

## 10.4 DAMPING OF OSCILLATIONS AND ATTENUATION OF SEISMIC WAVES

Vibrating solids dissipate mechanical energy by a number of processes, known collectively as *internal friction*. The standard laboratory techniques for measuring internal friction are to observe the damping of oscillatory motions that apply cyclic stresses to a solid under investigation, commonly a resonant oscillation of the solid itself, or to measure the attenuation of acoustic or ultrasonic waves. For the Earth the equivalent observations are damping of free oscillations and seismic waves. We also consider the extent to which the mantle may be responsible for tidal friction and damping of the Chandler wobble. There are numerous mechanisms of internal friction and we have no certain knowledge which of them is most important in the Earth, but for most purposes it suffices to describe damping in terms of empirical attenuation factors.

Attenuation of a sinusoidal oscillation or wave is represented by a parameter $Q$, defined in terms of the fractional loss of energy per cycle, $-\Delta E/E$:

$$\frac{2\pi}{Q} = -\frac{\Delta E}{E} \tag{10.21}$$

Equation 10.21 is only meaningful if $(-\Delta E/E) \ll 1$, but is readily generalized to include arbitrarily small $Q$ by representation in differential form:

$$-\frac{2\pi}{Q} = \frac{dE}{E} \cdot \frac{\tau}{dt} = \frac{\tau}{E}\frac{dE}{dt} \tag{10.22}$$

where $\tau$ is the period of the wave or oscillation. This integrates to give the exponential decay of $E$

$$E = E_0 \exp\left(-\frac{2\pi}{Q}\frac{t}{\tau}\right) \tag{10.23}$$

and, since energy is proportional to the square of wave amplitude, $A$:

$$A = A_0 \exp\left\{-\frac{\pi}{Q}\frac{t}{\tau}\right\} = A_0 \exp\left\{-\frac{\omega t}{2Q}\right\} \tag{10.24}$$

where $\omega/2\pi$ is the frequency. In Equations 10.23, 10.24, $t$ is the time since an oscillation had amplitude $A_0$ or energy $E_0$, but we are also interested in the case of a continuous travelling wave to which these equations apply if we neglect the diminution in amplitude by geometrical spreading, $t$ being then the time of travel of the wave from its source or from a reference point where the amplitude is $A_0$. Expressed in terms of distance of travel, $x$, Equation 10.24 is

$$A = A_0 \exp\left\{-\frac{\pi}{Q}\frac{x}{\lambda}\right\} = A_0 \exp(-\alpha x) \tag{10.25}$$

where $\lambda$ is the wavelength and $\alpha$ is referred to as the attenuation coefficient. To allow for geometrical spreading of a spherical wavefront of radius $r$, a factor $(r_0/r)^2$ should be included in Equation 10.23 and a factor $(r_0/r)$ in Equations 10.24 and 10.25.

$Q$ may also be defined in terms of the spectrum of the forced vibration of a geometrically simple body of resonant frequency $f_0$. If a plot of amplitude as a function of frequency for constant excitation gives a spectral line of width $\Delta f$ at the "half power" points, that is, at frequencies $f_0 \pm \frac{1}{2}\Delta f$ the amplitude is reduced to $1/\sqrt{2}$ of the resonant value, then the body has a $Q$ given by

$$Q = \frac{f_0}{\Delta f} \tag{10.26}$$

Equations 10.21 and 10.26 are equivalent in the sense that one follows from the other. The mathematics is identical to the problem of the series L-C-R circuit in electromagnetism, in which the symbol $Q$ (quality factor) originated. Equation 10.26 is of particular interest in connection with the Chandler wobble (Section 3.3) which may be excited by random impulses and therefore in effect by a source with a "white" spectrum with all frequencies equally represented. If this is so then the observed wobble spectrum, which by Equation 10.26 has a spread corresponding to $Q_W \approx 72$ or 600 (according to different estimates in Section 3.4), reflects the response of the Earth to this excitation and therefore indicates quite heavy damping of the wobble according to Equation 10.21. The nature of this damping has been problematical. It is suggested later in this section that the oceans may suffice.

The values of $Q$ for compressional waves, $Q_P$, are consistently higher than for shear waves, $Q_S$, and it appears possible that attenuation is due virtually entirely to the shear component of strain, even in compressional waves. Thus values of $Q$ for fluids, including sea water, are exceedingly high, reflecting the fact that the stresses associated with wave motion in fluids are purely compressional; similarly $Q$'s for the purely compressional radial modes of free oscillation of the Earth are also very high* (Slichter et al., 1966). We therefore expect the ratio $Q_P/Q_S$ to be close to the ratio of total strain energy to shear strain energy in a compressional wave

$$\frac{Q_P}{Q_S} \approx \frac{K + \frac{4}{3}\mu}{\frac{4}{3}\mu} = \frac{3}{2}\frac{(1-\nu)}{(1-2\nu)} \approx 2.4 \tag{10.27}$$

$\nu \approx 0.27$ being Poisson's ratio for the upper mantle, where anelastic losses are greatest. Reported observations have generally given somewhat smaller ratios. Kanamori (1967) estimated $Q_P/Q_S = 1.90$ for the mantle, with $\bar{Q}_S = 230$, averaged over the whole mantle. Sacks (1971) quoted values of $Q_P$, $Q_S$ for different paths giving an average $Q_P/Q_S =$ 2.15. The attenuation coefficients $\alpha_P$ and $\alpha_S$ of $P$ and $S$ waves of the same frequency (defined by Eq. 10.25) differ by a larger factor, because $S$ waves, being slower, have shorter wavelengths and therefore more

*Some damping must occur in inhomogeneous or polycrystalline material under hydrostatic compression, by virtue of stresses between grains of different elasticities.

stress cycles in any given wave path. Thus

$$\frac{\alpha_S}{\alpha_P} = \frac{Q_P}{Q_S} \cdot \frac{V_P}{V_S} \approx 3.6 \tag{10.28}$$

Estimates of $Q$ for the interior of the Earth appeal to the assumption that it is not intrinsically dependent on frequency. Frequency independence of $Q$ in rocks is well supported by observations (Knopoff, 1964a; Gordon and Davis, 1968) although a study of the mechanisms of internal friction (Jackson and Anderson, 1970) suggests that they are all basically relaxation phenomena, which give frequency dependent $Q$. The implication is that rocks contain components giving such a wide range of relaxation times that frequency dependence is smeared out and Mason (1969) suggested that frequency independence indicated damping by dislocation movements. For laboratory samples of rock and at least in the shallower levels of the Earth's crust $Q$ is quite low (usually 50 to 300) and is evidently dominated by "friction" at grain boundaries. Pandit and Tozer (1970) and Tittman et al. (1973) found that rocks subjected to high vacuum which removed volatiles, especially water, from internal grain surfaces gave much higher $Q$'s, in the range 1000 to 2000, which are comparable to values for individual mineral crystals. For this reason the surface layers of the Moon have very high $Q$. Water is important in the surface layers of the Earth and probably a fluid component due to partial melting has a similar effect in the upper mantle, but in the deep mantle grain boundary friction is probably less significant and consequently $Q$ is higher. But this means that, since the internal friction mechanisms in the lower mantle may not be the same as those in the upper mantle, the presently available evidence does not allow us to regard the extrapolation of frequency independence of $Q$ to the deep interior as a secure assumption.

Another assumption implicit in most determinations of $Q$ is that the dissipation of acoustic or seismic energy is linear in the sense that it can be calculated in terms of independent dissipations of the Fourier components of a complex waveform or pulse. This is a questionable assumption (Stacey et al., 1975) but is the best available until we have an adequate fundamental theory of attenuation of waves in terms of the stress-strain behavior of rocks (Fig. 10.19).

The broad picture of $Q$ in the mantle and core has been obtained by two independent methods, using body waves and free oscillations, which have different frequency ranges and therefore provide some test of the frequency independence assumption. By applying Equation 10.24 to two (or more) different frequency components in a single-wave arrival (e.g., PKP) a spectral ratio is obtained in terms of the source spectrum and the value of $\int Q^{-1}\,dt$ for the wave path. Comparison of the

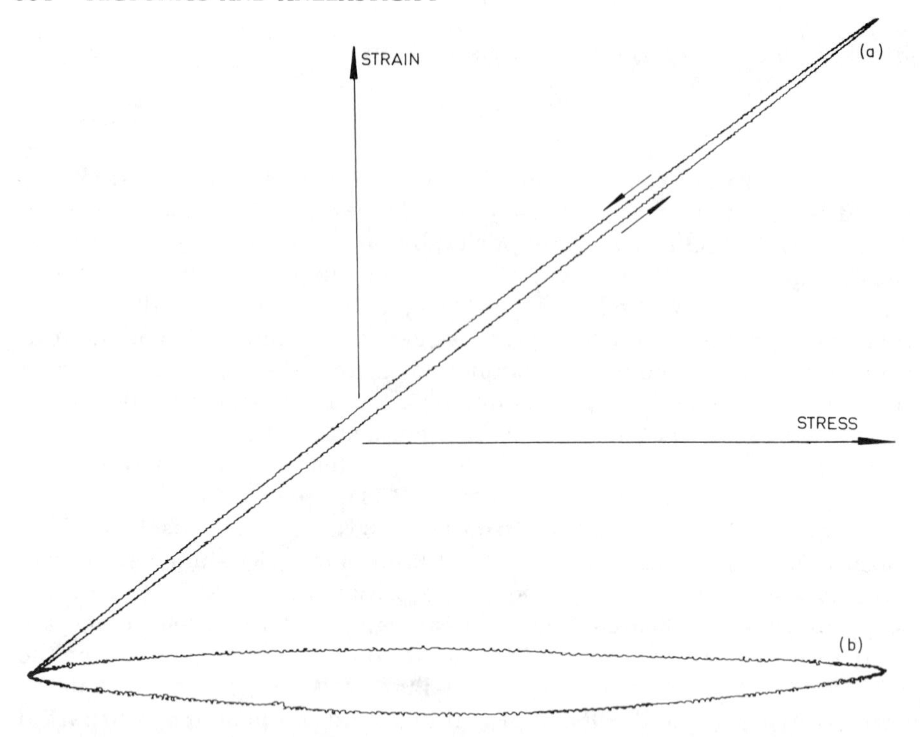


**Figure 10.19.** Stress-strain hysteresis of a sandstone sample sinusoidally loaded to a strain amplitude of $3.5 \times 10^{-5}$ (*a*). The loop area represents the energy loss per cycle. Also displayed is an expanded form of the loop shape (*b*). After McKavanagh and Stacey (1974).

spectra of two arrivals then relates the $Q$'s for the two paths without appeal to the source spectrum. The two arrivals may be records of the same earthquake at different stations or of different waves, for example, *P* and *PP* at the same station, but the method is particularly powerful when used with two wave paths that are very similar except for the addition of a particular component to one of them, for example, *PKP* and *PKIKP*. This allows the $Q$ for that component of the path to be determined. A variant of the spectral ratio method, which has been used on a small scale, is the pulse rise time or pulse width method. The rise time of an acoustic pulse (Fig. 10.20) increases with travel time in rock according to the empirical relationship (Gladwin and Stacey, 1974b).

$$\tau - \tau_0 = 0.5 \int_0^t Q^{-1}\, dt \qquad (10.29)$$

Although this is an experimental relationship it is consistent with the

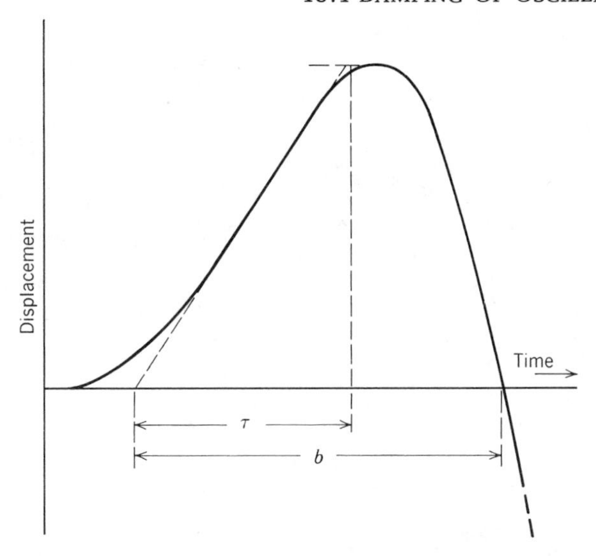


**Figure 10.20.** Definitions of pulse rise time $\tau$ and width $b$ for the first arrival of an acoustic or seismic pulse. Both increase linearly with the time of flight of a pulse through rock and with $Q^{-1}$ for the path.

hypothesis of linearity of the attenuation mechanism if appropriate velocity dispersion and frequency-dependent $Q$ are assumed (see Stacey et al., 1975).

Free oscillations decay with different $Q$'s because the stresses are differently distributed. The $Q$ for each mode is a differently weighted harmonic mean of the $Q$'s of rocks within the Earth. Thus comparison of the decays of the different modes of free oscillation allows the $Q$ profile of the Earth to be deduced (Anderson and Archambeau, 1964). The method cannot discern fine structure in the deep interior, which is only penetrated significantly by the lowest order modes, but the general agreement of the free oscillation and body wave data is important. An approximate $Q$ profile of the Earth is given in Figure 10.21.

The contribution of the mantle to tidal friction may be estimated from the lunar tidal strain amplitude, $\epsilon_T$, and the mean effective $Q$. Assuming a value for the tidal Love number, $k_2 = 0.27$, we have

$$\epsilon_T = \frac{k_2}{(C/Ma^2)}\frac{m}{M}\left(\frac{a}{R}\right)^3 = 4.65 \times 10^{-8} \qquad (10.30)$$

where $M$, $a$, $C$ are the mass, radius, and moment of inertia of the Earth and $m$, $R$ are the mass and distance of the Moon. This is one of the two mutually perpendicular shear strains. Thus the strain energy is

$$E_T = 2.\tfrac{1}{2}\bar{\mu}\,\epsilon_T^2 V = 3.4 \times 10^{17}\,\text{J} \qquad (10.31)$$

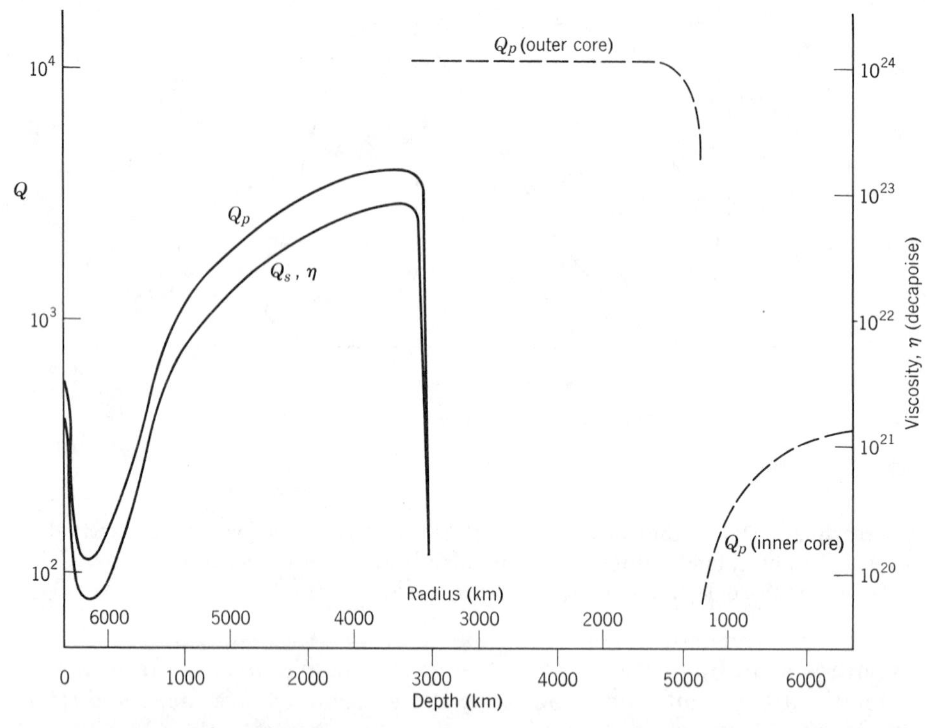


**Figure 10.21.** Profile of the anelastic parameter $Q$ within the Earth. Although $Q$ is not uniquely or necessarily related to viscosity or creep strength, the $Q$ profile coincides with an appropriately normalized viscosity profile within the uncertainties in these parameters for the mantle. $Q$ values for the core have been given by Sacks (1970a,b) and Qamar and Eisenberg (1974). Viscosity of the outer core is negligible (see, for example, Gans, 1972).

where $\bar{\mu} = 1.73 \times 10^{11}$ Pa is the mean rigidity of the mantle and $V = 9.06 \times 10^{20}$ m$^3$ is its volume. Taking $\bar{Q}_S = 230$ and the lunar tidal period $\tau_T = 4.46 \times 10^4$ sec (12 hours, 39 minutes) the rate of tidal energy dissipation by internal friction in the mantle is by Equation 10.22

$$-\frac{dE_T}{dt} = \frac{2\pi}{Q_S}\frac{E_T}{\tau_T} = 2.1 \times 10^{11}\ \mathrm{W} \qquad (10.32)$$

which is only 4% of the total tidal friction (Section 4.5), the balance being attributed to the seas.

A similar argument applies to the Chandler wobble for which the elastic strain energy is given by Equation 3.47. In effect there are two alternating elastic strains of this amplitude $\tau_W/4$ out of phase in time and mutually perpendicular, so that the anelastic dissipation of the total wobble energy, $E_W$ (Eq. 3.48), is

$$\left(-\frac{dE_W}{dt}\right)_{\text{mantle}} = \frac{2\pi}{Q_S} \cdot \frac{2E_{el}}{\tau_W} = 1.1 \times 10^{17}\,\alpha^2\,\text{W} \qquad (10.33)$$

where $\tau_W = 3.72 \times 10^7$ sec (431 days) is the wobble period. But the total rate of energy loss by wobble is

$$-\frac{dE_W}{dt} = \frac{2\pi}{Q_W} \cdot \frac{E_W}{\tau_W} = 1.6 \times 10^{18}\,\alpha^2\,\text{W} \quad \text{or} \quad 2 \times 10^{17}\,\alpha^2\,\text{W} \qquad (10.34)$$

according to the two estimates of wobble $Q$ ($Q_W = 72$ or 600) in Section 3.4. If the second figure is accepted the mantle appears close to adequate as a sink of wobble energy. But the wobble is also associated with a marine tide (the pole tide) and if this is more effective in damping the wobble by the same factor ($\sim 25$) as in the case of lunar tidal friction then dissipation by the pole tide suffices to explain the first of the alternative estimates in Equation 10.34. This is a very simple-minded approach to tidal damping, but clearly the role of the pole tide will require closer scrutiny (Wunsch, 1974).

# APPENDIX A

# Orbital Dynamics (Kepler's Laws) and Planetary Parameters

Although for most purposes in this text planetary and satellite orbits are assumed to be circular as a sufficient approximation, it has been known since the analysis by Johannes Kepler (1571–1630) of planetary observations by Tycho Brahe, that the orbits are ellipses. Apart from minor planets (and comets) the orbital ellipticities, $e$, are greatest for Mercury (0.2056) and Pluto (0.250). Kepler summarized his conclusions in three empirical laws:

1. The orbit of each planet is an ellipse with the Sun at one focus.
2. The line between a planet and the Sun sweeps our equal areas in equal times.
3. The square of the orbital period of a planet is proportional to the cube of its mean distance from the Sun (more correctly the semimajor axis of its orbit).

As Isaac Newton showed, these laws are consequences of the inverse square law of gravitational attraction; the second is simply a statement of conservation of angular momentum.

We should first note that a planetary orbit must lie in a plane because the gravitational force is central, that is, along the planet-Sun line, $\mathbf{r}$. This line and the instantaneous planetary velocity, $\mathbf{v}$, define the plane of the orbit since there is no initial velocity component and no force on the planet perpendicular to this plane and so the planet cannot leave it. It is, therefore, most convenient to represent the path of the planet in plane $(r, \theta)$ coordinates with its origin at the center of mass of the planet-Sun system (which must remain stationary or at least move with constant velocity, which can be subtracted). But the displacements of the Sun and planet from the center of mass are necessarily opposite and in a fixed ratio, being inversely proportional to their masses, so that whatever path the planet follows, the Sun follows its converse, appropriately diminished in size. Thus it suffices to consider the case of a planet of mass $m$ so slight compared with that of the Sun that we can assume the Sun to be stationary at the coordinate center.

The angular momentum of the planet about the Sun is:

$$L = mr^2\omega = mr^2\frac{d\theta}{dt} = 2m\frac{dS}{dt} \tag{A.1}$$

where $\omega$ is angular velocity about the Sun, $\theta$ is the angle of $\mathbf{r}$ with respect to a fixed line in the orbital plane, and $S$ is the area swept out by $\mathbf{r}$. Thus

$$\tfrac{1}{2}r^2\frac{d\theta}{dt} = \frac{dS}{dt} = \frac{L}{2m} = \text{constant} \tag{A.2}$$

which is Kepler's second law.

The total energy, being the sum of planetary kinetic energy and gravitational potential energy, is also conserved:

$$E = \tfrac{1}{2}mv^2 - \frac{GMm}{r} = \text{constant} \tag{A.3}$$

where $M$ is the solar mass and $v$ is planetary velocity, which may be written in terms of its radial and circumferential components:

$$\begin{aligned} v^2 &= \left(\frac{dr}{dt}\right)^2 + \left(r\frac{d\theta}{dt}\right)^2 = \left(\frac{dr}{d\theta}\cdot\frac{d\theta}{dt}\right)^2 + r^2\left(\frac{d\theta}{dt}\right)^2 \\ &= \left(\frac{d\theta}{dt}\right)^2\left[\left(\frac{dr}{d\theta}\right)^2 + r^2\right] \end{aligned} \tag{A.4}$$

Substituting for $(d\theta/dt)$ from Equation A.2 and then using $v^2$ in Equation A.3 and rearranging, we obtain the differential equation for $r(\theta)$

$$\frac{dr}{d\theta} = r\left(\frac{2mE}{L^2}r^2 + \frac{2GMm^2}{L^2}r - 1\right)^{1/2} \tag{A.5}$$

As may be verified by differentiation and substitution, the solution is the equation of an ellipse with one focus at the origin (the Sun):

$$r = \frac{p}{1 + e\cos(\theta + C)} \tag{A.6}$$

where

$$p = \frac{L^2}{GMm^2} \tag{A.7}$$

and

$$(e^2 - 1) = \frac{2EL^2}{G^2M^2m^3} \tag{A.8}$$

and $C$ is the constant of integration, being the phase angle of the orbit, and may be chosen to be zero if $r = r_{\min} = p/(1+e)$ at $\theta = 0$. This demonstrates Kepler's first law.

We may see what happens physically by comparing Equations A.3 and A.8. If $E < 0$ then, by Equation 3, the kinetic energy does not suffice to overcome the gravitational potential energy and the planet is held in orbit. By Equation A.8

$e^2 < 1$ and the orbit is elliptical. If $E = 0$, the planet can just escape, $e^2 = 1$ and the orbit is parabolic. If $E > 0$ the planet escapes on a hyperbolic orbit ($e^2 > 1$).

The closest approach distance (perigee), at $\theta = 0$, is

$$r_p = \frac{p}{1+e} \tag{A.9}$$

and the most remote point of the orbit (apogee), at $\theta = 180°$, is

$$r_a = \frac{p}{1-e} \tag{A.10}$$

so that the semimajor axis of the ellipse is

$$a = \tfrac{1}{2}(r_a + r_p) = \frac{p}{1-e^2} \tag{A.11}$$

Combining Equations A.7, A.8, and A.11 we obtain the simple result

$$E = -\frac{GMm}{2a} \tag{A.12}$$

that is, the total energy depends only on the semimajor axis of the orbit, independently of its eccentricity. Then combining Equations A.12 and A.3, we obtain the vis-viva equation for orbital velocity in terms of radius:

$$v^2 = GM\left(\frac{2}{r} - \frac{1}{a}\right) \tag{A.13}$$

Kepler's third law is obtained by integrating the second law (Eq. A.2):

$$S = \frac{L}{2m}T \tag{A.14}$$

$T$ being the period of the orbit and $S$ is its area:

$$S = \int_0^{2\pi} \tfrac{1}{2}r^2\, d\theta = \frac{p^2}{2}\int_0^{2\pi} \frac{d\theta}{(1+e\cos\theta)^2} = \frac{\pi p^2}{(1-e^2)^{3/2}} \tag{A.15}$$

which simplifies by Equation A.11 to give

$$S = \pi p^{1/2} a^{3/2} \tag{A.16}$$

Then substituting for $L/m$ from Equation A.7 and $S$ from A.16 we have Kepler's third law

$$T^2 = \frac{4\pi^2}{GM}a^3 \tag{A.17}$$

It is also useful to keep in mind the relationship for the semiminor axis, $b$, of an ellipse in terms of the semimajor axis or of $p$*

$$b^2 = a^2(1-e^2) = \frac{p^2}{(1-e^2)} = ap \tag{A.18}$$

*See also footnote on p. 54.

**Table A.1** Planetary Parameters.

| $n$ | Planet | Orbit Radius[a] (A.U.) $(R_n/R_3)$ | $\frac{R_n}{R_{n-1}}$ | Mass (in Earth Masses) $m_n/m_3$ | Mean Radius[b] (Earth Radii) | Mean Density ($10^3$ kg m$^{-3}$) | Estimated Density at Zero Pressure ($10^3$ kg m$^{-3}$) | Number of Known Satellites |
|---|---|---|---|---|---|---|---|---|
| 1 | Mercury | 0.387 | — | 0.05527 | 0.3828 | 5.44 | 5.3 | 0 |
| 2 | Venus | 0.723 | 1.86 | 0.8150 | 0.9500 | 5.24 | 3.9 | 0 |
| 3 | Earth | 1.000 | 1.38 | 1.00000 | 1.0000 | 5.515 | 4.04 | 1 |
| | Moon | | | 0.01230 | 0.2728 | 3.341 | 3.3 | |
| | Earth + Moon | | | 1.01230 | | | 4.03 | |
| 4 | Mars | 1.524 | 1.52 | 0.1074 | 0.5321 | 3.93 | 3.7 | 2 |
| (5) | Asteroids | Mean about 2.7 | (1.77) | — | | 3.9[c] | 3.9 | |
| 6 | Jupiter | 5.203 | (1.92) | 317.89 | 11.20 | 1.25 | Largely Gaseous (Jupiter–Neptune) | 12 |
| 7 | Saturn | 9.539 | 1.83 | 95.18 | 9.06 | $0.70_6$ | | 10 |
| 8 | Uranus | 19.18 | 2.00 | 14.60 | 3.95 | 1.31 | | 5 |
| 9 | Neptune | 30.06 | 1.56 | 17.23 | 3.78 | 1.76 | | 2 |
| | Pluto | ~40 (eccentric orbit) | | ~0.09 | ~0.5 | ~4 | ~4 | 0 |

NOTE. Recent data are summarized by J. D. Anderson (1974), with updating for Mercury by Howard et al. (1974). Lunar data are reviewed by Kaula (1969a).

[a]Semimajor axis of orbital ellipse.

[b]Radius of a sphere of equal volume.

[c]Average for total mass of meteorite falls.

# APPENDIX B

# The Roche Limit for Gravitational Stability of the Moon

The tide raised in one astronomical body by another is the result of a balance between the gravitational gradient across it, due to the second body, and its own self-gravitation. If the attracting body has mass $M$ and its distance $R$ from the one in which we are considering the tide is large, then the tidal bulge takes the form of a prolate ellipsoid with its long axis directed toward the attracting body and having an elongation varying approximately as $MR^{-3}$. Relevant equations are given in Section 4.4, with special reference to the tides raised in the Earth by the Moon and Sun. Interest here is in the much greater elongations that may occur in bodies that are relatively much closer and, in particular, the elongation of a close satellite in the gravitational field of a planet substantially more massive than itself. If it comes closer than a certain critical distance, known as the Roche limit, after E. Roche whose 1850 paper first studied the problem in detail, the satellite has no equilibrium ellipticity; its self-gravitation cannot hold it together against the disrupting force of the planet. The Roche limit for the Moon with respect to the Earth imposes a boundary condition on the origin of the Moon, that is, if it had ever come within the Roche limit it would have broken up.

The geometrical situation is represented in Figure B.1, in which the Moon is a prolate ellipsoid of mass $m$, with its longer ($c$) axis directed toward the Earth, mass $M$. Since $M \approx 80\,m$, the tidal distortion of the Earth is slight compared with the distortion of the Moon. We therefore approximate the gravitational field of the Earth at the Moon to that of a sphere without significant error. The Moon is highly elliptical and for points close to it, in particular on its surface, the self-gravitational potential cannot be adequately approximated by the equations of Section 3.1, which refer to ellipsoids of small ellipticity. The general equation for the potential of a uniform (solid) prolate ellipsoid is given in convenient form by MacMillan (1958, p. 63). Referred to axes, $x$, $y$, $z$ with the $c$ (long) axis of the

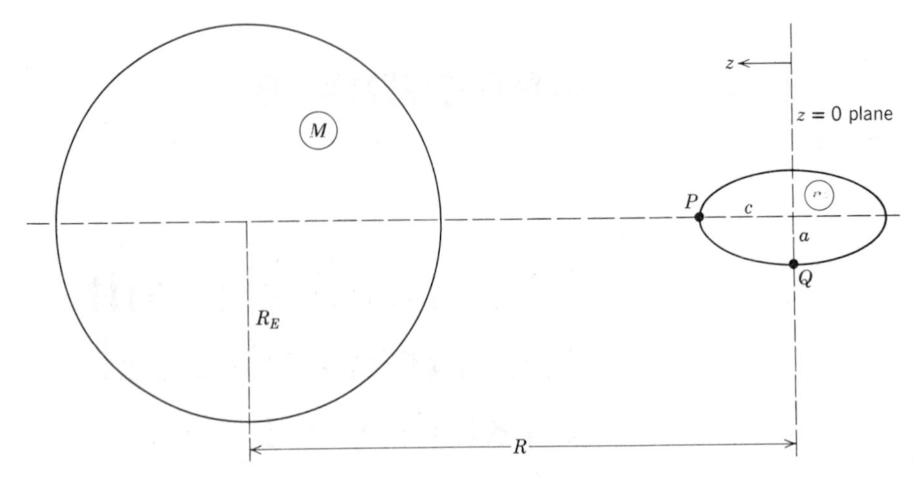


**Figure B.1.** Geometry of the calculation of the critical distance $R$ between a planet and its satellite for gravitational stability of the satellite.

ellipsoid aligned with $z$ the potential is, in the present notation,

$$V = -G\pi\rho a^2 c\left(1 + \frac{x^2 + y^2 - 2z^2}{2(c^2 - a^2)}\right)\frac{2}{\sqrt{c^2 - a^2}}\sinh^{-1}\sqrt{\frac{c^2 - a^2}{a^2 + \kappa}}$$

$$+ \frac{G\pi\rho a^2 c\sqrt{c^2 + \kappa}}{c^2 - a^2}\cdot\frac{x^2 + y^2}{a^2 + \kappa} - \frac{G\pi\rho a^2 c}{c^2 - a^2}\cdot\frac{2z^2}{\sqrt{x^2 + \kappa}} \tag{B.1}$$

where $\kappa$ satisfies the equation

$$\frac{x^2 + y^2}{a^2 + \kappa} + \frac{z^2}{c^2 + \kappa} = 1$$

We are interested in the potentials at two particular points, $P(x = y = 0,\ z = c)$ and $Q(x^2 + y^2 = a^2;\ z = 0)$ in Figure B.1, at both of which $\kappa = 0$ and Equation B.1 reduces to

$$V_P = G\cdot\frac{2\pi\rho a^4 c}{(c^2 - a^2)^{3/2}}\sinh^{-1}\left(\frac{c^2 - a^2}{a^2}\right)^{1/2} - G\cdot\frac{2\pi\rho a^2 c^2}{c^2 - a^2} \tag{B.2}$$

$$V_Q = -G\cdot\frac{\pi\rho a^2 c(2c^2 - a^2)}{(c^2 - a^2)^{3/2}}\sinh^{-1}\left(\frac{c^2 - a^2}{a^2}\right)^{1/2} + G\cdot\frac{\pi\rho a^2 c^2}{c^2 - a^2} \tag{B.3}$$

The tidal ellipticity is determined by equating the total potentials at $P$ and $Q$; thus to each of $V_P$ and $V_Q$ we add the gravitational potentials due to the Earth ($M$) and rotational potentials of the two points about the center of mass of the Earth-Moon system, assuming the Moon to keep a constant face toward the Earth, that is, the axial rotation rate of the Moon is equal to its orbital rotation rate,

$$V_P - \frac{GM}{R - c} - \tfrac{1}{2}\omega^2\left(\frac{M}{M + m}R - c\right)^2 = V_Q - \frac{GM}{R} - \tfrac{1}{2}\omega^2\left(\frac{M}{M + m}R\right)^2 \tag{B.4}$$

where $\omega$ is given by Equation 4.41

$$\omega^2 = \frac{G(M+m)}{R^3}$$

Equation B.4 simplifies to

$$V_P - V_Q = \frac{3}{2}\frac{GMc^2}{R^3}\left[\frac{R\left(1+\frac{1}{3}\frac{m}{M}\right) - \frac{1}{3}c\left(1+\frac{m}{M}\right)}{R-c}\right] = \frac{3}{2}\frac{GM^*c^2}{R^3} \tag{B.5}$$

where we may regard

$$M^* = M\left[\frac{\left(1+\frac{1}{3}\frac{m}{M}\right) - \frac{1}{3}\frac{c}{R}\left(1+\frac{m}{M}\right)}{1-\frac{c}{R}}\right] \tag{B.6}$$

as an "effective" mass of the Earth. $M^*$ and $M$ differ only to the extent that the mass of the Moon is not negligible with respect to that of the Earth and its semiaxis $c$ with respect to the Earth-Moon distance, $R$.

From Equations B.2, B.3 we also have

$$V_P - V_Q = \frac{G\pi\rho a^2 c}{(c^2-a^2)^{3/2}}(2c^2+a^2)\sinh^{-1}\left(\frac{c^2-a^2}{a^2}\right)^{1/2} - \frac{G\,.\,3\pi\rho a^2c^2}{c^2-a^2} \tag{B.7}$$

which may therefore be equated to Equation B.5. For this purpose it is convenient to represent the dimensions of the Moon in terms of the radius of a sphere of equal volume, $r_0 = (a^2c)^{1/3}$ and the ellipticity $e = \sqrt{(c^2-a^2)/a^2}$, so that

$$c = r_0(1+e^2)^{1/3} \qquad a = r_0(1+e^2)^{-1/6}$$

Then equating B.5 and B.7 gives

$$\frac{3}{2\pi\rho}\cdot\frac{M^*}{R^3} = 2\frac{r_0^3}{m}\cdot\frac{M^*}{R^3} = \frac{1}{e^3}\left[\frac{3+2e^2}{(1+e^2)^{1/2}}\sinh^{-1}e - 3e\right] \tag{B.8}$$

[It is of interest at this point to note that expressing Equation B.7 in terms of $r_0$, $e$ and expanding for small values of $e$, we obtain

$$V_P - V_Q = \frac{4\pi}{15}G\rho r_0^2 e^2 = \frac{1}{5}\frac{Gm}{r_0}e^2$$

which coincides with the result obtained from the approximation of Equation 3.15 with the substitutions

$$C = 0.4\,ma^2 \qquad A = 0.2\,m(c^2+a^2)$$

appropriate to a uniform ellipsoid.]

Now at the Roche limit the ellipticity is at the point of instability, that is, a small decrement in $R$ causes the moon to break up or, in other words, its ellipticity grows indefinitely. Thus we can determine the Roche limit from the condition $de/dR \to -\infty$ or $dR/de \to 0$ applied to Equation B.8, which means that the differential with respect to $e$ of the right-hand side of B.8 is equated to zero.

This gives

$$(4e^4 + 14e^2 + 9)\sinh^{-1} e = (9e + 8e^3)(1 + e^2)^{1/2} \tag{B.9}$$

the numerical solution of which is $e = 1.676_{554}$. Substituting this value into B.8 gives

$$\frac{r_0^3}{m} \cdot \frac{M^*}{R^3} = 0.07031 \tag{B.10}$$

or

$$R = 2.42_3 \left(\frac{M^*}{m}\right)^{1/3} r_0 = 2.42 \left(\frac{M^*}{M}\right)^{1/3} \left(\frac{\rho_E}{\rho}\right)^{1/3} R_E \tag{B.11}$$

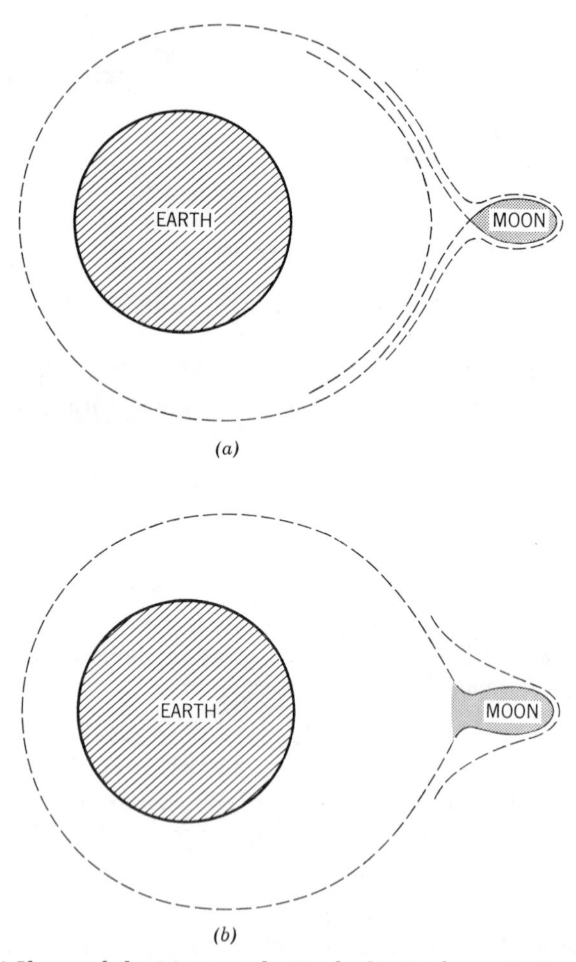


**Figure B.2.** (*a*) Shape of the Moon at the Roche limit of gravitational stability. (*b*) Inside the Roche limit there is no bounding equipotential surface and the Moon breaks up. For even closer approach the equipotential surface opens on the remote side of the Moon also. (Broken lines represent equipotential surfaces.)

where $\rho_E$, $R_E$ are the density and radius of the Earth.

Ignoring for the moment the factor $(M^*/M)^{1/3}$, we obtain, with $\rho_E = 5517$ kg m$^{-3}$, $\rho = 3340$ kg m$^{-3}$, the result $R = 2.86_4 R_E$. This allows an estimate of the term

$$\frac{c}{R} = \frac{c}{r_0} \cdot \frac{r_0}{R_E} \cdot \frac{R_E}{R} = 0.1435$$

which is the principal contribution to the "correction" factor $\left(\frac{M^*}{M}\right)^{1/3} = 1.037$, so that the "final" result is

$$R = 2.97 R_E \tag{B.12}$$

The correction for the finite mass and size of the Moon does not allow for the ellipticity of the Earth by virtue of the Moon's gravity, but the ellipticity of the Earth is less than 1% of that of the Moon and even at $3R_E$ its effect on the gravity field of the Earth is quite negligible. We can also readily admit as satisfactory the assumption that the Moon is homogeneous; the moment of inertia of the Moon is sufficiently close to that of a body of uniform density that this cannot be a significant error. More difficult to assess is the assumption that the Moon remains ellipsoidal near to the Roche limit. Since at the Roche limit the point $P$ in Figure B.1 is a neutral point in the field, the potential lines must cross there, so that although Figure B.1 properly represents the deformation of the Moon outside the Roche limit, at the limit itself the Moon must take the form of Figure B.2*a*. With further decrease in separation the potential surface bounding the Moon opens, allowing material to escape, as in Figure B.2*b*. However the error in assuming the self-potential of the Moon at the point $P$ to be that of an ellipsoid does not appear to be great.

# APPENDIX C

# Spherical Harmonic Functions

The principal purpose of this summary is to avoid repetition. Spherical harmonic analysis is used or referred to in Chapters 3, 4, 6 and 8 and is basic to several problems in geophysics. As far as possible, a common notation for the harmonic coefficients is used in the sections on gravity, geomagnetism, and free oscillations, although independent treatments of these subjects have used different notations. For a statement of the properties of spherical harmonic functions, refer to Chapter 3 of Sneddon (1961) or Chapter 2 of Kaula (1968), who considers the particular application to gravity. The discussion by Chapman and Bartels (1940, Vol. 2) is also useful, primarily as a specification for procedure in harmonic analysis of the geomagnetic field.

The gravitational and magnetic fields of the Earth as a whole are analyzed in terms of the potentials $V$ of the fields, which satisfy Laplace's equation at all points external to the Earth:

$$\nabla^2 V = \frac{\partial^2 V}{\partial x^2} + \frac{\partial^2 V}{\partial y^2} + \frac{\partial^2 V}{\partial z^2} = 0 \tag{C.1}$$

The natural boundaries that impose boundary conditions on solutions of Laplace's equation can be taken as spherical in most whole-Earth problems and it is therefore convenient to express the equation in spherical polar coordinates, $(r, \theta, \lambda)$:

$$\nabla^2 V = \frac{1}{r^2}\frac{\partial}{\partial r}\left(r^2 \frac{\partial V}{\partial r}\right) + \frac{1}{r^2 \sin\theta}\frac{\partial}{\partial \theta}\left(\sin\theta \frac{\partial V}{\partial \theta}\right) + \frac{1}{r^2 \sin^2\theta}\frac{\partial^2 V}{\partial \lambda^2} = 0 \tag{C.2}$$

Here $\theta$ is the angle to the axis of the coordinate system and is thus the colatitude (90° minus latitude) if the axis is chosen to be the Earth's rotational axis (or geomagnetic latitude if the magnetic axis is chosen), and $\lambda$ is longitude measured from a convenient reference. These are the commonly selected axes, but others may be used according to the symmetry of the problem.

The wave equation for propagation of seismic (or any other) waves has a similar geometrical form:

$$\frac{\partial^2 V}{\partial t^2} = c^2 \nabla^2 V \tag{C.3}$$

where $c$ is wave velocity and $V$ is the potential whose derivative in any direction gives the velocity in that direction. Thus the geometrical solutions of Equation C.2 appear also in the dynamic problem of free oscillations of the Earth (Section 6.3).

As may be verified by differentiation and substitution, Equation C.2 has solutions of the form

$$V = [r^l, r^{-(l+1)}](\cos m\lambda, \sin m\lambda)P_l^m(\cos\theta) \tag{C.4}$$

where $[r^l, r^{-(l+1)}]$ and $(\cos m\lambda, \sin m\lambda)$ represent alternative solutions, $l$ and $m$ are integers with $m \leqslant l$, and $P_l^m(\mu)$ satisfies the equation

$$(1-\mu^2)\frac{d^2P}{d\mu^2} - 2\mu\frac{dP}{d\mu} + \left[l(l+1) - \frac{m^2}{1-\mu^2}\right]P = 0 \tag{C.5}$$

This is Legendre's associated equation, which reduces to Legendre's equation for $m = 0$. Considering first the special case $m = 0$, we can verify, again by differentiation and substitution, that

$$P_l^0(\mu) = \frac{1}{2^l l!}\frac{d^l}{d\mu^l}[(\mu^2-1)^l] \tag{C.6}$$

As a solution of Equation C.5, the constant factor $(2^l l!)^{-1}$ is arbitrary, but has been chosen so that $P_l^0(1) = 1$. It should also be noted that $(\mu^2 - 1)^l$ is negative for odd $l$ since $\mu = \cos\theta \leqslant 1$, but that this makes $P_l^0(1) = +1$ rather than $(-1)$. The functions $P_l^0(\mu)$, normally written with the superscript 0 omitted, are the Legendre polynomials $P_l(\cos\theta)$ given in the $m = 0$ column of Table C.1. In some problems it is convenient to use latitude $\phi$ rather than colatitude $\theta$ by substituting $\sin\phi$ for $\cos\theta$.

The Legendre polynomials give solutions to Laplace's equation that have rotational symmetry, since by putting $m = 0$ in Equation C.4, dependence of $V$ on longitude is excluded. These are the zonal harmonics, functions of latitude only. A potential that can be expressed in terms of zonal harmonics only can be written as a sum of powers of distance $r$ from the origin of the coordinate system, with the Legendre polynomials appearing in the coefficients and giving the variation of potential with latitude. In geophysical problems it is convenient to make the coefficients dimensionally uniform by relating $r$ to the Earth radius $a$:

$$V = \frac{1}{a}\sum_{l=0}^{\infty}\left[C_l\left(\frac{a}{r}\right)^{l+1} + C_l'\left(\frac{r}{a}\right)^l\right]P_l(\cos\theta) \tag{C.7}$$

where $C_l$ are constant coefficients arising from sources internal to the surface considered and $C_l'$ are coefficients arising from external sources. Equation C.7 is simply a sum of terms of the form of Equation C.4 with $m = 0$.

The form of Equation C.7 is directly derivable by generalizing the method used in Section 3.1 to obtain MacCullagh's formula for gravitational potential. If, instead of terminating the expansion of Equation 3.8 at terms in $1/r^2$, it is continued to higher powers in $1/r$, the coefficients are higher-order Legendre polynomials:

$$\left[1 + \left(\frac{s}{r}\right)^2 - 2\frac{s}{r}\cos\psi\right]^{-1/2} = \sum_{l=0}^{\infty}\left(\frac{s}{r}\right)^l P_l(\cos\psi) \tag{C.8}$$

We can now consider the general case of Equations C.4 and C.5 in which $m \neq 0$. By writing

$$\left.\begin{aligned} P_l^m(\mu) &= (1-\mu^2)^{\frac{1}{2}m} \frac{d^m}{d\mu^m}[P_l^0(\mu)] \\ &\text{or} \\ P_l^m(\cos\theta) &= \sin^m\theta \frac{d^m}{d(\cos\theta)^m}[P_l^0(\cos\theta)] \end{aligned}\right\} \quad \text{(C.9)}$$

and carrying out the differentiation, we can verify that the function $P_l^m(\mu)$ is a solution of Equation C.5. This therefore gives the form of the solutions (C.4) in which a longitude variation is allowed. By writing $P_l^0(\mu)$ in the form of Equation C.6 we obtain

$$P_l^m(\mu) = \frac{(1-\mu^2)^{\frac{1}{2}m}}{2^l l!} \frac{d^{l+m}}{d\mu^{l+m}}[(\mu^2-1)^l] \quad \text{(C.10)}$$

At this point a departure from the sign convention used to normalize $P_l^0(\mu)$ is introduced because with odd $m$, $(\mu^2-1)^{m/2}$ would obviously be inconvenient. The form of Equations C.9 and C.10 is referred to as Ferrer's function (Sneddon, 1961). Since the highest power of $\mu$ in the expression to be differentiated is $\mu^{2l}$, it follows that differentiation gives a polynomial in $\mu$ of order $(l-m)$, being zero if $m > l$. Thus the distribution on a spherical surface of a potential with the form of Equation C.4 has $(l-m)$ nodal lines of latitude, arising from the $P_l^m(\cos\theta)$ term and $2m$ nodal lines of longitude, arising from the $(\cos m\lambda, \sin m\lambda)$ term. The polynomial form of the associated Legendre function is:

$$P_l^m(\cos\theta) = \frac{\sin^m\theta}{2^l} \sum_{t=0}^{\frac{1}{2}(l-m)} \frac{(-1)^t(2l-2t)!}{t!(l-t)!(l-m-2t)!}(\cos\theta)^{l-m-2t} \quad \text{(C.11)}$$

Examples are given in Figure C.1 and Table C.1.

The general expression for a function $V$ as a sum of spherical harmonics is thus

$$V = \frac{-1}{a} \sum_{l=0}^{\infty} \sum_{m=0}^{l} \left\{ \begin{aligned} &\left[C_l^m\left(\frac{a}{r}\right)^{l+1} + C_l'^m\left(\frac{r}{a}\right)^l\right]\cos m\lambda \\ &+\left[S_l^m\left(\frac{a}{r}\right)^{l+1} + S_l'^m\left(\frac{r}{a}\right)^l\right]\sin m\lambda \end{aligned} \right\} P_l^m(\cos\theta) \quad \text{(C.12)}$$

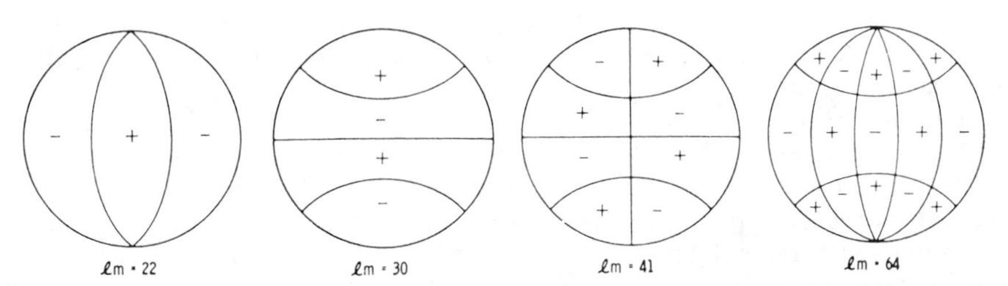


**Figure C.1.** Examples of spherical harmonics. $m=0$ gives zonal harmonics, $m=l$ gives sectoral harmonics and the general case $0<m<l$ are known as tesseral harmonics. Reproduced, by permission, from Kaula (1968).

**Table C.1** Legendre Polynomials $P_l(\cos\theta)$ and Associated Polynomials $P_l^m(\cos\theta)$[a]

| | $m=0$ | $m=1$ |
|---|---|---|
| $l=0$ | $1\ (1)$ | — |
| 1 | $\cos\theta\ (\sqrt{3})$ | $\sin\theta\ (\sqrt{3})$ |
| 2 | $\frac{1}{2}(3\cos^2\theta-1)\ (\sqrt{5})$ | $3\cos\theta\sin\theta\ (\sqrt{5/3})$ |
| 3 | $\frac{1}{2}(5\cos^3\theta-3\cos\theta)\ (\sqrt{7})$ | $\frac{3}{2}(5\cos^2\theta-1)\sin\theta\ (\sqrt{7/6})$ |
| 4 | $\frac{1}{8}(35\cos^4\theta-30\cos^2\theta+3)(\sqrt{9})$ | $\frac{5}{2}(7\cos^3\theta-3\cos\theta)\sin\theta\ (\sqrt{9/10})$ |

| | $m=2$ | $m=3$ | $m=4$ |
|---|---|---|---|
| $l=0$ | — | — | — |
| 1 | — | — | — |
| 2 | $3\sin^2\theta\ (\sqrt{5/12})$ | — | — |
| 3 | $15\cos\theta\sin^2\theta\ (\sqrt{7/60})$ | $15\sin^3\theta\ (\sqrt{7/360})$ | — |
| 4 | $\frac{15}{2}(7\cos^2\theta-1)\sin^2\theta\ (\sqrt{1/20})$ | $105\cos\theta\sin^3\theta$ $(\sqrt{1/280})$ | $105\sin^4\theta\ (\sqrt{1/2240})$ |

[a]Factors in brackets convert $P_l^m$ to $p_l^m$ by Equation C.14.

In problems of the gravitational or magnetic potential external to the Earth the coefficients $C'$ and $S'$ vanish, being incompatible with finite $V$ at $r\to\infty$. It must also be noted that the solutions considered here are restricted to those which remain finite everywhere on a spherical boundary.

Legendre and associated polynomials are orthogonal functions, which means that the integral over a sphere of a product,

$$\int_{-1}^{1} P_l^m(\mu)P_{l'}^{m'}(\mu)\,d\mu = 0 \tag{C.13}$$

unless both $l=l'$ and $m=m'$. This has an important consequence. If in a particular analysis of a function, only a few spherical harmonic terms are taken, the sum of these terms may poorly represent the function but, nevertheless, the values of the coefficients are not affected by taking additional terms in a more complete analysis. This is a property also of the terms in a Fourier series and we can regard spherical harmonic analysis as a development of Fourier analysis to spherical surfaces.

In some treatments of Legendre polynomials the subscript $n$ is used in place of $l$. Here $n$ is reserved for a further development that appears in the study of free oscillations—a harmonic radial variation. Free oscillations are classified in Section 6.3 according to the values of three integers; thus ${}_nS_l^m$ and ${}_nT_l^m$ denote spheroidal and torsional oscillations, respectively, where $l$, $m$ represent variations on a spherical surface, as for $P_l^m(\cos\theta)\cos m\lambda$ and $n$ is the number of internal spherical surfaces that are nodes of the motion.

The numerical factors in the associated polynomials defined by C.9 to C.11 increase rapidly with $m$; in order that the coefficients in a harmonic analysis relate more nearly to the physical significance of the terms they represent, various normalizing factors are used. The one that has been employed in most recent analyses of the geoid and must be favored for general adoption is the "fully normalized" function

$$p_l^m(\cos\theta) = \left[(2-\delta_{m,0})(2l+1)\frac{(l-m)!}{(l+m)!}\right]^{1/2} P_l^m(\cos\theta) \tag{C.14}$$

which is so defined that

$$\frac{1}{4\pi}\int_0^{2\pi}\int_{-1}^{1} \{p_l^m(\cos\theta)[\sin m\lambda, \cos m\lambda]\}^2\, d(\cos\theta)\, d\lambda = 1 \tag{C.15}$$

that is, the mean square value over a spherical surface is unity. Note that the factor $(2-\delta_{m,0})$ in Equation C.14 is unity if $m=0$, but $(2-\delta_{m,0})=2$ if $m \neq 0$ because the factor $[\sin m\lambda, \cos m\lambda]^2$ in Equation C.15 introduces a factor $\frac{1}{2}$. (There is no alternative $\sin m\lambda$ term if $m=0$.) The coefficients of a spherical harmonic expansion referred to the normalized coefficients, $p_l^m$ are distinguished by a bar: $\bar{C}_l^m$, $\bar{S}_l^m$. Thus

$$V = -\frac{1}{a}\sum_{l=0}^{\infty}\sum_{m=0}^{l} \left\{ \begin{aligned} &\left[\bar{C}_l^m\left(\frac{a}{r}\right)^{l+1} + \bar{C}_l'^m\left(\frac{r}{a}\right)^{l}\right]\cos m\lambda \\ +&\left[\bar{S}_l^m\left(\frac{a}{r}\right)^{l+1} + \bar{S}_l'^m\left(\frac{r}{a}\right)^{l}\right]\sin m\lambda \end{aligned} \right\} p_l^m(\cos\theta) \tag{C.16}$$

A different normalizing factor, $[2(l-m)!/(l+m)!]^{1/2}$, has been widely used in geomagnetism (e.g., see Chapman and Bartels, 1940). Considerable care is required in checking both normalization and signs of coefficients when using spherical harmonic data.

Of some interest is the spherical harmonic expansion of the equation for the surface of an oblate ellipsoid of equatorial radius $a$

$$r = a\left(1 + \frac{e^2}{1-e^2}\sin^2\phi\right)^{-1/2} \tag{C.17}$$

Expanding in powers of ellipticity $e = (1-c^2/a^2)^{1/2}$ to order $e^6$ or flattening $f = (1-c/a)$ to order $f^3$ and zonal harmonics to $P_6^0$ we have

$$\begin{aligned} \frac{r}{a} &= \left(1 - \frac{e^2}{6} - \frac{11}{20}e^4 - \frac{103}{1680}e^6\right) + \left(-\frac{e^2}{3} - \frac{5}{42}e^4 - \frac{3}{56}e^6\right)P_2^0 \\ &\quad + \left(\frac{3}{35}e^4 + \frac{57}{770}e^6\right)P_4^0 - \frac{5}{231}e^6 P_6^0 \end{aligned} \tag{C.18}$$

$$\begin{aligned} \frac{r}{a} &= \left(1 - \frac{f}{3} - \frac{f^2}{5} - \frac{13}{105}f^3\right) + \left(-\frac{2}{3}f - \frac{1}{7}f^2 + \frac{1}{21}f^3\right)P_2^0 \\ &\quad + \left(\frac{12}{35}f^2 + \frac{96}{385}f^3\right)P_4^0 - \frac{40}{231}f^3 P_6^0 \end{aligned} \tag{C.19}$$

Note that the $P_2^0$ term alone does not represent an ellipsoidal surface, although for the Earth the ellipticity is sufficiently slight that expansion to $e^4$, $f^2$, $P_4^0$ suffices.

# APPENDIX D

# Solution of the Integral Equation for the Velocity Profile in a Spherically Layered Earth

Equation 6.31 is rewritten

$$\Delta = \int_p^{\eta_0} \frac{2p}{r(\eta^2 - p^2)^{1/2}} \frac{dr}{d\eta} \, d\eta \tag{D.1}$$

the limits being $\eta_0$ and $\eta' = p$. Now $\eta_1$ is the value of $\eta$ corresponding to a radius $r_1$, between $r_0$ and $r'$, and it must be assumed that $\eta$ decreases monotonically with decreasing $r$ so that $\eta_1 > \eta'$, the value at maximum penetration of the ray. Also $\Delta_1$ is the value of $\Delta$ for the ray whose deepest point is at $r_1$. Multiply both sides of Equation D.1 by $(p^2 - \eta_1^2)^{-1/2}$ and integrate with respect to $p$ from $\eta_1$ to $\eta_0$:

$$\int_{\eta_1}^{\eta_0} \frac{\Delta \, dp}{(p^2 - \eta_1^2)^{1/2}} = \int_{\eta_1}^{\eta_0} \left\{ \int_p^{\eta} \frac{2p}{r[(p^2 - \eta_1^2)(\eta^2 - p^2)]^{1/2}} \frac{dr}{dp} \, d\eta \right\} dp \tag{D.2}$$

Now change the order of integration on the right-hand side, the limits being then $\eta_1$ to $\eta$ for $p$ and $\eta_1$ to $\eta_0$ for $\eta$:

$$\int_{\eta_1}^{\eta_0} \frac{\Delta \, dp}{(p^2 - \eta_1^2)^{1/2}} = \int_{\eta_1}^{\eta_0} \left\{ \int_{\eta_1}^{\eta} \frac{2p}{r[(p^2 - \eta_1^2)(\eta^2 - p^2)]^{1/2}} \frac{dr}{d\eta} \, dp \right\} d\eta \tag{D.3}$$

The left-hand side can be integrated by parts and the integration with respect to $p$ on the right-hand side gives a considerable simplification (for $\eta > \eta_1$) because

$$\int_{\eta_1}^{\eta} \frac{p \, dp}{[(p^2 - \eta_1^2)(\eta^2 - p^2)]^{1/2}} = \frac{\pi}{2} \tag{D.4}$$

so that

$$\left[\Delta \cosh^{-1}\left(\frac{p}{\eta_1}\right)\right]_{\eta_1}^{\eta_0} - \int_{\eta_1}^{\eta_0} \frac{d\Delta}{dp} \cosh^{-1}\left(\frac{p}{\eta_1}\right) dp = \pi \int_{\eta_1}^{\eta_0} \frac{1}{r} \frac{dr}{d\eta} d\eta \tag{D.5}$$

The first term on the left-hand side is zero because $\Delta = 0$ at $p = \eta_0$ and $\cosh^{-1}(p/\eta_1) = 0$ for $p = \eta_1$. Thus

$$\int_0^{\Delta_1} \cosh^{-1}\left(\frac{p}{n_1}\right) d\Delta = \pi \ln\left(\frac{r_0}{r_1}\right) \tag{D.6}$$

Since $\eta$ is the value of $(r/V)$ at the point of deepest penetration of a ray which travels a distance $\Delta$, it is convenient to put $\eta_1 = p_1$, as in Equation 6.32. $p(\Delta)$ is tabulated from the observed travel times, $T(\Delta)$, since $p = dT/d\Delta$, and thus $(r/V)$ is tabulated against $(r/r_0)$ by Equation 6.32.

# APPENDIX E

# The Thermodynamic Grüneisen Parameter

In discussions of the thermodynamic properties of the Earth, or of any materials at very high pressures, it is convenient to refer to a dimensionless parameter, $\gamma$, defined as follows

$$\gamma = \frac{\alpha K_T}{\rho C_V} = \frac{\alpha K_S}{\rho C_P} \tag{E.1}$$

where $\alpha$ is volume coefficient of expansion, $K_T$, $K_S$ are the isothermal and adiabatic incompressibilities, $\rho$ is density, and $C_V$, $C_P$ are specific heats at constant volume and constant pressure, respectively. Whereas $\alpha$ gives the change in volume by heating at constant pressure, $\gamma$ is a measure of the change in pressure on heating at constant volume (see Eq. E.3). Numerical values of $\gamma$ are close to unity for virtually all materials*, including liquids and gases, and for solids above about half of their Debye temperatures it appears to be virtually independent of temperature. For solids under high compression, $\gamma$ decreases somewhat, but usually less strongly than $1/\rho$. Thus it is much more nearly constant over the range of physical conditions in the Earth than the component parameters $\alpha$, $K$ or $\rho$. We refer to $\gamma$, defined as in Equation E.1, as the thermal or thermodynamic Grüneisen parameter. If necessary to distinguish it from related, but not necessarily identical parameters with different definitions, the symbol $\gamma_T$ may be used.

In his studies of crystal lattice dynamics, Grüneisen (1926) defined the parameter

$$\gamma_i = \frac{d(\ln \nu_i)}{d(\ln V)} \tag{E.2}$$

where $\nu_i$ is the frequency of the $i$th mode of lattice vibration and $V$ is volume.

*O. L. Anderson (1966, 1974) has tabulated values for geologically interesting materials.

Under the conditions that all $\nu_i$ have the same volume dependence and are independent of temperature, $\gamma_i$, defined by Equation E.2, can be shown to be formally equivalent to $\gamma$ as defined by Equation E.1 (see, for example, Knopoff, 1963). Since the geophysical interest is in high temperatures, at which atomic vibrations can be treated classically, in the context of the present interest it is permissible to ignore the quantization of lattice vibrations and treat the oscillations of atoms or bonds as independent of one another.

A simple physical picture of the process of thermal expansion is seen by referring to Figure E.1, which shows the approximate form of the mutual potential energy $E(r)$ of two neighboring atoms as a function of their separation, $r$. The potential well is asymmetrical, that is, it is easier to pull the atoms apart from the equilibrium separation ($a$) than to push them together. At high temperature the atoms vibrate and the bond oscillates in length, but the oscillation is not harmonic (sinusoidal) because the potential well is not parabolic, that is, restoring force is not proportional to displacement from equilibrium. For a particular energy of oscillation, as indicated, the bond extension is greater than the bond compression. The bond is not only extended

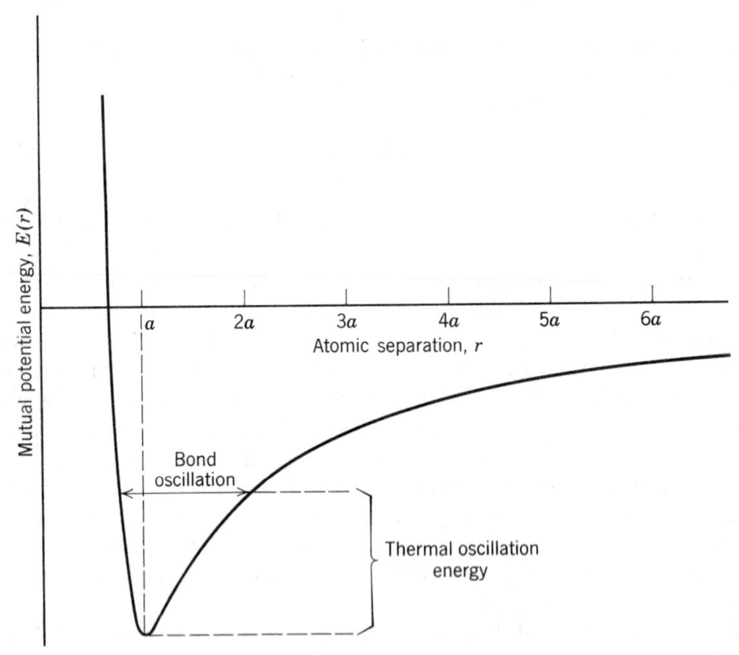


**Figure E.1.** Mutual potential energy $E(r)$ of neighboring atoms as a function of their separation $r$. The equilibrium separation at zero pressure is $a$. This is termed an anharmonic potential. Its asymmetry means that thermal oscillation of the bond is anharmonic or nonsinusoidal. An oscillation of constant energy causes greater extension than compression of the bond and an average extension. Thus anharmonicity causes thermal expansion.

further than it is compressed, but spends more time in the extended state and the time-averaged extension is the linear thermal expansion.

Various properties of a material described by a simple central force atomic potential function, such as that in Figure E.1, can be represented in terms of $E(r)$ and its derivatives (Eqs. 6.48 to 6.52). Thus when the material is compressed the pressure, $P$, or force pushing the atoms together, is expressed in terms of $dE/dr$. Incompressibility or bulk modulus, $K$, is the rate of increase in $P$ with compression and so is determined by $d^2E/dr^2$. The strongly increasing atomic repulsion with decreasing $r$ means that $K$ is not constant but increases with compression. Then $dK/dP$ is given in terms of $d^3E/dr^3$ (and lower derivatives). Now this third derivative is a measure of the asymmetry or departure of the potential well from parabolic form, which is responsible for thermal expansion. Thus thermal expansion can also be represented by $d^3E/dr^3$ and lower derivatives. It follows that thermal expansion $\alpha$ can be represented in terms of $dK/dP$, $K$, $P$ without any explicit assumption about the form of the function $E(r)$ and when the substitutions are made it is seen that $\alpha$ is so related to the function $K(P)$ and the thermal energy (or specific heat) that what we are actually obtaining is an expression for $\gamma$ (Eq. E.1) in terms of $dK/dP$, $K$ and $P$. The principle has been recognized by many authors (e.g., Slater, 1939; Dugdale and MacDonald, 1953; Vashchenko and Zubarev, 1963), but the alternative presentations have led to differences in detail and interpretation (e.g., Gilvarry, 1956b; Pastine, 1965; Altschuler, 1965; Knopoff and Shapiro, 1969). The geophysical significance is that since $K_S(P)$ is a well-documented function for the interior of the Earth (Appendix G and Fig. 6.17) we have an important constraint on the thermodynamic parameters.

In the simple, classical thermodynamic problem of a material that is heated to temperature $T$ (in principle from 0 K) and then recompressed to its original volume, the total pressure $P$ can be represented by an initial ambient pressure $P_0$ plus a thermal pressure $P_T$ which is given by the volume expansion coefficient, $\alpha$, and incompressibility, $K_T$. Then use of Equation E.1 introduces $\gamma$:

$$P = P_0 + P_T = P_0 + \alpha K_T T = P_0 + \gamma \rho C_V T \tag{E.3}$$

Equation E.3 expresses the classical definition of Grüneisen's ratio as the ratio of thermal pressure to thermal energy per unit volume. This is the starting point for the calculation of a relationship between $\gamma$ and $K(P)$ in terms of the thermal pressure resulting from classical atomic vibrations with a three-dimensional but arbitrarily asymmetrical (anharmonic) atomic potential $E(r)$ from which Irvine and Stacey (1975) obtained the expression

$$\gamma = \frac{\frac{1}{2}\frac{dK}{dP} - \frac{5}{6} + \frac{2}{9}\frac{P}{K} - \frac{f}{18K} + \frac{1}{6}\frac{df}{dP}}{1 - \frac{4}{3}\frac{P}{K} + \frac{f}{3K}} \tag{E.4}$$

where $f$ is a correction factor arising from noncentral atomic forces. Equation E.4 with $f = 0$ was derived in a quite different manner by Vashchenko and Zubarev (1963).

Comparison of Equation E.4 with laboratory measurements on materials for which the assumption of central atomic forces appears to be a reasonably good one (e.g., many metals and ionic crystals) is straightforward because $f$ can be neglected. However, this is unsatisfactory for covalent materials in which atomic bonds are rigid, and especially for tetrahedrally bonded crystals. Diamond is the simplest example of this type of bonding, but the Si–O bond in silicates, in the low pressure forms prevailing in the Earth's crust and upper mantle, is also tetrahedrally coordinated. These materials have an intrinsic bond rigidity, arising from inflexibility of the angles between bonds, which has the effect of restricting atomic vibrations, thereby reducing the thermal pressure and Grüneisen's ratio. However through much of the Earth it appears reasonable to assume that $f \approx 0$. In the outer core this is readily justified by the fluidity (but at such high pressures, even in a fluid, the instantaneous atomic structure must closely resemble close packing and each atom effectively "sees" a lattice in which it vibrates, creating a thermal pressure exactly as in a solid). In the lower mantle the compression of silicates to close-packed structures greatly reduces the effectiveness of the bond rigidity (because the bond angles cannot assume the preferred values). Thus for tetrahedrally bonded (low pressure) $\alpha$-quartz, $\gamma = 0.746$, requiring a value of $f/3K$ about 2, whereas for stishovite, the densest (high pressure) form of quartz, which is commonly taken to be representative of the crystal structures of the lower mantle, $\gamma = 1.5$, requiring $f/3K \approx 0$ to 0.3.

Estimation of $\gamma$ for the interior of the Earth from Equation E.4 requires extrapolation of the seismological $K(P)$ data to the zero-temperature condition, but the correction terms are not large. Representative values of $\gamma$ for important regions of the Earth are:

| | |
|---|---|
| Upper mantle | 0.8 |
| Lower mantle | 1.0 |
| Outer core | 1.4 |

It may be noted that this is not the trend of the pressure dependence of $\gamma$ for any particular material at constant phase, which shows a decrease with pressure.

# APPENDIX F

# Numerical Data of Geophysical Interest

## PHYSICAL CONSTANTS*

| | |
|---|---|
| Speed of light | $c = 2.997925 \times 10^{8}$ m sec$^{-1}$ |
| Electronic charge | $-e = 1.60219 \times 10^{-19}$ C |
| Electron rest mass | $m = 9.1096 \times 10^{-31}$ kg |
| Proton mass | $M_P = 1.6726 \times 10^{-27}$ kg |
| Mass of hydrogen atom ($H^1$) | $M_H = 1.6735 \times 10^{-27}$ kg |
| Planck's constant | $h = 6.6262 \times 10^{-34}$ J sec |
| Boltzmann's constant | $k = 1.3806 \times 10^{-23}$ J deg$^{-1}$ |
| Stefan-Boltzmann constant | $\sigma = 5.670 \times 10^{-8}$ W m$^{-2}$ deg$^{-4}$ |
| Gas constant | $R = 8.314$ J mole$^{-1}$ deg$^{-1}$ |
| | $= 8.314 \times 10^{3}$ J deg$^{-1}$ per kg mole |
| Avogadro's number | $N = 6.022094 \times 10^{23}$ mole$^{-1}$ |
| | $= 6.022094 \times 10^{26}$ per kg mole |
| Wiedemann-Franz constant | $L = 2.45 \times 10^{-8}$ W $\Omega$ deg$^{-2}$ |
| Bohr magneton | $\mu_B = 9.274 \times 10^{-24}$ A turn m$^2$ (J T$^{-1}$) |
| Permittivity of free space | $\epsilon_0 = 8.854 \times 10^{-12}$ F m$^{-1}$ |
| Permeability of free space | $\mu_0 = 4\pi \times 10^{-7}$ H m$^{-1}$ |
| Gravitational constant | $G = 6.6732 \times 10^{-11}$ m$^3$ kg$^{-1}$ sec$^{-2}$ (N m$^2$ kg$^{-2}$) |

## DIMENSIONS AND PROPERTIES OF THE EARTH

| | |
|---|---|
| Equatorial radius | $a = 6.378139 \times 10^{6}$ m |
| Polar radius | $c = 6.35675 \times 10^{6}$ m |
| Volume | $V = 1.083 \times 10^{21}$ m$^3$ |
| Radius of sphere of equal volume | $6.3708 \times 10^{6}$ m |

*Factors for conversion to alternative units are given later in this appendix.

| | |
|---|---|
| Flattening | $f=\dfrac{a-c}{a}=3.35282\times10^{-3}=1/298.256$ |
| Surface areas | |
| land | $1.48\times10^{14}$ m$^2$ |
| sea | $3.62\times10^{14}$ m$^2$ |
| total | $A=5.100\times10^{14}$ m$^2$ |
| Mass | $M=5.973_2\times10^{24}$ kg |
| Gravitational const. × mass (including atmosphere) | $GM=3.986005\times10^{14}$ m$^3$ sec$^{-2}$ |
| Mean density | $\rho=5.515\times10^3$ kg m$^{-3}$ |
| Moments of inertia | |
| About polar axis | $C=8.0378\times10^{37}$ kg m$^2$ |
| About equatorial axis | $A=8.0115\times10^{37}$ kg m$^2$ |
| Core | $C_c=0.920\times10^{37}$ kg m$^2$ |
| Atmosphere | $C_A=1.38\times10^{32}$ kg m$^2$ |
| Dynamical ellipticity | $H=\dfrac{C-A}{C}=3.2732\times10^{-3}=1/305.51$ |
| Ellipticity coefficient | $J_2=\dfrac{C-A}{Ma^2}=1.08264\times10^{-3}$ |
| Coefficient of moment of inertia | $\dfrac{J_2}{H}=\dfrac{C}{Ma^2}=0.33076$ |
| Solar day | 86,400 sec |
| Sidereal day | 86,164 sec |
| Rotational angular velocity | $\omega=7.292115\times10^{-5}$ sec$^{-1}$ |
| Obliquity of ecliptic | $\theta=23°27'8''.26=23°.473$ |
| Equatorial gravity (see Eq. 4.13 for latitude variation) | $g_e=9.780317$ m sec$^{-2}$ |
| Geoid potential | $W_0=6.26368\times10^7$ m$^2$ sec$^{-2}$ |
| Ratio $\dfrac{\text{centrifugal force}}{\text{equatorial gravity}}$ | $m=\dfrac{\omega^2a}{g_e}=3.46775\times10^{-3}$ |
| Semimajor axis of orbit | $r_E=1.4959789\times10^{11}$ m = 1 Astronomical Unit (AU) |
| Mean orbital velocity | $2.977\times10^4$ m sec$^{-1}$ |
| Ratio $\dfrac{\text{mass of Sun}}{\text{mass of Earth}}$ | 332,946.8 |
| Solar constant | $S=1360$ watt m$^{-2}$ |
| Mean Earth-Moon distance | $R=3.8440_5\times10^8$ m = 60.34 Earth radii |
| Lunar orbital angular velocity | $\omega_L=2.6844\times10^{-6}$ sec$^{-1}$ |
| Ratio $\dfrac{\text{mass of Earth}}{\text{mass of Moon}}$ | $\mu=\dfrac{M}{m}=81.302$ |
| Rate of precession of equinox | $\omega_P=50''.37$ year$^{-1}=7.738\times10^{-12}$ rad sec$^{-1}$ |
| Period of precession of equinox | 25,730 years $=8.120\times10^{11}$ sec |
| Total geothermal flux | $(3.14\pm0.17)\times10^{13}$ W |
| Mean geothermal flux | $(6.15\pm0.34)\times10^{-2}$ W m$^{-2}$ |

## DIMENSIONS AND PROPERTIES OF THE EARTH *(Contd.)*

| | |
|---|---|
| Magnetic dipole moment (1975) | $m = 7.94 \times 10^{22}$ A m$^2$ |
| Mass of atmosphere | $5.1 \times 10^{18}$ kg |
| Mass of oceans | $1.4 \times 10^{21}$ kg |
| Mass of crust | $2.6 \times 10^{22}$ kg |
| Mass of mantle | $4.0 \times 10^{24}$ kg |
| Mass of outer core | $1.85 \times 10^{24}$ kg |
| Mass of inner core | $9.7 \times 10^{22}$ kg |

## PROPERTIES OF GEOLOGICAL MATERIALS

### Chemical Compositions of Representative Igneous Rocks (percent by mass)

| | Tholeiitic Basalt | Andesite | Granite |
|---|---|---|---|
| $SiO_2$ | 50 | 58 | 74 |
| MgO | 6 | 3 | 0.3 |
| $FeO + Fe_2O_3$ | 12 | 6 | 2 |
| $Al_2O_3$ | 14 | 18 | 13 |
| CaO | 10 | 7 | 1.0 |
| $Na_2O$ | 2 | 4 | 3 |
| $K_2O$ | 0.8 | 2 | 5 |
| $H_2O$ | 1.0 | 0.3 | 0.7 |
| Balance | 4.2 | 1.7 | 1 |

## DENSITIES AND ELASTICITIES

| | Density $\rho(10^3$ kg m$^{-3})$ | Incompressibility $K\,(10^{10}$ Pa) | Rigidity $\mu(10^{10}$ Pa) |
|---|---|---|---|
| Granite | 2.7 | 5.5 | 2 |
| Basalt | 2.9 | 7.5 | 3 |
| Eclogite | 3.4 | 9 | 5 |
| Dunite | 3.3 | 10 | 5.5 |
| Iron 20°C | 7.87 | 16 | 8 |
| Liquid iron (MP) | 7.01 | $\sim 9$ | – |
| Sea water | 1.025 | 0.2 | – |

## THERMAL PROPERTIES

| | | Igneous Rock | Iron | Sea Water |
|---|---|---|---|---|
| At 20°C | Conductivity ($W\ m^{-1}\ deg^{-1}$) | 2.5 | 67 | 0.6 |
| | Specific heat ($J\ kg^{-1}\ deg^{-1}$) | 710 | 447 | 4180 |
| | Diffusivity ($m^2\ sec^{-1}$) | $1.2_6 \times 10^{-6}$ | $1.8 \times 10^{-5}$ | $1.4 \times 10^{-7}$ |
| | Volume expansion coefficient ($deg^{-1}$) | $1.8 \times 10^{-5}$ | $3.6 \times 10^{-5}$ | zero at 2°C; $15 \times 10^{-5}$ at 15°C |
| Latent heat of melting ($J\ kg^{-1}$) | | $4.2 \times 10^5$ | $2.74 \times 10^5$ | $3.35 \times 10^5$ |
| Melting point (K) | | ~1500 | 1805 | 273 (pure water) |

## ELECTRIC AND MAGNETIC PROPERTIES

| | Resistivity (ohm m) | Dielectric Constant | Magnetic Susceptibility[a] |
|---|---|---|---|
| Granite (dry, 20°C) | $3 \times 10^{10}$[b] | 8 | $\sim 2.5 \times 10^{-3}$ |
| Basalt (dry, 20°C) | $1.5 \times 10^{9}$[b] | 20 | $\sim 6 \times 10^{-2}$ |
| Iron (20°C) | $9.8 \times 10^{-8}$ | — | $10^2$ to $10^4$ |
| Sea water | 0.23 | 80 | $10^{-5}$ |

[a]Divide values by $4\pi$ for electromagnetic units.
[b]Values for mineral conduction with all water expelled.

## CONVERSION OF UNITS

### SI (rationalized mks) Units and cgs Electromagnetic Units

| | SI Unit | Equivalent in emu |
|---|---|---|
| Mass | 1 kilogram (kg) | $10^3$ grams (g or gm) |
| Length | 1 meter (m) | $10^2$ centimeters (cm) |
| Time | 1 second (s or sec) | 1 sec |
| Force | 1 newton (N) | $10^5$ dynes |
| Pressure | 1 Pascal (Pa) $\equiv 1\ Nm^{-2}$ | 10 dyne $cm^{-2}$ |
| Energy | 1 joule (J) | $10^7$ ergs |

## CONVERSION OF UNITS (*Contd.*)

| | | |
|---|---|---|
| Power | 1 watt (W) | $10^7$ erg $sec^{-1}$ |
| Electric Current | 1 ampere (A or amp) | 0.1 emu (ab amp) |
| Potential difference (voltage) | 1 volt (V) | $10^8$ emu (ab volt) |
| Electric field | 1 volt per meter ($Vm^{-1}$) | $10^6$ emu |
| Magnetic flux | 1 weber (Wb) | $10^8$ maxwells |
| Magnetic intensity (B) | 1 Tesla (T) $\equiv$ 1 $Wbm^{-2}$ | $10^4$ gauss |
| Magnetic field strength (H) | 1 ampere turn $m^{-1}$ ($Am^{-1}$) | $4\pi \times 10^{-3}$ oersted |
| Magnetic moment | 1 ampere turn $m^2$ ($Am^2$) | $10^3$ gauss $cm^3$ |
| Magnetization (magnetic moment/unit volume) | 1 ampere turn $m^{-1}$ ($Am^{-1}$) | $10^{-3}$ emu |
| Resistance | 1 ohm ($\Omega$) | $10^9$ emu |
| Resistivity | 1 ohm meter ($\Omega$m) | $10^{11}$ emu |
| Inductance | 1 henry (H) | $10^9$ emu |
| Capacitance | 1 farad (F) | $10^{-9}$ emu |

## OTHER CONVERSIONS

| | |
|---|---|
| 1 statute mile | = 1609 m |
| 1 nautical mile | = 1852 m |
| 1 pound (lb) | = 0.4536 kg |
| 1 ton (U.S.A.) (2000 lb) | = 907.2 kg |
| 1 ton (Imperial) (2240 lb) | = 1016 kg |
| 1 tonne | = 1000 kg |
| 1 year | $= 3.15567 \times 10^7$ sec |
| 1 kg wt | $= 9.807$ N $= 9.807 \times 10^5$ dyn |
| 1 atmosphere of pressure (supports 76 cm of mercury) | $= 1.013$ bar $= 1.013 \times 10^5$ Pa |
| 1 bar ($10^6$ dyne $cm^{-2}$) | $= 10^5$ Pa |
| 1 kg wt $cm^{-2}$ | $= 0.9807$ bar $= 9.807 \times 10^4$ Pa |
| 1 calorie | = 4.184 J |
| 1 Heat Flux Unit (1 $\mu$ cal $cm^{-2}$ $sec^{-1}$) | $= 4.184 \times 10^{-2}$ W $m^{-2}$ |
| 1 electron volt (ev) | $= 1.602 \times 10^{-19}$ J |
| 1 radian | $= 57^\circ.30 = 2.063 \times 10^5$ arc sec |
| 1 arc sec | $= 4.848 \times 10^{-6}$ rad |
| 1 gamma | $= 10^{-5}$ G = 1 nano tesla (nT) ($\equiv 10^{-9}$ Wb $m^{-2}$) |
| 1 gal (1 cm $sec^{-2}$) | $= 10^{-2}$ m $sec^{-2}$ |
| 1 milligal | $= 10^{-5}$ m $sec^{-2}$ |

# APPENDIX G

# Earth Model

**Table G.1** Parameters of Parametric Earth Models (PEM's) by Dziewonski et al. (1975) as Functions of Normalized Radius $R$ (i.e., radius/6371 km)

<table>
<tr><th>Region</th><th>Radius Range (km)</th><th>Density ($10^3$ kg m$^{-3}$)</th><th>$V_P$(km sec$^{-1}$)</th><th>$V_S$(km sec$^{-1}$)</th></tr>
<tr><td>Inner core</td><td>0–1217.1</td><td>$13.01219 - 8.45292R^2$</td><td>$11.24094 - 4.09689R^2$</td><td>$3.56454 - 3.45241R^2$</td></tr>
<tr><td>Outer core</td><td>1217.1–3485.7</td><td>$12.58416 - 1.69929R - 1.94128R^2 - 7.11215R^3$</td><td>$10.03904 + 3.75665R - 13.67046R^2$</td><td>0</td></tr>
<tr><td>Lower mantle</td><td>3485.7–5701.0</td><td>$6.81430 - 1.66273R - 1.18531R^2$</td><td>$16.69287 - 6.38826R + 4.68676R^2 - 5.30512R^3$</td><td>$9.20501 - 6.85512R + 9.39892R^2 - 6.25575R^3$</td></tr>
<tr><td>Transition zone</td><td>5701.0–5951.0</td><td>$11.11978 - 7.87054R$</td><td>$21.05692 - 12.31433R$</td><td>$15.04371 - 10.69726R$</td></tr>
<tr><td colspan="5">Oceanic Structure</td></tr>
<tr><td>Below LVZ</td><td>5951.0–6151.0</td><td rowspan="3">$7.15855 - 3.85999R$</td><td>$30.00026 - 22.53683R$</td><td>$15.87214 - 11.86483R$</td></tr>
<tr><td>LVZ</td><td>6151.0–6311.0</td><td>7.87320</td><td>4.33450</td></tr>
<tr><td>Above LVZ</td><td>6311.0–6360.0</td><td>7.90000</td><td>4.55000</td></tr>
</table>

Table G.1 (*Contd.*)

| Region | Radius Range (km) | Density ($10^3$ kg m$^{-3}$) | $V_P$(km sec$^{-1}$) | $V_S$(km sec$^{-1}$) |
|---|---|---|---|---|
| **Oceanic Structure** (*Contd.*) | | | | |
| Crust | 6360.0–6366.0 | 2.85000 | 6.40000 | 3.70000 |
| Sediments | 6366.0–6367.0 | 1.50000 | 2.00000 | 1.00000 |
| Ocean | 6367.0–6371.0 | 1.03000 | 1.50000 | 0 |
| **Continental Structure** | | | | |
| Below LVZ | 5951.0–6151.0 | 7.15855 | 25.60797 $-17.63609R$ | 13.52229 $-9.32106R$ |
| LVZ | 6151.0–6251.0 | $-3.85999R$ | 7.84750 | 4.45860 |
| Above LVZ | 6251.0–6336.0 | | 8.02000 | 4.69000 |
| Lower crust | 6336.0–6351.0 | 2.92000 | 6.50000 | 3.75000 |
| Upper crust | 6351.0–6371.0 | 2.72000 | 5.80000 | 3.45000 |
| **Average Structure** | | | | |
| Below LVZ | 5951.0–6151.0 | 7.15855 | 28.48832 $-20.90003R$ | 15.09536 $-11.01544R$ |
| LVZ | 6151.0–6291.0 | $-3.85999R$ | 7.89520 | 4.34060 |
| Above LVZ | 6291.0–6352.0 | | 7.93420 | 4.65400 |
| Lower crust | 6352.0–6357.0 | 2.90200 | 6.50000 | 3.75000 |
| Upper crust | 6357.0–6368.0 | 2.80200 | 6.00000 | 3.55000 |
| Ocean | 6368.0–6371.0 | 1.03000 | 1.50000 | 0 |

**Table G.2** Data for Parametric Earth Models (PEM's) by Dziewonski et al. (1975).

| Radius (km) | Density ($10^3$ kg $m^{-3}$) | Pressure ($10^{11}$ Pa ≡ M bar) | Gravity (m $sec^{-2}$) | Incompressibility ($10^{11}$ Pa) | Rigidity ($10^{11}$ Pa) |
|---|---|---|---|---|---|
| 0.0 | 13.012 | 3.6324 | 0.000 | 14.237 | 1.653 |
| 100.0 | 13.010 | 3.6300 | 0.364 | 14.233 | 1.652 |
| 200.0 | 13.004 | 3.6230 | 0.727 | 14.220 | 1.649 |
| 300.0 | 12.993 | 3.6112 | 1.090 | 14.200 | 1.643 |
| 400.0 | 12.979 | 3.5947 | 1.452 | 14.170 | 1.636 |
| 500.0 | 12.960 | 3.5735 | 1.813 | 14.133 | 1.627 |
| 600.0 | 12.937 | 3.5477 | 2.174 | 14.087 | 1.615 |
| 700.0 | 12.910 | 3.5173 | 2.533 | 14.033 | 1.602 |
| 800.0 | 12.879 | 3.4823 | 2.891 | 13.971 | 1.586 |
| 900.0 | 12.844 | 3.4428 | 3.247 | 13.901 | 1.569 |
| 1000.0 | 12.804 | 3.3989 | 3.601 | 13.822 | 1.550 |
| 1100.0 | 12.760 | 3.3507 | 3.953 | 13.736 | 1.529 |
| 1200.0 | 12.712 | 3.2981 | 4.302 | 13.642 | 1.506 |
| 1217.1 | 12.704 | 3.2887 | 4.362 | 13.625 | 1.502 |
| 1217.1 | 12.139 | 3.2887 | 4.362 | 12.773 | 0.000 |
| 1300.0 | 12.096 | 3.2436 | 4.613 | 12.674 | 0.000 |
| 1400.0 | 12.042 | 3.1861 | 4.918 | 12.538 | 0.000 |
| 1500.0 | 11.984 | 3.1252 | 5.226 | 12.384 | 0.000 |
| 1600.0 | 11.922 | 3.0609 | 5.533 | 12.210 | 0.000 |
| 1700.0 | 11.857 | 2.9933 | 5.841 | 12.019 | 0.000 |
| 1800.0 | 11.789 | 2.9224 | 6.147 | 11.810 | 0.000 |
| 1900.0 | 11.716 | 2.8484 | 6.451 | 11.584 | 0.000 |
| 2000.0 | 11.639 | 2.7713 | 6.752 | 11.341 | 0.000 |
| 2100.0 | 11.558 | 2.6913 | 7.051 | 11.082 | 0.000 |
| 2200.0 | 11.473 | 2.6084 | 7.347 | 10.808 | 0.000 |
| 2300.0 | 11.383 | 2.5227 | 7.639 | 10.520 | 0.000 |
| 2400.0 | 11.288 | 2.4345 | 7.926 | 10.218 | 0.000 |
| 2500.0 | 11.189 | 2.3439 | 8.210 | 9.903 | 0.000 |
| 2600.0 | 11.084 | 2.2509 | 8.488 | 9.577 | 0.000 |
| 2700.0 | 10.974 | 2.1558 | 8.762 | 9.239 | 0.000 |
| 2800.0 | 10.859 | 2.0587 | 9.030 | 8.892 | 0.000 |
| 2900.0 | 10.738 | 1.9598 | 9.292 | 8.536 | 0.000 |
| 3000.0 | 10.611 | 1.8592 | 9.548 | 8.173 | 0.000 |
| 3100.0 | 10.478 | 1.7572 | 9.798 | 7.804 | 0.000 |
| 3200.0 | 10.340 | 1.6540 | 10.040 | 7.430 | 0.000 |
| 3300.0 | 10.195 | 1.5497 | 10.276 | 7.052 | 0.000 |
| 3400.0 | 10.043 | 1.4446 | 10.504 | 6.671 | 0.000 |
| 3485.7 | 9.909 | 1.3540 | 10.693 | 6.345 | 0.000 |
| 3485.7 | 5.550 | 1.3540 | 10.693 | 6.582 | 2.911 |
| 3500.0 | 5.543 | 1.3455 | 10.672 | 6.559 | 2.904 |
| 3600.0 | 5.496 | 1.2870 | 10.537 | 6.396 | 2.852 |
| 3700.0 | 5.449 | 1.2296 | 10.421 | 6.233 | 2.800 |

Table G.2 (*Contd.*)

| Radius (km) | Density ($10^3$ kg $m^{-3}$) | Pressure ($10^{11}$ Pa ≡ M bar) | Gravity (m $sec^{-2}$) | Incompressibility ($10^{11}$ Pa) | Rigidity ($10^{11}$ Pa) |
|---|---|---|---|---|---|
| 3800.0 | 5.401 | 1.1734 | 10.323 | 6.069 | 2.747 |
| 3900.0 | 5.352 | 1.1181 | 10.240 | 5.905 | 2.694 |
| 4000.0 | 5.303 | 1.0637 | 10.169 | 5.741 | 2.641 |
| 4100.0 | 5.253 | 1.0102 | 10.111 | 5.576 | 2.587 |
| 4200.0 | 5.203 | 0.9574 | 10.063 | 5.412 | 2.533 |
| 4300.0 | 5.152 | 0.9054 | 10.025 | 5.248 | 2.478 |
| 4400.0 | 5.101 | 0.8541 | 9.994 | 5.084 | 2.423 |
| 4500.0 | 5.049 | 0.8035 | 9.971 | 4.921 | 2.367 |
| 4600.0 | 4.996 | 0.7534 | 9.954 | 4.758 | 2.310 |
| 4700.0 | 4.943 | 0.7040 | 9.942 | 4.596 | 2.252 |
| 4800.0 | 4.889 | 0.6551 | 9.936 | 4.435 | 2.194 |
| 4900.0 | 4.834 | 0.6068 | 9.934 | 4.275 | 2.135 |
| 5000.0 | 4.779 | 0.5591 | 9.935 | 4.116 | 2.075 |
| 5100.0 | 4.724 | 0.5119 | 9.940 | 3.958 | 2.015 |
| 5200.0 | 4.668 | 0.4652 | 9.947 | 3.802 | 1.953 |
| 5300.0 | 4.611 | 0.4190 | 9.957 | 3.647 | 1.891 |
| 5400.0 | 4.553 | 0.3734 | 9.969 | 3.494 | 1.828 |
| 5500.0 | 4.496 | 0.3282 | 9.982 | 3.343 | 1.765 |
| 5600.0 | 4.437 | 0.2836 | 9.996 | 3.194 | 1.701 |
| 5701.0 | 4.377 | 0.2391 | 10.012 | 3.045 | 1.636 |
| 5701.0 | 4.077 | 0.2391 | 10.012 | 2.480 | 1.220 |
| 5751.0 | 4.015 | 0.2188 | 10.006 | 2.414 | 1.165 |
| 5801.0 | 3.953 | 0.1989 | 10.000 | 2.348 | 1.111 |
| 5851.0 | 3.892 | 0.1793 | 9.993 | 2.284 | 1.060 |
| 5901.0 | 3.830 | 0.1600 | 9.985 | 2.220 | 1.010 |
| 5951.0 | 3.768 | 0.1411 | 9.976 | 2.157 | 0.961 |
| | | ***Continental Structure*** | | | |
| 5951.0 | 3.553 | 0.1407 | 9.976 | 1.865 | 0.823 |
| 6001.0 | 3.523 | 0.1231 | 9.957 | 1.794 | 0.792 |
| 6051.0 | 3.492 | 0.1057 | 9.939 | 1.724 | 0.761 |
| 6101.0 | 3.462 | 0.0884 | 9.921 | 1.656 | 0.731 |
| 6151.0 | 3.432 | 0.0713 | 9.904 | 1.590 | 0.702 |
| 6151.0 | 3.432 | 0.0713 | 9.904 | 1.203 | 0.682 |
| 6201.0 | 3.402 | 0.0544 | 9.887 | 1.193 | 0.676 |
| 6251.0 | 3.371 | 0.0377 | 9.870 | 1.182 | 0.670 |
| 6251.0 | 3.371 | 0.0377 | 9.870 | 1.179 | 0.741 |
| 6301.0 | 3.341 | 0.0211 | 9.854 | 1.169 | 0.734 |
| 6336.0 | 3.320 | 0.0097 | 9.842 | 1.161 | 0.730 |
| 6336.0 | 2.920 | 0.0097 | 9.842 | 0.686 | 0.410 |
| 6351.0 | 2.920 | 0.0053 | 9.833 | 0.686 | 0.410 |
| 6351.0 | 2.720 | 0.0053 | 9.833 | 0.483 | 0.323 |
| 6371.0 | 2.720 | 0.0000 | 9.806 | 0.483 | 0.323 |

**Table G.2** (*Contd.*)

| Radius (km) | Density ($10^3$ kg m$^{-3}$) | Pressure ($10^{11}$ Pa ≡ M bar) | Gravity (m sec$^{-2}$) | Incompressibility ($10^{11}$ Pa) | Rigidity ($10^{11}$ Pa) |
|---|---|---|---|---|---|
| | | ***Oceanic Structure*** | | | |
| 5951.0 | 3.553 | 0.1411 | 9.976 | 1.758 | 0.815 |
| 6001.0 | 3.523 | 0.1235 | 9.957 | 1.674 | 0.776 |
| 6051.0 | 3.492 | 0.1061 | 9.939 | 1.593 | 0.740 |
| 6101.0 | 3.462 | 0.0888 | 9.921 | 1.514 | 0.704 |
| 6151.0 | 3.432 | 0.0717 | 9.904 | 1.438 | 0.669 |
| 6151.0 | 3.432 | 0.0717 | 9.904 | 1.267 | 0.644 |
| 6161.0 | 3.426 | 0.0683 | 9.901 | 1.265 | 0.643 |
| 6231.0 | 3.383 | 0.0448 | 9.877 | 1.249 | 0.635 |
| 6271.0 | 3.359 | 0.0314 | 9.864 | 1.240 | 0.631 |
| 6311.0 | 3.335 | 0.0182 | 9.850 | 1.231 | 0.626 |
| 6311.0 | 3.335 | 0.0182 | 9.850 | 1.160 | 0.690 |
| 6336.0 | 3.320 | 0.0101 | 9.842 | 1.155 | 0.687 |
| 6360.0 | 3.305 | 0.0022 | 9.835 | 1.150 | 0.684 |
| 6360.0 | 2.850 | 0.0022 | 9.835 | 0.647 | 0.390 |
| 6366.0 | 2.850 | 0.0006 | 9.830 | 0.647 | 0.390 |
| 6366.0 | 1.500 | 0.0006 | 9.830 | 0.040 | 0.015 |
| 6367.0 | 1.500 | 0.0004 | 9.829 | 0.040 | 0.015 |
| 6367.0 | 1.030 | 0.0004 | 9.829 | 0.023 | 0.000 |
| 6371.0 | 1.030 | 0.0000 | 9.820 | 0.023 | 0.000 |
| | | ***Average Structure*** | | | |
| 5951.0 | 3.553 | 0.1411 | 9.976 | 1.762 | 0.820 |
| 6001.0 | 3.523 | 0.1234 | 9.957 | 1.683 | 0.784 |
| 6051.0 | 3.492 | 0.1060 | 9.939 | 1.606 | 0.749 |
| 6101.0 | 3.462 | 0.0887 | 9.921 | 1.532 | 0.715 |
| 6151.0 | 3.432 | 0.0716 | 9.904 | 1.459 | 0.682 |
| 6151.0 | 3.432 | 0.0716 | 9.904 | 1.277 | 0.646 |
| 6186.0 | 3.411 | 0.0598 | 9.892 | 1.269 | 0.642 |
| 6221.0 | 3.389 | 0.0480 | 9.884 | 1.261 | 0.638 |
| 6256.0 | 3.368 | 0.0363 | 9.869 | 1.253 | 0.634 |
| 6291.0 | 3.347 | 0.0248 | 9.857 | 1.245 | 0.630 |
| 6291.0 | 3.347 | 0.0248 | 9.857 | 1.140 | 0.724 |
| 6321.0 | 3.329 | 0.0149 | 9.847 | 1.134 | 0.721 |
| 6352.0 | 3.310 | 0.0048 | 9.837 | 1.127 | 0.716 |
| 6352.0 | 2.902 | 0.0048 | 9.837 | 0.681 | 0.408 |
| 6357.0 | 2.902 | 0.0033 | 9.834 | 0.681 | 0.408 |
| 6357.0 | 2.802 | 0.0033 | 9.834 | 0.537 | 0.353 |
| 6368.0 | 2.802 | 0.0003 | 9.826 | 0.537 | 0.353 |
| 6368.0 | 1.030 | 0.0003 | 9.826 | 0.023 | 0.000 |
| 6371.0 | 1.030 | 0.0000 | 9.819 | 0.023 | 0.000 |

# APPENDIX H

## Radioactive Elements

**Table H.1** Radioactive Elements Used in Dating.

| Element | Radio-isotope | Percent of Natural Element | Decay Mechanism | Half-Life (years) | Decay Constant ($year^{-1}$) | Stable Daughter Product |
|---|---|---|---|---|---|---|
| Uranium | $U^{238}$ | 99.2743[a] | $(8\alpha + 6\beta)$ series decay; $5.4 \times 10^{-5}$% spontaneous fission | $4.468 \times 10^{9}$ | $1.55125 \times 10^{-10}$ | Lead, $Pb^{206}$ |
| | $U^{235}$ | 0.7201[a] | $(7\alpha + 4\beta)$ series decay (also neutron-induced fission) | $7.038 \times 10^{8}$ | $9.8485 \times 10^{-10}$ | Lead, $Pb^{207}$ |
| Thorium | $Th^{232}$ | Nearly 100[b] | $(6\alpha + 4\beta)$ series decay | $1.401 \times 10^{10}$ | $4.9475 \times 10^{-11}$ | Lead, $Pb^{208}$ |
| Rubidium | $Rb^{87}$ | 27.85 | $\beta$ | $4.88 \times 10^{10}$ | $1.42 \times 10^{-11}$ | Strontium, $Sr^{87}$ |
| Potassium | $K^{40}$ | 0.01167 | 10.5% electron capture; 89.5% $\beta$ | $1.250 \times 10^{9}$ | $5.544 \times 10^{-10}$[c] | 10.5% Argon, $Ar^{40}$; 89.5% Calcium, $Ca^{40}$ |
| Rhenium | $Re^{187}$ | 66 | $\beta$ | $4.3 \times 10^{10}$ | $1.6 \times 10^{-11}$ | Osmium, $Os^{187}$ |
| Carbon | $C^{14}$ | $1.6 \times 10^{-10}$ (in atmospheric $CO_2$) | $\beta$ | $5.73 \times 10^{3}$ | $1.21 \times 10^{-4}$ | Nitrogen, $N^{14}$ |

[a] Not quite 100% taken together because $U^{234}$, a shorter-lived intermediate product in $U^{238}$ decay, coexists with it. The atomic ratio $U^{238}/U^{235}$ is 137.88.

[b] A short-lived thorium isotope ($Th^{230}$) occurs in the decay series of $U^{238}$.

[c] $\lambda_{Ar} = 0.5811 \times 10^{-10}$, $\lambda_{Ca} = 4.963 \times 10^{-10}$ $year^{-1}$.

**Table H.2** Some Extinct Isotopes.

| Element | Isotope | Decay Mechanism | Half-Life (years) | Decay Constant (year$^{-1}$) | Stable Daughter Product |
|---|---|---|---|---|---|
| Aluminium | $Al^{26}$ | 85% positron emission<br>15% electron capture | $8.8 \times 10^5$ | $7.9 \times 10^{-7}$ | $Mg^{26}$ |
| Iodine | $I^{129}$ | $\beta$ | $1.64 \times 10^7$ | $4.2 \times 10^{-8}$ | $Xe^{129}$ |
| Plutonium | $Pu^{244}$ | 99.7% $\alpha$ (and following series)<br>0.3% spontaneous fission | $8.2 \times 10^7$ | $8.5 \times 10^{-9}$ | 99.7% $Pb^{208}$ (via $Th^{232}$)<br>0.3% fission products, notably Xe isotopes |
| Curium | $Cm^{248}$ | 93% $\alpha$ (and following series)<br>7% spontaneous fission | $3.7 \times 10^5$ | $1.87 \times 10^{-6}$ | 93% $Pb^{208}$ (via $Pu^{244}$, $Th^{232}$)<br>7% fission products, including Xe isotopes |

**Table H.3** Energy Output of Radioactive Elements.

| Isotope | Energy/Distintegration (MeV) | Power ($\mu$W/kg of isotope) | Power ($\mu$W/kg of element) | Total Heat in Earth[a] ($10^{12}$ W) | Total Heat $4.5 \times 10^9$ yr ago ($10^{12}$ W) |
|---|---|---|---|---|---|
| $U^{238}$ | 47.7 | 95.0 | 94.35 } 98.4 | 11.33 | 22.77 |
| $U^{235}$ | 43.9 | 562.0 | 4.05 } 98.4 | 0.486 | 40.8 |
| $Th^{232}$ | 40.5 | 26.6 | 26.6 | 11.18 | 13.97 |
| $K^{40}$ | 0.71 | 30.0 | 0.0035 | 8.41 | — |
| | | | | 31.4 | |

Primary data from Hamza and Beck (1972) (see also Birch, 1954). Values are locally absorbed energies and assume complete escape of neutrinos.

[a]These data assume a total heat generation of $31.4 \times 10^{12}$ W from an earth with proportions of radioactive elements characteristic of carbonaceous chondrites, Th/U = 3.5, K/U = $2 \times 10^4$ by mass. Note however the uncertainty in the terrestrial K/U ratio, which is considered in Section 2.3.

# APPENDIX I

# The Geological Time Scale[a]

| Era | Period | Epoch | Time before present ($10^6$ years) |
|---|---|---|---|
| Cenozoic (mammals) | Quaternary | Recent (Holocene) | 0–0.01 |
| | | Pleistocene | 0.01–1.8 |
| | Tertiary | Pliocene | 1.8–5.6 |
| | | Miocene | 5.6–23 |
| | | Oligocene | 23–36.5 |
| | | Eocene | 36.5–54.5 |
| | | Paleocene | 54.5–65 |
| Mesozoic (reptiles) | Cretaceous | | 65–143 |
| | Jurassic | | 143–212 |
| | Triassic | | 212–247 |
| Paleozoic (invertebrates) | Permian | | 247–289 |
| | Carboniferous | | 289–367 |
| | Devonian | | 367–416 |
| | Silurian | | 416–446 |
| | Ordovician | | 446–509 |
| | Cambrian | | 509–575 |
| Pre-Cambrian | | | 575– |

[a]Based on reviews of data by Harland et al. (1964), Berggren (1969) and Armstrong and McDowall (1975).

# APPENDIX J

# Problems

**1.1** Consider a spherical black body at 1 AU from the Sun to be rotating and tumbling in such a way as to equalize the temperature over its whole surface. What is that temperature? (Use the value of the solar constant in Appendix F.)

**1.2** What would be the equilibrium temperature of the subsolar point on the body in Problem 1.1 if it presented a constant face to the Sun? (Neglect conduction within the body.)

**1.3** What is the equivalent black-body temperature for the surface of the Sun (radius $6.96 \times 10^8$ m)?

**1.4** How does the equilibrium temperature of a planet depend on the radius of its orbit?

**1.5** By virtue of the rotation of the Sun, radiation from opposite (approaching and receding) limbs is seen to be Doppler shifted (to blue and red respectively). Thus the radiation field carries away some of the solar angular momentum. What is the fractional rate of slowing of the Sun's rotation by this effect? (Assume the moment of inertia of the Sun to be $0.3\,MR^2$.)

**1.6** Suppose that collisions in the asteroidal belt are continuously producing interplanetary dust with a distribution of sizes such that the number of particles $dN$ within any mass range $m$ to $(m + dm)$ given by $dN \propto m^{-n}\,dm$, where $n > 1$. (According to Dohnanyi, 1968, $n \approx 1.83$.)
Once produced, the dust spirals toward the Sun by the Poynting-Robertson effect (Section 1.4). What is the distribution of particle masses intercepted by the Earth? Assume the mass intercepted to be only a small fraction of the total and to be determined by residence time near to the Earth's orbit and that all particles have equal densities.

**1.7** Assuming meteorites to be of two kinds, irons that are 100% metal and chondrites that average 10% metal and that their total metal content is the same as that of the Earth, what fraction of the total mass of meteorites is in the chondrites?

**1.8** Consider the Moon to be momentarily directly between the Sun and Earth (during a solar eclipse) and calculate the ratio of its gravitational attractions to the Sun and Earth. Since the solar gravity is stronger, explain how the Moon remains in orbit about the Earth instead of going into an independent orbit about the Sun.

**1.9** The uncompressed density of the planet Mercury is estimated to be 5300 kg $m^{-3}$. If it consists of a core and mantle with compositions similar to those of the Earth, what is the ratio of core radius to total radius?

**1.10** What is the average speed of arrival on the Earth of meteoroids falling freely from outside the solar system? How does it compare with the escape speed from the Earth?

**2.1** Derive Equations 1.4 and 1.5.

**2.2** Derive Equation 2.2

**2.3** Show that for atoms of a radioactive species with decay constant $\lambda$, the mean life is $\lambda^{-1}$.

**2.4** Consider a rock that yields the following isochron data:

$$\frac{d(\mathrm{Ar}^{40})}{d(\mathrm{K}^{40})} = 0.098 \pm 0.005$$

$$\frac{d(\mathrm{Sr}^{87}/\mathrm{Sr}^{86})}{d(\mathrm{Rb}^{87}/\mathrm{Sr}^{86})} = 0.0215 \pm 0.0007$$

$$\frac{d(\mathrm{Pb}^{207}/\mathrm{Pb}^{204})}{d(\mathrm{Pb}^{206}/\mathrm{Pb}^{204})} = 0.090 \pm 0.005$$

(a) Calculate "ages" from each of these isochrons with the uncertainties inferred from the isochron data.

(b) Suggest reasons for the discrepancies between the alternative "ages" and a probable history.

**2.5** Of the hydrogen atoms in ordinary terrestrial water (in the atmosphere, sea, etc.) approximately 0.014% are of the heavy isotope, D or $H^2$. Juvenile water that emerges from the interior of the earth in the outgassing process is about 10% less rich in D. Assuming all terrestrial water to be an outgassing product, the compositional difference can be explained by dissociation of water in the upper atmosphere and selective loss of the light isotope $H^1$ to space. How much atmospheric oxygen would that leave? How does this compare with the mass of oxygen in the atmosphere? [Note: atmospheric oxygen is continuously lost by oxidation of freshly produced minerals in the crust, e.g., magnetite ($Fe_3O_4$) to hematite ($Fe_2O_3$), and volcanic gases are deficient in oxygen and even contain hydrogen. Thus it is apparent that without the selective escape of hydrogen to space there could not be a significant oxygen content in the atmosphere.]

**2.6** If the process of nuclear synthesis produced $I^{127}$ and $I^{129}$ in equal atomic abundances and proceeded at a uniform rate for a time $\tau$, and did not operate either before or after this, and if accumulation of $Xe^{129}$ in a particular meteorite commenced at time $t$ after the cessation of nuclear synthesis, obtain the expression relating the ratio ($Xe^{129}/I^{127}$) in the meteorite to $\tau$, $t$, and the decay constant $\lambda$ for $I^{129}$. Consider further the special cases $\tau \ll \lambda^{-1}$ and $\tau \gg \lambda^{-1}$.

**2.7** If the sodium occurring as NaCl in sea water amounts to 1.07% of the mass of the sea and has accumulated by solution from eroding rocks that on average contribute 1% of their mass as sodium, how much rock has been eroded over geological time? What fraction of the present crust is this? (Estimated rates of erosion and of accumulation of salt in the sea have been used to infer an approximate age of the Earth.)

**2.8** Why is the $K^{40} \rightarrow Ca^{40}$ decay not usable as a means of dating rocks?

**3.1** Consider the Earth to be a sphere of mass $M$ and radius $R$ consisting of two zones each of uniform density, a mantle of thickness $R/2$ and a core of radius $R/2$ having a density $f$ times the mantle density. The moment of inertia is $\frac{1}{3}MR^2$. What is the value of $f$?

**3.2** (a) Mars rotates axially at an angular rate $\omega = 7.0882 \times 10^{-5}$ rad sec$^{-1}$ and has an ellipticity coefficient of gravitational potential, $J_2 = 0.001972$. Assuming it to be in hydrostatic equilibrium throughout, estimate the surface flattening and the coefficient of the moment of inertia, $C/Ma^2$. (Other relevant data may be obtained from Table A.1 and Appendix F.)

(b) What is the rate of precession of Mars, due to solar torques, from these values, if the inclination of its equator to the orbital plane is 24°.

**3.3** What are the values for mass and moment of inertia of the inner core according to the density relationship in Table G.1? What fractions of the whole Earth values are these?

**3.4** (a) If the Greenland ice sheet (at 75°N) decreases by the melting of 100 km$^3$ of ice annually, distributing melt water uniformly over the surface of the Earth with no other disturbance or adjustment, estimate:

(i) The angular migration of the pole.

(ii) The associated slowing of the Earth's rotation rate.

(b) What influence would isostatic rebound have on your conclusions to part (a)?

**3.5** What is the effect on the Earth's rotation of the diurnal development of traffic in North America by virtue of the fact that by driving on the right it causes a net circulation of mass? (Guesses are required for the mass of traffic, its mean speed and the effective width of highways.)

**3.6** Consider two concentric spheres that are constrained to rotate about axes inclined to one another by an angle $\phi$. If a thin layer of viscous fluid between them exerts a uniform frictional drag on the differential motion and the rotation of the outer sphere is maintained at an angular velocity $\omega_0$, what is the rotational speed of the inner one? If the fluid layer has radius $r$, thickness $d \ll r$ and viscosity $\eta$ what is the power dissipation?
What limit must be imposed on the viscosity of an outer core boundary layer 20 km thick, if the core rotates about an axis at an angle of $1.7 \times 10^{-5}$ rad to the mantle axis?

**3.7** Show that the critical speed of rotational instability of a spherical planet of mean density $\bar{\rho}$ (i.e., the speed at which equatorial gravity becomes zero) is

$$\omega_{crit} = \sqrt{\tfrac{4}{3}\pi\bar{\rho}G}$$

and the corresponding angular momentum (for uniform density) is

$$L_{\text{crit}} = 0.32\, G^{1/2} M^{5/3} \bar{\rho}^{-1/6}$$

where $M$ is the planetary mass.

A better calculation considers the ellipticity of the planet, using a general equation for potential or gravity of an ellipsoid as in Appendix B.

(*Note*: Observed planetary and asteroidal angular momenta follow quite closely a law of this form but with a numerical constant of 0.07. This suggests that processes of accretion and collision lead to rotational speeds about 25% of the speed of rotational instability, which, by the first of the above equations, is roughly constant.)

**4.1** (a) The following series of gravity measurements was made at noon each day on a ship sailing due north at constant speed. At what time on what day did it cross the equator and what was its speed?

| Day No. | $g$ | Day No. | $g$ |
|---|---|---|---|
| 1 | 9.80222 | 9 | 9.78050 |
| 2 | 9.79805 | 10 | 9.78033 |
| 3 | 9.79415 | 11 | 9.78083 |
| 4 | 9.79059 | 12 | 9.78197 |
| 5 | 9.78747 | 13 | 9.78374 |
| 6 | 9.78484 | 14 | 9.78609 |
| 7 | 9.78278 | 15 | 9.78897 |
| 8 | 9.78132 | 16 | 9.79232 |

(b) At latitude $\phi$ the ship turns due East. What is the apparent difference in gravity (measured while the ship is in motion) due to this changed direction of motion?

**4.2** If 30% of the "excess" ellipticity is to be accounted for in terms of a delay in the response of the Earth to slowing rotation, what is the time lag?

**4.3** Show that for a fluid earth of uniform density the potential Love number is $k_2 = \frac{3}{2}$. (Relate the potential on the surface of a uniform prolate ellipsoid by Eq. 3.15 to the exciting potential by Eq. 4.43.)

**4.4** Substituting Eq. 4.37 in 4.35 we find a $\cos \psi$ term in $W$, that is, the tidal potential appears asymmetrical with respect to the center of the Earth. Why does the asymmetrical term not appear in Equation 4.42?

**4.5** What is the approximate amplitude of the tide raised by the Sun in the planet Mercury? Is this sufficient to exceed the probable yield point?

**4.6** Suppose that the primeval earth had a rotation period one quarter of the present value and that subsequent slowing has been due entirely to angular momentum exchange with the Moon, all motions being coplanar.

(a) If the Moon was formed in a circular orbit, what was the orbital radius?

(b) If the Moon was captured, what was the distance of closest approach?

**4.7** Consider the marine tide in a hypothetical 5-km deep channel around the equator. It is resonant when the speed of a free shallow water wave coincides with the tidal speed. What are the angular velocities of the earth's rotation and of the Moon in orbit, and the Moon's distance from the Earth when this condition is satisfied? Assume conservation of angular momentum in the Earth-Moon system. (The wave speed is given by Eq. 5.47.)

**4.8** What would be the effect on the Earth's rotation if tidal power stations were used to harness $10^{12}$ W of tidal energy (about one sixth of world energy requirements)?

**4.9** Show that the gravitational field of a uniform sheet of material is independent of distance from it for distances small compared with the linear dimensions of the sheet. Why can the correct gravitational field not be obtained by considering a spherical shell of indefinitely large radius?

**4.10** If isostatic balance is maintained during deposition of sediments ($\rho = 2000$ kg m$^{-3}$) on the sea floor by asthenospheric flow at a depth at which $\rho = 3400$ kg m$^{-3}$, what depression of basement rocks results from 1 km of sediment? (Take the density of sea water to be 1000 kg m$^{-3}$.)

**4.11** If all harmonic terms in gravitational potential of the same degree, $l$, are summed we obtain the degree amplitudes, $U_l$, given by Equation 4.26. (Coefficients are summed as squares because spherical harmonics are orthogonal.) Values of $U_l$ from the coefficients of Rapp (1973), with minor correction for bias introduced by uncertainties in the coefficients, are

| $l$ | $U_l(10^{-6})$ | $l$ | $U_l(10^{-6})$ |
|---|---|---|---|
| 2 | 5.702* | 11 | 0.295 |
| 3 | 2.970 | 12 | 0.203 |
| 4 | 1.487* | 13 | 0.283 |
| 5 | 1.186 | 14 | 0.172 |
| 6 | 0.888 | 15 | 0.183 |
| 7 | 0.737 | 16 | 0.157 |
| 8 | 0.470 | 17 | 0.134 |
| 9 | 0.424 | 18 | 0.184 |
| 10 | 0.382 | 19 | 0.177 |
| | | 20 | 0.150 |

If these amplitudes are to be explained in terms of undulations of a nearly spherical internal boundary of density contrast $\Delta\rho$ and radius $r < a$, where $a$ is Earth radius, then the undulation amplitude for each harmonic degree is $h_l$, given by $U_l$ as in Equation 4.24.

(a) Neglecting $U_2$ and $U_3$ (which probably include major contributions from incomplete recovery from former glaciations) find the value of $(r/a)$ for which $h_l$ is as nearly as possible independent of $l$. This gives the depth $(a - r)$ of a boundary that must explain the gravity field, with the assumption that the spatial spectrum of the undulations is white.

*Equilibrium ellipticity subtracted.

(A much better fit to the data can be obtained by considering two boundaries with a partial negative correlation, both in the upper mantle.)

(b) How much additional ice on each polar cap would bring $U_2$ into line with the other data? (suppose that it extends with uniform thickness down to latitude 60°N and S.)

**5.1** If Equations 5.1 and 5.3 are strictly correct, how does the duration of a seismic wave train depend on the magnitude or energy of the earthquake causing it?

**5.2** By comparing Equations 5.1 and 5.12 (noting that in 5.1 $a$ represents amplitude of ground motion in microns and in 5.12 $a$ is ground acceleration in m $\sec^{-2}$), show that if $I$ is the intensity of an earthquake at distance $\Delta$, its magnitude is

$$M = \frac{I}{3} + \log_{10} T + f(\Delta, h) + C + 1.9$$

What can you conclude about the relationship between dominant wave period, $T$, and magnitude by comparing this result with Equation 5.13?

**5.3** If earthquakes on a transcurrent fault have stress release of 100 bars ($10^7$Pa) and the fault is moving at a long-term average rate of 5 cm/year, estimate approximately the strength of the heat flow anomaly along it.

**6.1** Consider a laminated medium composed of alternate layers of materials in which the $P$-wave velocities (in bulk) are $V_1$, $V_2$. The layers are of equal thicknesses and the two materials have the same density, but different elasticities, subject to the constraint that Poisson's ratio $\nu$ is the same for both. Show that, for $P$ waves of wavelength long compared with the thicknesses of the layers (but much shorter than the overall dimension of the medium), the velocities perpendicular and parallel to the layering are

$$V_\perp = \frac{\sqrt{2}\, V_1 V_2}{(V_1^2 + V_2^2)^{1/2}}$$

$$V_{/\!/} = \left\{\tfrac{1}{2}(V_1^2 + V_2^2)\left[1 - \frac{(V_1^2 - V_2^2)^2}{(V_1^2 + V_2^2)^2} \cdot \frac{\nu^2}{(1-\nu)^2}\right]\right\}^{1/2}$$

What values of $V_1/V_2$ (or of $V_2/V_1$ since this is equivalent) are required for velocity anisotropies of 1%, 5%, that is, $V_{/\!/}/V_\perp = 1.01,\ 1.05$?

[*Hint*. To calculate the effective elastic modulus parallel to the layers write down a set of equations in the form of Eq. 6.5, with the constraints that the total (combined) strain perpendicular to an axial applied stress is zero. Assume a representative value of $\nu$.]

**6.2** Deduce the thicknesses and velocities of plane horizontal layers which would give the following (hypothetical) first arrivals:

| $S$(km) | $T$ (sec) |
|---|---|
| 1 | 0.33 |
| 3 | 1.00 |
| 5 | 1.53 |
| 10 | 2.53 |
| 20 | 4.53 |
| 30 | 6.53 |
| 50 | 10.53 |
| 60 | 12.38 |

| $S$(km) | $T$ (sec) |
|---|---|
| 80 | 15.45 |
| 100 | 18.53 |
| 120 | 21.61 |
| 140 | 24.62 |
| 160 | 27.05 |
| 200 | 31.93 |
| 250 | 38.03 |
| 300 | 44.13 |

**6.3** Consider a highly simplified model of the Earth, which has three uniform layers only. The total radius is 6400 km with upper mantle 600 km deep, the total mantle depth 2900 km and a core of radius 3500 km. $V_P = 9$ km/sec in the upper mantle and core and $V_P = 12$ km/sec in the lower mantle. What are the epicentral ranges of the core shadow zone and the triplication of arrivals due to the mantle transition?

**6.4** Assuming the Murnaghan equation 6.43 show that

$$P = \frac{K_0}{K_0'}\left[\left(\frac{\rho}{\rho_0}\right)^{K_0'} - 1\right]$$

**6.5** Show that $(\partial P/\partial T)_V = \gamma\rho C_V$, where $\gamma$ is defined by Equation E.1 (Appendix E). (This the differential form of the Mie-Grüneisen equation.)

**6.6** Derive Equation 6.38. Also show that the factor $(1 - g^{-1}\, d\phi/dr)$, used by K. E. Bullen as a test for homogeneity within the Earth, is formally equivalent to $dK/dP$ for a homogeneous adiabatic layer.

**6.7** Show that Equation 6.57 requires
(a) $\alpha = K_0'$
(b) $\rho K' = \rho_0 K_0'$

**6.8** What is the central pressure in a planet of uniform density in terms of its total mass $M$ and radius $R$?

**6.9** (a) Deduce expressions for Poisson's ratio, $\nu$, in terms of $K/\mu$ and $V_P/V_S$.
(b) How does $\nu$ vary with depth according to the model data in Appendix G? On a sketch graph of the radial variation of $\nu$ for the mantle, identify the lithosphere, asthenosphere and mesosphere (which may crudely be identified as the rigid upper layer, or plate, the weak layer and the more rigid lower mantle, respectively; see Fig. 10.3).

**6.10** Consider waves incident normally on a plane boundary between media of densities $\rho_1$, $\rho_2$ in which the velocities are $V_1$, $V_2$. By imposing boundary conditions that the total wave amplitudes must be the same in the two media and the flux of wave energy approaching the boundary must be equal to that leaving it, show that the ratio of reflected to incident amplitudes is

$$\frac{V_2\rho_2 - V_1\rho_1}{V_2\rho_2 + V_1\rho_1}$$

What is the meaning of the possible alternative signs of this quantity?

**6.11** The specific heats of metals at very low temperatures are dominated by the effect of conduction electrons, being proportional to absolute temperature. For iron at ordinary pressures the numerical relationship for electronic specific heat is

$$C_e = 0.089\ T\ \mathrm{J\ kg^{-1}\ deg^{-1}}$$

(see Kittel, 1971, p. 252). However, the numerical factor is proportional to the density of states at the Fermi level $D$, which decreases with pressure, and an electronic Grüneisen ratio $\gamma_e$ may be defined in terms of this variation:

$$\gamma_e = \frac{\partial \ln D}{\partial \ln V}$$

For iron $\gamma_e = 2.1$ at laboratory pressures but probably decreases somewhat with pressure (but not approaching the free electron value $\gamma_e = \frac{2}{3}$).

(a) Assume $\gamma_e = 2$ as an approximation for the core and estimate $C_e$ for iron at $T = 4000$ K and $P = 2$ Mbar ($2 \times 10^{11}$ Pa).

(b) What is the electron contribution to thermal pressure (see Eq. E.3).

(c) How does this affect the interpretation of outer core densities?

[Notes: (i) The classical lattice contribution to the specific heat of iron (3R/mole) is

$$C_V = 447\ \mathrm{J\ kg^{-1}\ deg^{-1}}$$

which is completely dominant at $T \simeq 300$ K but not at core temperatures.

(ii) $C_e$ cannot increase indefinitely according to the above equation but would approach a saturation value of $\frac{3}{2}k$ per available electron if temperatures of several tens of thousands of degrees could be applied.

(iii) The role of electron pressure is discussed by Zharkov and Kalinin (1971).]

**6.12** Show that the thermodynamic Grüneisen ratio according to the Vashchenko-Zubarev relationship (Eq. E.4 with $f = 0$) has the constant value $(n+2)/6$ at arbitrary pressure for a material described by the Born-Mie atomic potential function (Eq. 6.46) with $m = 1$.

**7.1** The equation for thermal diffusion in one dimension (along the $z$ axis) in a material with no internal heat source is

$$\frac{\partial T}{\partial t} = \eta \frac{\partial^2 T}{\partial z^2}$$

where $\eta = K/\rho C$ is the thermal diffusivity. Show that the temperature at depth $z$ from a free surface ($z = 0$) at which there is an imposed temperature $T_0 \sin \omega t$ follows the equation given in the footnote on p. 184.

**7.2** Show that the amplitude of oscillation of the local temperature gradient at depth $z$ in the Earth due to penetration of a surface temperature oscillation $T_0 \sin \omega t$ is

$$\left(\frac{dT}{dz}\right)_{\max} = \sqrt{\frac{\omega}{\eta}}\, T_0 \exp\left(-\sqrt{\frac{\omega}{2\eta}} \cdot z\right)$$

**7.3** Show that the temperature difference over the lower half of a borehole of depth $z$, due to penetration of the thermal wave considered in Problems 7.1 and 7.2 oscillates with amplitude $fT_0$, where

$$f = \left[e^{-x}\left(e^{-x} + 1 - 2e^{-x/2}\cos\frac{x}{2}\right)\right]^{1/2}$$

and $x = \dfrac{z}{z^*} = \sqrt{\dfrac{\omega}{2\eta}} \cdot z$

$z^* = \sqrt{\dfrac{2\eta}{\omega}}$ being the effective penetration depth of the wave

Show further that $f$ has a maximum value of 0.347 at $x = 1.335$. If $z = 1000$ m and $\eta = 1.26 \times 10^{-6}\ \text{m}^2\ \text{sec}^{-1}$ what is the corresponding period of the thermal wave?

**7.4** Consider a plane earth in which the rate of heat generation per unit volume, $q$, varies with depth as

$$q = \frac{Q}{z_0} e^{-z/z_0}$$

where $Q$ is the equilibrium surface heat flux per unit area and $z_0$ is the scale depth of the exponential distribution of sources and is assumed to be much less than the radius of the Earth.

(a) Verify that the total flux integrates to $Q$.

(b) What is the temperature profile in diffusive equilibrium with constant conductivity, $K$?

(c) What is the asymptotic temperature at great depth in terms of the gradient near to the surface $(dT/dz)_0$?

**7.5** What heat flux would you expect from the Moon if it is in equilibrium with internal heat sources and either

(a) the proportions of radioactive elements are the same (by mass) as in the Earth,

(b) the proportions are the same as for the Earth's crust and mantle only. (An assumption about potassium in the core is involved here.)

Accepting as valid averages the observed lunar heat flux of about $0.019\ \text{Wm}^{-2}$ and the mean K/U ratios in Figure 7.3 are representative of the Moon, the crust-mantle mix (terrestrial data) and the whole Earth (carbonaceous chondrite data) and assuming a constant ratio Th/U = 3.5

throughout, what are the enrichment or depletion factors for K and U for the Moon relative to the crust-mantle mix, and relative to carbonaceous chondrites?

**7.6** Consider a planetary model in which the planets have different radii $R$ but similar heat generation per unit volume, $q$, uniform throughout, and temperature profiles in equilibrium with the heat generation. The surfaces are maintained at constant temperature $T_0$. Obtain expressions for the variation of central temperature with radius for each of three alternative assumptions about heat transfer:

(i) Heat transfer is by conduction only with constant conductivity, $K$.

(ii) Heat transfer is purely radiative with constant opacity, that is, $K = AT^3$, where $A$ is a constant.

(iii) Convection maintains an adiabatic gradient throughout. Solve this problem approximately by assuming initially a uniform density to obtain the central pressure, as by Problem 6.8. Then assume compression by Murnaghan's equation (6.39) and obtain $T_{central}/T_0$, assuming constant Grüneisen ratio. (But see Problem 7.7).

**7.7** Strictly, the assumption of constant Grüneisen ratio should be used with a finite strain equation of the form Equation 6.46 with $m = 1$ (see Problem 6.12). Find $\rho/\rho_0$ and hence $(T/T_0)_{adiabatic}$ as functions of $P/K_0$ for material satisfying this equation.

**7.8** What is the total heat capacity of the core if we assume the classical specific heat, $3R$ and a mean atomic weight of 50? What is the mean cooling rate if $5 \times 10^{12}$ W of core heat is accounted for by cooling? (Neglect latent heat of solidification.)

**7.9** If the entropy of melting is a constant and the inner core solidified at a mean temperature of 4400 K what is the total latent heat release by solidification of the inner core? What is the rate of release (in watts) if it has occurred steadily for $4.5 \times 10^9$ years?

**7.10** What is the temperature drop of the core associated with the migration of the inner core boundary to its present position from the center of the Earth? (Assume validity of Eq. 7.25.) Combining solutions to Problems 7.8 and 7.9, what are the values of total heat release and the average rate over $4.5 \times 10^9$ years in this process?

**7.11** If the density increment on solidification of core material is 2.5%, estimate the gravitational energy release in the whole Earth due to solidification of the inner core. What is the corresponding mean temperature rise?

**7.12** What is the gravitational energy release resulting from settling of a core of density $2\bar{\rho}$ and radius $0.55R$ from an initially uniform Earth of density $\bar{\rho}$ and radius $R$, assuming as an approximation that the density of each element of material is unaffected? Substitute numerical values for the Earth. What average temperature rise would this energy release cause?

**8.1** Considering the core to be a sphere with a uniform parallel internal magnetic field, driven by an appropriate pattern of currents which are confined to the core, show that the magnetic field energy external to the core is half of the field energy within the core.

**8.2** If the total kinetic energy of core motion is equal to the total magnetic energy of the dipole field according to the assumption of Problem 8.1, what is the rms speed of core fluid?

**8.3** Consider the model of the core rotating coherently about an axis at a *small* angle $\phi$ to the mantle rotation axis, as in Problem 3.6. What is the value of $\phi$ required to explain the geomagnetic westward drift?

**8.4** Verify the equation for dipole moment given in the footnote on p. 239. What is the corresponding power dissipation?

**8.5** Show that if the core is a sphere of radius $R$ and uniform electrical resistivity $\rho$, carrying a current which is a simple circulation about an axis, then the power dissipated $P$ is related to dipole moment $m$ by:

(a) $P = \dfrac{256}{3\pi^3}\dfrac{m^2\rho}{R^5}$ if the current density is uniform

(b) $P = \dfrac{15}{2\pi}\dfrac{m^2\rho}{R^5}$ if the current density is proportional to radial distance from the axis

Case (b) is the minimum dissipation condition considered by Parker (1972).

**8.6** What is the power dissipation by differential rotation of core and mantle at the westward drift rate, $\Delta\omega = 10^{-10}$ rad sec$^{-1}$, if the electromagnetic coupling coefficient $K_R$, estimated from Equation 3.53, is used?

**8.7** What is the electromagnetic skin depth of the core for a magnetic fluctuation with a 1000-year period?
(Since major components of the secular variation have periods of this order, the estimate gives an extreme upper limit to the thickness of a core boundary layer that is not participating in dynamo action.)

**9.1** Consider the directions of magnetization in a body of rock that was formed over a sufficient time interval to allow the secular variation to be averaged out. Its paleomagnetic colatitude is $\theta$ and the corresponding magnetic inclination is $I_0$, as given by Equation 8.9. But the measured values of inclination are scattered by secular variation so that the individual directions form a cone about the mean. If magnetic colatitudes are calculated for each sample and averaged, the mean colatitude is not equal to the colatitude corresponding to the mean direction because of the nonlinearity of Equation 8.9. A bias is introduced. Calculate the bias in apparent $\theta$ obtained by averaging pole positions determined from inclinations $(I_0 + 15°)$ and $(I_0 - 15°)$ and show that it has a maximum of nearly 4° at $\theta = 36°$. Could this effect explain the "far sidedness" of paleomagnetic poles noted by R. L. Wilson (1972)?

**9.2** If the fraction of intrusive rocks that have spontaneously reversed their magnetizations is $f_1$ and the self-reversed fraction of intruded (rebaked and remagnetized) country rocks is $f_2$, show that the fraction of corresponding pairs that disagree in polarity is $F=(f_1+f_2-2f_1f_2)$ and that this has a maximum value of 0.5 if $f_1=f_2=0.5$. Observations indicate that $F=0.02$; assuming $f_1=f_2=f$, what value of $f$ does this imply?

**10.1** Consider two seismic stations, one very close to an earthquake and another on the opposite side of the Earth. In a particular frequency range the first station sees a "white" spectrum from the *P*-wave arrival and the record from the remote one gives a spectrum such that a plot of ln (amplitude) versus frequency (hertz) has a gradient of $-6$.

(a) What is the effective $Q_P$ for the trans-Earth path?

(b) What is the effective value of $Q_P$ for the mantle if the core is assumed to have infinite $Q_P$?

(c) What approximate value for $Q_s$ of the mantle do these data suggest?

(d) What is the corresponding gradient of the ln (amplitude) versus frequency plot for the *S*-wave spectrum?

(Mantle and core travel times for paths along a diameter may be obtained from data in Fig. 6.5.)

**10.2** Using the data in Problem 10.1 with Equation 10.29 what are the minimum rise times for *P* and *S* pulses at $\Delta = 180°$? What high frequency cutoff could be used with a seismometer without significant loss of these signals?

**10.3** If the effective anelastic $Q$ for the planet Mercury is 200, what is the time for "destruction" of its rotation by tidal friction? (Amplitude of the tidal strain is given by Problem 4.5.

**10.4** With the simplifying assumption that stress-strain loops of rocks are elliptical, what is the resolution in strain measurement required to determine to 10% the anelastic $Q$ of a rock of order 100 from a strain cycle of peak-to-peak amplitude $10^{-5}$?

**10.5** If the rising and falling limbs in the pattern of mantle convection have effective thicknesses $t$, horizontal lengths much greater than $t$ (i.e., on the surface they appear as linear features) and depth $z$, what is the maximum average temperature difference between them permitted by the mean convective stress $\sigma_c$ inferred from the thermodynamic argument of Sections 7.4 and 10.3? Consider cases $t = 100$ km or 1000 km and $z = 700$ km or 2800 km. (Thermal expansion coefficient as a function of depth within the mantle may be obtained from the Grüneisen parameter, as in Appendix E, with $K$ and $\rho$ from Earth model data and a classical specific heat assumed.)

# Bibliography

Acuna, M. H., and Ness, N. F. (1975) Jupiter's main magnetic field measured by Pioneer 11. *Nature*, **253**, 327.

Ade-Hall, J. M., and Watkins, N.D. (1970) Absence of correlations between opaque petrology and natural remanence polarity in Canary Island lavas. *Geophys. J.*, **19**, 351.

Ade-Hall, J. M., and Wilson, R. L. (1963) Petrology and natural remanence of the Mull lavas. *Nature*, **198**, 659.

Aggarwal, Y. P., Sykes, L. R., Armbruster, J., and Sbar, M. L. (1973) Premonitory changes in seismic velocities and prediction of earthquakes. *Nature*, **241**, 101.

Ahrens, L. H. (ed.) (1968) *Origin and distribution of the elements*. Oxford: Pergamon.

Aitken, M. J. (1970) Dating by archaeomagnetic and thermoluminescent methods. *Phil. Trans. Roy. Soc. A*, **269**, 77.

Aitken, M. J., and Weaver, G. H. (1965) Recent archaeomagnetic results in England. *J. Geomag. Geolect.*, **17**, 391.

Aki, K. (1966) Generation and propagation of *G* waves from the Niigata earthquake of June 16, 1964. Part 2. Estimation of earthquake moment, released energy, and stress-strain drop from the *G*-wave spectrum. *Bull. Earthquake Res. Inst., Tokyo*, **44**, 73.

Aki, K. (1972) Scaling law of earthquake source time-function. *Geophys. J.*, **31**, 3.

Alexander, E. C., Lewis, R. S., Reynolds, J. H., and Michel, M. C. (1971) Plutonium-244: confirmation as an extinct radioactivity. *Science*, **172**, 837.

Alfvén, H. (1954) *The origin of the solar system*. Oxford: Clarendon Press.

Alfvén, H. (1965) Origin of the Moon. *Science*, **148**, 476.

Allan, D. W. (1958) Reversals of the Earth's magnetic field. *Nature*, **182**, 469.

Alldredge, L. R., and Hurwitz, L. (1964) Radial dipoles as the sources of the Earth's main magnetic field. *J. Geophys. Res.*, **69**, 2631.

Alldredge, L. R., and Stearns, C. O. (1969) Dipole model of the sources of the Earth's magnetic field and secular change. *J. Geophys. Res.*, **74**, 6583.

Alldredge, L. R., Van Voorhis, G. D., and Davis, T. M. (1963) A magnetic profile around the world. *J. Geophys. Res.*, **68**, 3679.

Allen, C. W. (1963) *Astrophysical quantities. (2nd ed.)* London: Athlone Press.

Alterman, Z., Jarosch, H., and Pekeris, C. L. (1959) Oscillations of the Earth. *Proc. Roy. Soc. A*, **252**, 80.

Al'tshuler, L. V. (1965) Use of shock waves in high pressure physics. *Soviet Physics Uspekhi*, **8**(1), 52.

Anders, E. (1962) Meteorite ages. *Revs. Mod. Phys.*, **34**, 287.

Anders, E. (1963) Meteorite ages. In Middlehurst and Kuiper (1963), p. 402.

Anders, E. (1964) Origin, age and composition of meteorites. *Space Science Revs.*, **3**, 583.

Anders, E., and Larimer, J. W. (1972) Extinct superheavy element in meteorites: attempted characterization. *Science*, **175**, 981.

Anders, E., Higuchi, H., Gros, J., Takahashi, H., and Morgan, J. W. (1975) Extinct superheavy element in the Allende meteorite. *Science*, **190**, 1262.

Anderson, D. L. (1967) A seismic equation of state. *Geophys. J.*, **13**, 9.

Anderson, D. L., and Archambeau, C. B. (1964) The anelasticity of the Earth. *J. Geophys. Res.*, **69**, 2071.

Anderson, D. L., and Jordan, T. (1970) The composition of the lower mantle. *Phys. Earth Plan. Int.*, **3**, 23.

Anderson, J. D. (1974) Geodetic and dynamical properties of planets. *EOS (Trans. Am. Geophys. Un.)*, **55**, 515.

Anderson, O. L. (1966) The use of ultrasonic measurements under modest pressure to estimate compression at high pressure. *J. Phys. Chem. Solids*, **27**, 547.

Anderson, O. L. (1974) The determination of the volume dependence of the Grüneisen parameter $\gamma$. *J. Geophys. Res.*, **79**, 1153.

Anderson, O. L., and Nafe, J. E. (1965) The bulk modulus-volume relationship for oxide compounds and related geophysical problems. *J. Geophys. Res.*, **70**, 3951.

Anderson, O. L., Schreiber, E., Liebermann, R. C., and Soga, N. (1968) Some elastic constant data on minerals relevant to geophysics. *Revs. Geophys.*, **6**, 491.

Andrews, D. J. (1975) A numerical investigation of the thermal state of the Earth's mantle. *Tectonophysics*, **25**, 177.

Archambeau, C. B. (1968) General theory of elastodynamic source fields. *Revs. Geophys.*, **6**, 241.

Armstrong, R. L., and McDowall, W. G. (1975) Proposed refinement of the phanerozoic time scale. *Geodynamic Highlights (Inter-Union Commission on Geodynamics Bulletin)*, No. 2 (February, 1975), 33.

Babcock, H. W., and Babcock, H. D. (1955) The Sun's magnetic field, 1952–1954. *Astrophys. J.*, **121**, 349.

Bailey, J. M. (1971a) Jupiter: its captured satellites. *Science*, **173**, 812.

Bailey, J. M. (1971b) Origin of the outer satellites of Jupiter. *J. Geophys. Res.*, **76**, 7827.

Banks, R. J. (1969) Geomagnetic variations and the electrical conductivity of the upper mantle. *Geophys. J.*, **17**, 457.

Banks, R. J. (1972) The overall conductivity distribution of the Earth. *J. Geomag. Geoelect.*, **24**, 337.

Barazangi, M., and Dorman, J. (1969) World seismicity map of ESSA Coast and Geodetic Survey epicentre data for 1961–1967. *Bull. Seism. Soc. Am.* **59**, 369.

Bardeen, J. (1938) Compressibilities of the alkali metals. *J. Chem. Phys.*, **6**, 372.

Barraclough, D. R., Harwood, J. M., Leaton, B. R. and Malin S. R. C. (1975) A model of the geomagnetic field at epoch 1975. *Geophys. J.*, **43**, 645.

Båth, M. (1966) Earthquake energy and magnitude. *Phys. and Chem. of Earth*, **7**, 115.

Baum, B. A., Gel'd, P. V., and Tyagunov, G. V. (1967) Resistivity of ferrosilicon alloys in the temperature range 800–1700°C. *Phys. Metals and Metallography*, **24**(1), 181.

Beck, A. E. (1970) Nonequivalence of oceanic and continental heat flows and other geothermal problems. *Comments on Earth Sciences: Geophysics*, **1**, 29.

Benioff, H. (1949) Seismic evidence for the fault origin of oceanic deeps. *Bull. Geol. Soc. Am.*, **60**, 1837.

Benioff, H. (1962) Movements on major transcurrent faults. In Runcorn (1962), p. 103.

Berggren, W. A. (1969) Cenozoic chronostratigraphy, planktonic foraminiferal zonation and the radiometric time scale. *Nature*, **224**, 1072.

Bernas, R., Gradsztajn, E., and Yanif, A. (1969) Isotopic composition of lithium in some meteorites and the role of neutrons in the nucleosynthesis of the light elements in the solar system. In Millman (1969), p. 123.

Bertojo, M., Chui, M. F., and Townes, C. H. (1974) Isotopic abundances and their variations within the galaxy. *Science*, **184**, 619.

Bingham, D. K., and Stone, D. B. (1972) Secular variation in the Pacific Ocean region. *Geophys. J.*, **28**, 337.

Birch, F. (1948) The effects of pleistocene climatic variations upon geothermal gradients. *Am. J. Sci.*, **246**, 729.

Birch, F. (1952) Elasticity and constitution of the Earth's interior. *J. Geophys. Res.*, **57**, 227.

Birch, F. (1954) Heat from radioactivity. In Faul (1954), p. 148.

Birch, F. (1958) Differentiation of the mantle. *Bull. Geol. Soc. Am.*, **69**, 483.

Birch, F. (1961) The velocity of compressional waves in rocks to 10 kilobars, Part 2. *J. Geophys. Res.*, **66**, 2199.

Birch, F. (1972) The melting relations of iron and temperatures in the Earth's core. *Geophys. J.*, **29**, 373.

Black, D. I. (1967) Cosmic ray effects and faunal extinctions at geomagnetic field reversals. *Earth and Plan. Sci. Letters*, **3**, 225.

Blanco, V. M., and McCuskey, S. W. (1961) *Basic physics of the solar system*. Reading, Mass.: Addison-Wesley.

Bodu, R., Bouzigues, H., Morin, N., and Pfiffelmann, J. (1972) Sur l'existence d'anomalies isotopiques rencountrées dans l'uranium du Gabon. *Compt. Rend. Acad. Sci. Paris D*, **275**, 1731.

Boettcher, A. L. (1974) Review of symposium on deep seated rocks and geothermometry. *EOS (Trans. Am. Geophys. Un.)*, **56**, 1068.

Boland, J. N., McLaren, A. C., and Hobbs, B. E. (1971) Dislocations associated with optical features in naturally deformed olivine. *Contr. Mineral. Petrol.*, **30**, 53.

Bolt, B. A. (1964) Recent information on the Earth's interior from studies of mantle waves and eigenvibrations. *Phys. and Chem. of the Earth*, **5**, 55.

Bolt, B. A., and Nuttli, O. W. (1966) P Wave residuals as a function of azimuth 1: Observations. *J. Geophys. Res.*, **71**, 5977.

Bomford, G. (1971) *Geodesy*. (3rd ed.) Oxford: Clarendon Press.

Boore, D. B., McEvilly, T. V., and Lindh, A. (1975) Quarry blast sources and earthquake prediction: the Parkfield, California, earthquake of June 28, 1966. *Pure Appl. Geophys.*, **113**, 293.

Boschi, E. (1974) Melting of iron. *Geophys. J.*, **38**, 327.

Boyd, F. R. (1973) A pyroxene geotherm. *Geochim. Cosmochim. Acta*, **37**, 2533.

Bradley, R. S. (ed) (1963) *High pressure physics and chemistry, Vol. 1.* New York: Academic Press.

Braginskiy, S. I., and Nikolaychik, V. V. (1973) Estimation of the electrical conductivity of the Earth's lower mantle from the lag of an electromagnetic signal. *Physics of the Solid Earth*, **1973**, 601.

Brennen, C. (1974) Isostatic recovery and the strain rate dependent viscosity of the Earth's mantle. *J. Geophys. Res.*, **79**, 3993.

Briden, J. C. (1966) Variation of the intensity of the geomagnetic field through geological time. *Nature*, **212**, 246.

Briden, J. C., and Irving, E. (1964) Palaeolatitude spectra of sedimentary palaeoclimatic indicators. In Nairn (1964).

Bridge, H. S., Lazarus, A. J., Snyder, C. W., Smith, E. J., Davis, L., Coleman, P. J., and Jones, D. E. (1967) Mariner V: plasma and magnetic fields observed near Venus. *Science*, **158**, 1669.

Bridgman, P. W. (1957) Effects of pressure on binary alloys VI: Systems for the most part of dilute alloys of high melting metals. *Proc. Am. Acad. Arts. Sci.*, **84**, 179.

Brune, J. N. (1964) Travel times, body waves and normal modes of the Earth. *Bull. Seism. Soc. Am.*, **54**, 2099.

Brune, J. N. (1968) Seismic moment, seismicity, and rate of slip along major fault zones. *J. Geophys. Res.*, **73**, 777.

Brune, J. N. (1970) Tectonic stress and the spectra of seismic shear waves from earthquakes. *J. Geophys. Res.*, **75**, 4997 (also *J. Geophys. Res.*; **76**, 5002, 1971.)

Brune, J. N., Henyey, T. L., and Roy, R. F. (1969) Heat flow, stress, and rate of slip along the San Andreas fault in California. *J. Geophys. Res.*, **74**, 3821.

Brune, J. N., and Oliver, J. (1959) The seismic noise of the Earth's surface. *Bull. Seism. Soc. Am.*, **49**, 349.

Bryson, R. A. (1974) A perspective on climatic change. *Science*, **184**, 753.

Bucha, V. (1965) Results of archaeomagnetic research in Czechoslovakia for the epoch from 4400 B.C. to the present. *J. Geomag. Geoelect.*, **17**, 407.

Bukowinski, M. S. T. (1976) On the electronic structure of iron core pressures. *Phys. Earth Planet. Int.*, **13**, 57.

Bukowinski, M. S. T. and Knopoff, L. (1976) Electronic structure of iron and the Earth's core. *Geophys. Res. Letters*, **3**, 45.

Bullard, E. C. (1949) The magnetic field within the Earth. *Proc. Roy. Soc. A*, **197**, 433.

Bullard, E. C. (1956) Edmond Halley (1656–1742). *Endeavour*, **15**, 189.

Bullard, E. C. (1960) Response to award of Arthur L. Day medal. *Proc. Vol. for 1959, Geol. Soc. Am.*, p. 92.

Bullard, E. C. (1965) Concluding remarks, symposium on continental drift. *Phil. Trans. Roy. Soc. A*, **258**, 322.

Bullard, E. C. (1967) The removal of trend from magnetic surveys. *Earth Plan. Sci. Lett.*, **2**, 293.

Bullard, E., Everett, J. E., and Smith, A. G. (1965) The fit of the continents around the Atlantic. *Phil. Trans. Roy. Roc. A*, **258**, 41.

Bullard, E. C., Freedman, C., Gellman, H., and Nixon, J. (1950) The westward drift of the Earth's magnetic field. *Phil. Trans. Roy. Soc. A*, **243**, 67.

Bullard, E. C., and Gellman, H. (1954) Homogeneous dynamos and terrestrial magnetism. *Phil. Trans. Roy. Soc. A*, **247**, 213.

Bullard, E. C., and Griggs, D. T. (1961) The nature of the Mohorovičić discontinuity. *Geophys. J.*, **6**, 118.

Bullen, K. E. (1954) *Seismology*. London: Methuen.

Bullen, K. E. (1963) *An introduction to the theory of seismology*. (3rd ed.) Cambridge: Cambridge University Press.

Bullen, K. E., and Haddon, R. A. W. (1967a) Earth oscillations and the Earth's interior. *Nature*, **213**, 574.

Bullen, K. E., and Haddon, R. A. W. (1967b) Derivation of an earth model from free oscillation data. *Proc. U.S. Nat. Acad. Sci.*, **58**, 846.

Bullen, K. E., and Haddon, R. A. W. (1973) The ellipticities of surfaces of equal density inside the Earth. *Phys. Earth Plan. Int.*, **7**, 199.

Burnett, D. S., and Wasserburg, G. J. (1967a) $^{87}$Rb–$^{87}$Sr ages of silicate intrusions in iron meteorites. *Earth and Plan. Sci. Letters*, **2**, 397.

Burnett, D. S., and Wasserburg, G. J. (1967b) Evidence for the formation of an iron meteorite at $3.8 \times 10^9$ years. *Earth and Plan. Sci. Letters*, **2**, 137.

Busse, F. H. (1974) A theory of a convection driven dynamo (Abstract). *EOS (Trans. Am. Geophys. Un.)*, **56**, 1108.

Byerlee, J. D. (1970) The mechanics of stick slip. *Tectonophysics*, **9**, 475.

Byerlee, J. D., and Brace, W. F. (1968) Stick slip, stable sliding and earthquakes—effect of rock type, pressure, strain rate and stiffness. *J. Geophys. Res.*, **73**, 6031.

Cain, J. C., Hendricks, S. J., Langel, R. A., and Hudson, W. V. (1967) A proposed model for the International Geomagnetic Reference Field—1965. *J. Geomag. Geoelect.*, **19**, 335.

Caldirola, P., and Knoepfel, H. (eds.) (1971) *Physics of high energy density (Proc. Int. School Phys. 'Enrico Fermi', Course 48)*. New York: Academic Press.

Cameron, A. G. W. (1970) Formation of the Earth-Moon system. *EOS (Trans. Am. Geophys. Un.*, **51**, 628.

Cameron, A. G. W. (1973) Abundances of the elements in the solar system. *Space Sci. Revs.*, **15**, 121.

Carmichael, C. M. (1967) An outline of the intensity of the paleomagnetic field of the Earth. *Earth Plan. Sci. Letters*, **3**, 351.

Carpenter, E. W. (1966) A quantitative evaluation of teleseismic explosion records. *Proc. Roy. Soc. A*, **290**, 287.

Ceplecha, Z. (1961) Multiple fall of Pribram meteorites photographed. *Bull. Astr. Inst. Czech.*, **12**, 21.

Chapman, S., and Bartels, J. (1940) *Geomagnetism*. (2 vols.) Oxford: Clarendon Press.

Chapman, S., and Lindzen, R. S. (1970) *Atmospheric tides*. Dordrecht: Reidel.

Chinnery, M. A. (1961) The deformation of the ground around surface faults. *Bull. Seism. Soc. Am.*, **51**, 355.

Chinnery, M. A., and North, R. G. (1975) The frequency of very large earthquakes. *Science*, **190**, 1197.

Chow, T. J., and Patterson, C. C. (1962) The occurrence and significance of lead isotopes in pelagic sediments. *Geochim. et Cosmochim. Acta*, **26**, 263.

Clark, S. P. (1957) Radiative transfer in the Earth's mantle. *Trans. Am. Geophys. Un.*, **38**, 931.

Clark, S. P., Turekian, K. K., and Grossman, L. (1972) Model for the early history of the Earth. In Robertson (1972), p. 1.

Clayton, R. N., Grossman, L., and Mayeda, T. K. (1973) A component of primitive nuclear composition in carbonaceous meteorites. *Science*, **182**, 485.

Clayton, R. N., Mayeda, T. K., and Onuma, N. (1975) An oxygen isotope classification of meteorites. *EOS (Trans. Am. Geophys. Un.)*, **56**, 392.

Clendenen, R. L., and Drickamer, H. G. (1964) The effect of pressure on the volume and lattice parameters of ruthenium and iron. *J. Phys. Chem. Solids*, **25**, 865.

Colburn, D. S. (1972) Lunar magnetic field measurements, electrical conductivity calculations and thermal profile inferences. In Runcorn and Urey (1972), p. 355.

Collinson, D. W., Creer, K. M., and Runcorn, S. K. (eds.) (1967) *Methods in paleomagnetism*. Amsterdam: Elsevier.

Colombo, G. (1965) Rotational period of the planet Mercury. *Nature*, **208**, 575.

Colombo, G., and Shapiro, I. I. (1966). The rotation of the planet Mercury. *Astrophys. J.*, **145**, 296.

Compston, W., Jeffrey, P. M., and Riley, G. H. (1960) Age of emplacement of granites. *Nature*, 186, 702.

Compston, W., Lovering, J. F., and Vernon, M. J. (1965) The rubidium-strontium age of the Bishopville aubrite and its component enstatite and feldspar. *Geochim. et Cosmochim. Acta*, **29**, 1085.

Cottrell, A. H. (1953) *Dislocations and plastic flow in crystals*. Oxford: Clarendon Press.

Cox, A. (1969) Geomagnetic reversals. *Science*, **163**, 237.

Cox, A. (ed.) (1973) *Plate tectonics and geomagnetic reversals*. San Francisco: Freeman.

Cox, A., Dalrymple, G. B., and Doell, R. R. (1967) Reversals of the Earth's magnetic field. *Scientific American*, **216**(2), 44.

Cox, A., and Doell, R. R. (1960) Review of paleomagnetism. *Bull. Geol. Soc. Am.*, **71**, 645.

Craig, H., Miller, S. L., and Wasserburg, G. J. (eds.) (1964) *Isotopic and cosmic chemistry*. Amsterdam: North Holland.

Crain, I. K. (1968) The glacial effect and the significance of continental terrestrial heat flow measurements. *Earth Plan. Sci. Letters*, **4**, 69.

Cramer, C. H., and Kovach, R. L. (1974) A search for teleseismic travel-time anomalies along the San Andreas fault zone. *Geophys. Res. Letters*, **1**, 90.

Currie, R. G. (1968) Geomagnetic spectrum of internal origin and lower mantle conductivity. *J. Geophys. Res.*, **73**, 2779.

Currie, R. G. (1974) Period and $Q_W$ of the Chandler wobble. *Geophys. J.*, **38**, 179.

Dalrymple, G. B., and Lanphere, M. A. (1969) Potassium-argon dating. Principles, techniques and applications to geochronology. San Francisco: Freeman.

Dalrymple, G. B., and Lanphere, M. A. (1971) $^{40}Ar/^{39}Ar$ technique of K-Ar dating: a comparison with the conventional technique. *Earth Plan. Sci. Letters*, **12**, 300.

Darbyshire, J. (1962) Microseisms. In Hill (1962), p. 700.

Darwin, G. H. (1962) *The tides and kindred phenomena in the solar system*. (Original edition 1898, reprinted 1962). San Francisco: Freeman.

Davies, D., and Sheppard, R. M. (1972) Lateral heterogeneity in the Earth's mantle. *Nature*, **239**, 318.

Davies, G. F. (1974a) Limits on the constitution of the lower mantle. *Geophys. J.*, **38**, 479.

Davies, G. F. (1974b) Elasticity, crystal structure and phase transitions. *Earth Plan. Sci. Letters*, **22**, 339.

Davies, G. F., and Dziewonski, A. M. (1975) Homogeneity and constitution of the Earth's lower mantle and outer core. *Phys. Earth Plan. Int.*, **10**, 336.

Davis, E. E., and Lister, C. R. B. (1974) Fundamentals of ridge crest topography. *Earth Plan. Sci. Letters*, **21**, 405.

Davis, P. M., and Stacey, F. D. (1972) Geomagnetic anomaly caused by a man-made Lake. *Nature*, **240**, 348.

Deacon, G. E. R. (1947) Relations between sea waves and microseisms. *Nature*, **160**, 419.

Denham, C. R. (1974) Counter-clockwise motion of paleomagnetic directions 24,000 years ago at Mono Lake, California. *J. Geomag. Geoelect.*, **26**, 487.

Dermott, S. F. (1968) On the origin of commensurabilities in the solar system—II the orbital period relation. *Mon. Not. R. Astr. Soc.*, **141**, 363.

Derr, J. S. (1969) Internal structure of the Earth inferred from free oscillations. *J. Geophys. Res.*, **74**, 5202.

Doell, R. R., and Cox, A. (1971) Pacific geomagnetic secular variation. *Science*, **171**, 248.

Doell, R. R., and Cox, A. (1972) The Pacific geomagnetic secular variation anomaly and the question of lateral uniformity in the lower mantle. In Robertson (1972), p. 245.

Dohnanyi, J. S. (1968) Collisional model of asteroids and their debris. In Kresák and Millman (1968), p. 486.

Dolginov, Sh. Sh., Yeroshenko, Ye. G., and Zhuzgov, L. N. (1973) Magnetic field in the very close neighborhood of Mars according to data from Mars 2 and Mars 3 spacecraft. *J. Geophys. Res.*, **78**, 4779.

Dorman, J. (1969) Seismic surface wave data on the upper mantle. In Hart (1969), p. 257.

Dorman, J., Ewing, M., and Oliver, J. (1960) Study of shear-velocity distribution in the upper mantle by Rayleigh waves. *Bull. Seism. Soc. Am.*, **50**, 87.

Dorman, L. M. and Lewis, B. T. R. (1972) Experimental isostasy 3. Inversion of the isostatic Green function and lateral density changes. *J. Geophys. Res.*, **77**, 3068.

Drickamer, H. G., and Frank, C. W. (1973) *Electronic transitions and the high pressure chemistry and physics of solids*. London: Chapman and Hall.

Du Fresne, E. R., and Anders, E. (1963) Chemical evolution of the carbonaceous chondrites. In Middlehurst and Kuiper (1963), p. 496.

Dugdale, J. S., and MacDonald, D. K. C. (1953) Thermal expansion of solids. *Phys. Rev.*, **89**, 832.

Dunn, J. R., Fuller, M., Ito, H., and Schmidt, V. A. (1971) Paleomagnetic study of a reversal of the Earth's magnetic field. *Science*, **172**, 840.

Durrani, S. A. (1971) Origin and ages of tektites. *Phys. Earth Plan. Int.*, **4**, 251.

Du Toit, A. L. (1937) *Our wandering continents*. Edinburgh: Oliver and Boyd.

Dyal, P., Parkin, C. W., and Sonett, C. P. (1970) Apollo 12 magnetometer: measurement of a steady magnetic field on the surface of the Moon. *Science*, **169**, 762.

Dyal, P., Parkin, C. W., and Daily, W. D. (1974) Magnetism and the interior of the Moon. *Revs. Geophys. Space Phys.*, **12**, 568.

Dziewonski, A. M., and Gilbert, F. (1972) Observations of normal modes from 84 recordings of the Alaskan earthquake of 1964 March 28. *Geophys. J.*, **27**, 393.

Dziewonski, A. M., and Gilbert, F. (1973) Observations of normal modes from 84 recordings of the Alaskan earthquake of 1964 March 28—II. Further remarks based on new spheroidal overtone data. *Geophys. J.*, **35**, 401.

Dziewonski, A. M., Hales, A. L., and Lapwood, E. R. (1975) Parametrically simple earth models consistent with geophysical data. *Phys. Earth Plan. Int.*, **10**, 12.

Eaton, J. P., Richter, D. H., and Ault, W. U. (1961) The tsunami of May 23, 1960, on the island of Hawaii. *Bull. Seism. Soc. Am.*, **51**, 135.

Eirich, F. R. (ed.) (1958) *Rheology; theory and applications*. (3 vols.) New York: Academic Press.

Eldridge, J. S., O'Kelley, G. D., and Northcutt, K. J. (1974) Primordial radioelement concentrations in rocks and soils from Taurus-Littrow. *Proc. Fifth Lunar Conference (Suppl. 5 Geochim. Cosmochim. Acta)*, **2**, 1025.

Elsasser, W. M. (1941) A statistical analysis of the Earth's magnetic field. *Phys. Rev.*, **60**, 876.

Elsasser, W. M. (1950) The Earth's interior and geomagnetism. *Revs. Mod. Phys.*, **22**, 1.

Elsasser, W. M. (1956) Hydromagnetic dynamo theory. *Revs. Mod. Phys.*, **28**, 135.

Elsasser, W. M. (1963) Early history of the Earth. In Geiss and Goldberg (1963), p. 1.

Elsasser, W. M. (1966) Thermal structure of the upper mantle and convection. In Hurley (1966), p. 461.

Elsasser, W. M. (1967) Interpretation of heat flow equality. *J. Geophys. Res.*, **72**, 4768.

Elsasser, W. M., and Isenberg, I. (1949) Electronic phase transition in iron at extreme pressures (abstract). *Phys. Rev.*, **76**, 469.

Emiliani, C., and Shackleton, N. J. (1974) The Brunhes epoch: isotopic paleotemperatures and geochronology. *Science*, **183**, 511.

Evans, R., and Jain, A. (1972) Calculations of electrical transport properties of liquid metals at high pressures. *Phys. Earth Plan. Int.*, **6**, 141.

Evernden, J. F., Best, W. J., Pomeroy, P. W., McEvilly, T. V., Savino, J. M., and Sykes, L. R. (1971) Discrimination between small magnitude earthquakes and explosions. *J. Geophys. Res.*, **76**, 8042.

Ewing, W. M., Jardetsky, W. S., and Press, F. (1957) *Elastic waves in layered media.* New York: McGraw-Hill.

Faul, H. (ed.) (1954) *Nuclear geology.* New York: Wiley.

Faul, H. (1966) *Ages of rocks, planets and stars.* New York: McGraw-Hill.

Faure, G., and Powell, J. L. (1972) *Strontium isotope geology.* Berlin: Springer.

Fechtig, H., and Kalbitzer, S. (1966) The diffusion of argon in potassium-bearing solids. In Schaeffer and Zähringer (1966), p. 68.

Finch, H. F., and Leaton, B. R. (1957) The Earth's main magnetic field-epoch 1955.0. *Monthly Notices, R. Astr. Soc., Geophys. Suppl.*, **7**, 314.

Fisher, D. (1966) The origin of meteorites: space erosion and cosmic radiation ages. *J. Geophys. Res.*, **71**, 3251.

Fisher, D. E. (1975) Trapped helium and argon and the formation of the atmosphere by degassing. *Nature*, **256**, 113.

Fleischer, R. L., and Price, P. B. (1964) Techniques for geological dating of minerals by chemical etching of fission fragment tracks. *Geochim. et Cosmochim. Acta*, **28**, 1705.

Fleischer, R. L., Price, P. B., and Walker, R. M. (1965) Tracks of charged particles in solids. *Science*, **149**, 383.

Fleischer, R. L., Naesser, C. W., Price, P. B., and Walker, R. M. (1965) Cosmic ray exposure ages of tektites by the fission track technique. *J. Geophys. Res.*, **70**, 1491.

Fleischer, R. L., Price, P. B., Walker, R. M., and Maurette, M. (1967) Origins of fossil charged-particle tracks in meteorites. *J. Geophys. Res.*, **72**, 331.

Fleischer, R. L., Price, P. B., and Woods, R. T. (1970) A second tektite fall in Australia. *Earth Plan. Sci. Letters*, **7**, 51.

Foster, J. H., and Opdyke, N. D. (1970) Upper Miocene to recent magnetic stratigraphy in deep-sea sediments. *J. Geophys. Res.*, **75**, 4465.

Fowles, G. R. (1970) *Analytical mechanics* (second edition). New York: Holt, Rinehart and Winston.

Frank, F. C. (1965) On dilatancy in relation to seismic sources. *Revs. Geophys.*, **3**, 485.

Friedel, J. (1964) *Dislocations.* Oxford: Pergamon.

Fujisawa, H., Fujii, N., Mizutani, H., Kanamori, H., and Akimoto, S. (1968) Thermal diffusivity of $Mg_2SiO_4$, $Fe_2SiO_4$ and NaCl at high pressures and temperatures. *J. Geophys. Res.*, **73**, 4727.

Fukao, Y., Hitoshi, M., and Uyeda, S. (1968) Optical absorption spectra at high temperatures and radiative thermal conductivity of olivines. *Phys. Earth Planet. Interiors.*, **1**, 57.

Fuller, M. (1974) Lunar magnetism. *Revs. Geophys. Space Phys.*, **12**, 23.

Ganguly, J., and Kennedy, G. C. (1977) Solubility of K in Fe-S liquid, silicate-K-$(Fe\text{-}S)^{liq}$ equilibria, and their planetary implications. *Earth Plan. Sci. Letters*, **35**, 411.

Gans, R. F. (1972) Viscosity of the Earth's core. *J. Geophys. Res.*, **77**, 360.

Gapcynski, J. P., Blackshear, W. T., Tolson, R. H., and Compton, H. R. (1975) A determination of the lunar moment of inertia. *Geophys. Res. Letters*, **2**, 353.

Gaposchkin, E. M., and Lambeck, K. (1971) Earth's gravity field to sixteenth degree and station coordinates from satellite and terrestrial data. *J. Geophys. Res.*, **76**, 4855.

Gardiner, R. B., and Stacey, F. D. (1971) Electrical resistivity of the core. *Phys. Earth Plan. Int.* **4**, 406.

Garland, G. D. (1965) *The Earth's shape and gravity.* Oxford: Pergamon.

Gaskell, T. F. (ed.) (1967) *The Earth's mantle.* London: Academic Press.

Gast, P. W. (1960) Limitations on the composition of the upper mantle. *J. Geophys. Res.*, **65**, 1287.

Gast, P. W. (1962) The isotopic composition of strontium and the age of stone meteorites. *Geochim. et Cosmochim. Acta*, **26**, 927.

Gast, P. W. (1972) The chemical composition of the Earth, the Moon and chondritic meteorites. In Robertson (1972), p. 19.

Gastil, G. (1960) The distribution of mineral dates in time and space. *Amer. J. Sci.*, **258**, 1.

Geiss, J., and Goldberg, E. D. (eds.) (1963) *Earth science and meteoritics.* Amsterdam: North Holland.

Gentner, W., Glass, B. P., Storzer, D., and Wagner, G. A. (1970) Fission track ages and ages of deposition of deep sea microtektites. *Science*, **168**, 359.

Gilbert, F., and Dziewonski, A. M. (1975) An application of normal mode theory to the retrieval of structural parameters and source mechanisms from seismic spectra. *Phil. Trans. Roy. Soc. A*, **278**, 187.

Gilchrist, J., Thorpe, A. N., and Senftle, F. E. (1969) Infra red analysis of water in tektites and other glasses. *J. Geophys. Res.*, **74**, 1475.

Gilvarry, J. J. (1956a) The Lindemann and Grüneisen laws. *Phys. Rev.*, **102**, 308.

Gilvarry, J. J. (1956b) Grüneisen parameter for a solid under finite strain. *Phys. Rev.*, **102**, 331.

Gladwin, M. T., and Stacey, F. D. (1974a) Ultrasonic pulse velocity as a rock stress sensor. *Tectonophysics*, **21**, 39.

Gladwin, M. T., and Stacey, F. D. (1974b) Anelastic degradation of acoustic pulses in rock. *Phys. Earth Plan. Int.* **8**, 332.

Glaessner, M. F. (1966) Pre-Cambrian palaeontology. *Earth Sci. Revs.*, **1**, 29.

Glass, B., and Heezen, B. C. (1967) Tektites and geomagnetic reversals. *Nature*, **214**, 372.

Goettel, K. A. (1972) Partitioning of potassium between silicates and sulphide melts: experiments relevant to the Earth's core. *Phys. Earth Plan. Int.*, **6**, 161.

Goldich, S. S., Muehlberger, W. R., Lidiak, E. G., and Hedge, C. E. (1966) Geochronology of the midcontinent region, United States 1. Scope, methods and principles. *J. Geophys. Res.*, **71**, 5375.

Goldreich, P., and Peale, S. (1966) Spin-orbit coupling in the solar system. *Astronomical Journal*, **71**, 425.

Goldreich, P., and Toomre, A. (1969) Some remarks on polar wandering. *J. Geophys. Res.*, **74**, 2555.

Goldstein, J. I., and Ogilvie, R. E. (1965) A re-evaluation of the iron-rich portion of the Fe–Ni system. *Trans. Metal. Soc. AIME*, **233**, 2083.

Goldstein, J. I., and Short, J. M. (1967) Cooling rates of 27 iron and stony iron meteorites. *Geochim. et. Cosmochim. Acta*, **31**, 1001.

Goles, G. G. (1969) Cosmic abundances. In K. H. Wedepohl (ed.) *Handbook of Geochemistry, Vol. 1*. Berlin: Springer, p. 116.

Gordon, R. B., and Davis, L. A. (1968) Velocity and attenuation of seismic waves in imperfectly elastic rock. *J. Geophys. Res.*, **73**, 3917.

Gough, D. I. (1973a) The geophysical significance of geomagnetic variation anomalies. *Phys. Earth Plan. Int.*, **7**, 379.

Gough, D. I. (1973b) The interpretation of magnetometer array studies. *Geophys. J.*, **35**, 83.

Graber, M. A. (1976) Polar motion spectra based upon Doppler, I.P.M.S. and B.I.H. data. *Geophys. J.*, **46**, 75.

Gray, C. M., Papanastassiou, D. A., and Wasserburg, G. J. (1973) The identification of early condensates from the solar nebula. *Icarus*, **20**, 213.

Griggs, D. T., and Baker, D. W. (1968) The origin of deep-focus earthquakes. In Mark and Fernbech (1968), p. 23.

Grommé, S., and Vine, F. J. (1972) Paleomagnetism of Midway Atoll lavas and northward movement of the Pacific plate. *Earth Plan. Sci. Letters*, **17**, 159.

Grossman, L., and Larimer, J. W. (1974) Early chemical history of the solar system. *Revs. Geophys. Space Phys.*, **12**, 71.

Grover, R., Getting, I. C., and Kennedy, G. C. (1973) Simple compressibility relation for solids. *Phys. Rev. B*, **7**, 567.

Grüneisen, E. (1926) The state of a solid body. *Handbuch der Phys.*, **10**, 1. National Aeronautics and Space Administration Translation RE2-18-59W.

Gubbins, D. (1974) Theories of the geomagnetic and solar dynamos. *Revs. Geophys. Space Phys.*, **12**, 137.

Guinot, B. (1970) Work of the Bureau International de L'Heure on the rotation of the Earth. In Mansinha et al. (1970), p. 54.

Gupta, I. N. (1973) Premonitory variations in *S*-wave velocity anisotropy before earthquakes in Nevada. *Science*, **182**, 1129.

Gutenberg, B. (1958) Rheological problems of the Earth's interior. In Eirich (1958), Vol. 2, p. 401.

Gutenberg, B. (1959) *Physics of the Earth's interior*. New York: Academic Press.

Gutenberg, B., and Richter, C. F. (1954) *Seismicity of the Earth and associated phenomena.* (2nd ed.) Princeton, N.J.: Princeton University Press.

Gutenberg, B., and Richter, C. F. (1956) Earthquake magnitude, intensity, energy and acceleration. *Bull. Seism. Soc. Am.*, **46**, 105.

Haddon, R. A. W., and Cleary, J. R. (1974) Evidence for scattering of seismic PKP waves near the mantle-core boundary. *Phys. Earth Plan. Int.*, **8**, 211.

Hadley, K. (1975) $V_P/V_S$ anomalies in dilatant rock samples. *Pure Appl. Geophys.* **113**, 1.

Hall, H. T., and Murthy, V. R. (1971) The early chemical history of the Earth: some critical element fractionations. *Earth Plan. Sci. Letters*, **11**, 239.

Hamilton, E. I., and Farquhar, R. M. (eds.) (1968) *Radiometric dating for geologists.* New York, Wiley-Interscience.

Hamza, V. M., and Beck, A. E. (1972) Terrestrial heat flow, the neutrino problem, and a possible energy source in the core. *Nature*, **240**, 343.

Harland, W. B., Smith, A. G., and Wilcock, B. (eds.) (1964) Geological Society phanerozoic time scale 1964. *Quart J. Geol. Soc. Lond.*, **120S**, 260. (Supplement volume: The phanerozoic time scale.)

Harnwell, G. P. (1949) Principles of electricity and electromagnetism. (2nd ed.) New York: McGraw-Hill.

Harrison, C. G. A. (1968) Evolutionary processes and reversals of the Earth's magnetic field. *Nature*, **217**, 46.

Hart, P. J. (ed.) (1969) *The Earth's crust and upper mantle.* Geophysical Monograph **13**. Washington: American Geophysical Union.

Hart, R. A. (1973) Geochemical and geophysical implications of the reaction between sea water and the oceanic crust. *Nature*, **243**, 76.

Hart, S. R., Schilling, J. G., and Powell, J. L. (1973) Basalts from Iceland and along the Reykjanes Ridge: Sr isotope geochemistry. *Nature Physical Science*, **246**, 104.

Hartman, W. K., and Larson, S. M. (1967) Angular momenta of planetary bodies. *Icarus*, **7**, 257.

Hasselmann, K. (1963) A statistical analysis of the generation of microseisms. *Revs. Geophys.*, **1**, 177.

Haubrich, R. A., Munk, W. H., and Snodgrass, F. E. (1963) Comparative spectra of microseisms and swell. *Bull. Seism. Soc. Am.*, **53**, 27.

Hays, J. D., Saito, T., Opdyke, N. D., and Burckle, L. H. (1969) Pliocene-pleistocene sediments of the equatorial Pacific: their paleomagnetic, biostratigraphic and climatic record. *Geol. Soc. Am. Bull.*, **80**, 1481.

Healy, J. H., Rubey, W. W., Griggs, D. T., and Raleigh, C. B. (1968) The Denver earthquakes. *Science*, **161**, 1301.

Heard, H. C., Borg, I. Y., Carter, N. L., and Raleigh, C. B. (1972) *Flow and Fracture of rocks.* Geophysical Monograph 16. Washington: American Geophysical Union.

Hedge, C. E., Watkins, N. D., Hildreth, R. A., and Doering, W. P. (1973) $^{87}Sr/^{86}Sr$ ratios in basalts from islands in the Indian Ocean. *Earth Plan. Sci. Letters*, **21**, 29.

Heirtzler, J. R., Dickson, G. O., Herron, E. M., Pitman, W. C., and LePichon, X. (1968) Marine magnetic anomalies, geomagnetic field reversals and motions of the ocean floor and continents. *J. Geophys. Res.*, **73**, 2119.

Heirtzler, J. R., LePichon, X., and Baron, J. G. (1966) Magnetic anomalies over the Reykjannes ridge. *Deep Sea Res.*, **13**, 427.

Heiskanen, W. A., and Meinesz, F. A. V. (1958) *The Earth and its gravity field.* New York: McGraw-Hill.

Hensley, W. K., Bassett, W. A., and Huizenga, J. R. (1973) Pressure dependence of the radioactive decay constant of beryllium-7. *Science*, **181**, 1164.

Herrin, E. (1968) Introduction to "1968 Seismological tables for *P* phases." *Bull. Seism. Soc. Am.*, **58**, 1193. (Tabulations and discussion papers by several coauthors follow this introduction).

Hide, R. (1966) Free hydromagnetic oscillations of the Earth's core and the theory of the geomagnetic secular variation. *Phil. Trans. Roy. Soc.*, **A259**, 615.

Hide, R., and Roberts, P. H. (1961) The origin of the main geomagnetic field. *Phys. and Chem. of the Earth*, **4**, 27.

Higbie, J., and Stacey, F. D. (1970) Depth of density variations responsible for features of the satellite geoid. *Phys. Earth Plan. Int.*, **4**, 145.

Higbie, J. W., and Stacey, F. D. (1971) Interpretation of global gravity anomalies. *Nature Phys. Sci.*, **234**, 130.

Higgins, G. H., and Kennedy, G. C. (1971) The adiabatic gradient and the melting point gradient in the core of the Earth. *J. Geophys. Res.*, **76**, 1870.

Hill, M. N. (ed.) (1962) *The sea. Vol. 1: Physical oceanography.* New York: Wiley-Interscience.

Hindmarsh, W. R., Lowes, F. J., Roberts, P. H., and Runcorn, S. K. (eds.) (1967) *Magnetism and the cosmos.* Edinburgh: Oliver and Boyd.

Hobbs, B. E., McLaren, A. C., and Paterson, M. S. (1972) Plasticity of single crystals of synthetic quartz. In Heard et al. (1972), p. 29.

Hofmann, R. B. (1968) *Geodimeter fault movement investigations in California.* Bulletin 116-6, California Department of Water Resources.

Holmes, A. (1965) *Principles of physical geology.* (rev. ed.) London: Nelson.

Houben, H., Gierasch, P. J., and Turcotte, D. L. (1975) Can the geomagnetic dynamo be driven by the semidiurnal tides? (Abstract). *EOS (Trans. Am. Geophys. Un.)*, **56**, 356.

Howard, H. T. et al. (21 authors) (1974). Mercury: results on mass, radius, ionosphere, and atmosphere from Mariner 10 dual frequency radio signals. *Science*, **185**, 179.

Hoyle, F. (1960) On the origin of the solar nebula. *Quart. J. Roy. Astr. Soc.*, **1**, 28.

Hubbert, M. K., and Rubey, W. W. (1959) Role of fluid pressure in mechanics of overthrust faulting. *Bull. Geol. Soc. Am.*, **70**, 115.

Hudson, J. D. (1964) Sedimentation rates in relation to the Phanerozoic time scale. *Quart. J. Geol. Soc. London*, **120S**, 37.

Huey, J. M., and Kohman, T. P. (1973) $^{207}$Pb-$^{206}$Pb isochron and age of chondrites. *J. Geophys. Res.*, **78**, 3227.

Hurley, P. M. (ed.) (1966) *Advances in Earth science*. Cambridge, Mass.: M.I.T. Press.

Hurley, P. M., Hughes, H., Faure, G., Fairbairn, H. W., and Pinson, W. H. (1962) Radiogenic strontium-87 model of continent formation. *J. Geophys. Res.*, **67**, 5315.

Hurley, P. M., and Rand, J. R. (1970) Continental radiometric ages. In Maxwell (1970), Part 2, p. 575.

Hurwitz, L., Fabiano, E. B., and Peddie, N. W. (1974) A model of the geomagnetic field for 1970. *J. Geophys. Res.*, **79**, 1716.

International Union of Geodesy and Geophysics (1967) Resolution No. 1, XIV General Assembly. *Bulletin Geodesique*, **86**, 367.

IAGA Study Group (1976) International geomagnetic reference field 1975. *Geophys. J.*, **44**, 733.

Irvine, R. D., and Stacey, F. D. (1975) Pressure dependence of the thermal Grüneisen parameter, with application to the Earth's lower mantle and outer core. *Phys. Earth Plan. Int.*, **11**, 157.

Irving, E. (1964) *Paleomagnetism*. New York: Wiley.

Irving, E. (1966) Paleomagnetism of some carboniferous rocks from New South Wales and its relation to geological events. *J. Geophys. Res.*, **71**, 6025.

Isacks, B., Oliver, J., and Sykes, L. R. (1968) Seismology and the new global tectonics. *J. Geophys. Res.*, **73**, 5855.

Ishikawa, Y., and Syono, Y. (1963) Order-disorder transformation and reverse thermoremanent magnetism in the $FeTiO_3$-$Fe_2O_3$ system. *J. Phys. Chem. Solids*, **24**, 517.

Iyer, H. M., Evans, J. R., and Coakley, J. (1974) Teleseismic evidence for the existence of low velocity material deep into the upper mantle under Yellowstone caldera. *EOS (Trans. Am. Geophys. Un.)*, **55**, 1190.

Jacchia, L. G. (1963) Meteors, meteorites and comets: Interrelations. In Middlehurst and Kuiper (1963), p. 774.

Jackson, D. D., and Anderson, D. L. (1970) Physical mechanisms of seismic wave attenuation. *Revs. Geophys. Space Phys.*, **8**, 1.

Jain, A., and Evans, R. (1972) Calculation of the electrical resistivity of liquid iron in the Earth's core. *Nature, Physical Science*, **235**, 165.

Jain, A. V., Gordon, R. B., and Lipschutz, M. E. (1972) Hardness of kamacite and shock histories of 119 meteorites. *J. Geophys. Res.*, **77**, 6940.

Jastrow, R., and Cameron, A. G. W. (eds.) (1963) *Origin of the solar system*. New York: Academic Press.

Jeffreys, H. (1963) On the hydrostatic theory of the figure of the Earth. *Geophys. J.*, **8**, 196.

Jeffreys, H. (1970) *The Earth, its origin, history and physical constitution*. (5th ed.) Cambridge: Cambridge University Press.

Jeffreys, H., and Bullen, K. E. (1940) *Seismological tables*. (Reprint issued 1958.) London: British Association for the Advancement of Science.

Johnson, T. V., and Fanale, F. P. (1973) Optical properties of carbonaceous chondrites and their relationship to asteroids. *J. Geophys. Res.*, **78**, 8507.

Johnston, M. J. S., and Stacey, F. D. (1969a) Volcano-magnetic effect observed on Mt. Ruapehu, New Zealand. *J. Geophys. Res.*, **74**, 6541.

Johnston, M. J. S., and Stacey, F. D. (1969b) Transient magnetic anomalies accompanying volcanic eruptions in New Zealand. *Nature*, **224**, 1289.

Jordan, T. H. (1974) Mantle heterogeneity, plate motions and global tectonics (Abstract). *EOS* (*Trans. Am. Geophys. Un.*), **55**, 1102.

Jordan, T. H., and Anderson, D. L. (1974) Earth structure from free oscillations and travel times. *Geophys. J.*, **36**, 411.

Jurdy, D. M., and Van der Voo, R. (1975) Polar wander since the early Cretaceous. *Science*, **187**, 1193.

Kahle, A. B., Ball, R. H., and Cain, J. C. (1969) Prediction of geomagnetic secular change confirmed. *Nature*, **223**, 165.

Kanamori, H. (1967) Spectrum of short period core phases in relation to the attenuation in the mantle. *J. Geophys. Res.*, **72**, 2181.

Kanasewich, E. R. (1968) The interpretation of lead isotopes and their geological significance. In Hamilton and Farquhar (1968).

Kanasewich, E. R., Ellis, R. M., Chapman, C. H., and Gutowski, P. R. (1973) Seismic array evidence of a core boundary source for the Hawaiian linear volcanic chain. *J. Geophys. Res.*, **78**, 1361.

Kanasewich, E. R., and Gutowski, P. R. (1975) Detailed seismic analysis of a lateral mantle inhomogeneity. *Earth Plan. Sci. Letters*, **25**, 379.

Karnik, V. (1961) Seismicity of Europe. Progress Report II. *International Union of Geodesy and Geophysics Monograph*, **9**.

Kato, Y., and Utashiro, S. (1949) On the changes of terrestrial magnetic field accompanying the great Nankaido earthquake of 1946. *Sci. Repts. Tohoku Univ. Series 5, Geophysics*, **1**, 40.

Kaula, W. M. (1966) *Theory of satellite geodesy*. Waltham, Mass.: Blaisdell.

Kaula, W. M. (1968) *An introduction to planetary physics; The terrestrial planets*. New York: Wiley.

Kaula, W. M. (1969a) The gravitational field of the Moon. *Science*, **166**, 1581.

Kaula, W. M. (1969b) A tectonic classification of the main features of the Earth's gravitational field. *J. Geophys. Res.*, **74**, 4807.

Kawai, N. (1972) The magnetic control on the climate in the geological time. *Proc. Japan Acad.*, **48**, 687.

Kawai, N., Nakajima, T., Yasukawa, K., Hirooka, K., and Kobayashi, K. (1973) The oscillation of field in the Matuyama geomagnetic epoch. *Proc. Japan Acad.*, **49**, 619.

Keane, A. (1954) An investigation of finite strain in an isotropic material subjected to hydrostatic pressure and its seismological applications. *Austral. J. Phys.*, **7**, 323.

Keeler, R. N. (1971) Electrical conductivity of condensed media at high pressures. In Caldirola and Knoepfel (1971), p. 106.

Keeler, R. N., and Royce, E. B. (1971) Shock waves in condensed media. In Caldirola and Knoepfel (1971), p. 51.

Keen, C., and Tramontini, C. (1970) A seismic refraction survey on the Mid-Atlantic ridge. *Geophys. J.*, **20**, 473.

Kelvin, Lord (1899) The age of the Earth as an abode fitted for life. *Phil. Mag.*, **47**, 66.

Kennedy, G. C., and Higgins, G. H. (1972) Melting temperatures in the Earth's mantle. *Tectonophysics*, **13**, 221.

Kennett, J. P., and Watkins, N. D. (1970) Geomagnetic polarity change, volcanic maxima and faunal extinctions in the South Pacific. *Nature*, **227**, 930.

Khan, M. A. (1969) General solution of the problem of hydrostatic equilibrium of the Earth. *Geophys. J.*, **18**, 177.

Khan, M. A., and O'Keefe, J. A. (1974) Relation of the Antarctic gravity low to the Earth's equilibrium figure. *J. Geophys. Res.*, **79**, 3027.

King, D. A., and Ahrens, T. J. (1973) Shock compression of iron sulphide and the possible sulphur content of the Earth's core. *Nature, Phys. Sci.*, **243**, 82.

Kirby, S. H., and Raleigh, C. B. (1973) Mechanisms of high temperature, solid state flow in minerals and ceramics and their bearing on the creep behavior of the mantle. *Tectonophysics*, **19**, 165.

Kirshenbaum, A. D., and Cahill, J. A. (1962) The density of liquid iron from the melting point to 2500°K. *Trans. Metallurgical Soc. A.I.M.E.*, **224**, 816.

Kirsten, T., Steinbrunn, F., and Zähringer, J. (1971) Location and variation of trapped rare gases in Apollo 12 lunar samples. *Geochim. Cosmochim Acta Suppl. 2* (Proc. Second Lunar Science Conference) Vol. 2, 1651.

Kittel, C. (1949) Ferromagnetic domain theory. *Revs. Mod. Phys.*, **21**, 541.

Kittel, C. (1971) *Introduction to solid state physics* (4th ed.). New York: Wiley.

Knopoff, L. (1963) Equations of state of matter at moderately high pressures. In Bradley (1963), p. 227.

Knopoff, L. (1964a) *"Q." Revs. Geophys.*, **2**, 625.

Knopoff, L. (1964b) Earth tides as a triggering mechanism for earthquakes. *Bull. Seism. Soc. Am.*, **54**, 1865.

Knopoff, L. (1964c) The statistics of earthquakes in Southern California. *Bull. Seism. Soc. Am.*, **54**, 1871.

Knopoff, L., and Shapiro, J. N. (1969) Comments on the interrelationships between Grüneisen's parameter and shock and isothermal equations of state. *J. Geophys. Res.*, **74**, 1439.

Kobayashi, K. (1968) Paleomagnetic determination of the intensity of the geomagnetic field in the pre-Cambrian period. *Phys. Earth Planet. Int.*, **1**, 387.

Kohlstedt, D. L., and Goetze, C. (1974) Low stress, high temperature creep in olivine single crystals. *J. Geophys. Res.*, **79**, 2045.

Kolenkiewicz, R., Smith, D. E., and Dunn, P. J. (1973) A reevaluation of the tidal perturbation of the Beacon Explorer C spacecraft (Abstract). *EOS (Trans. Am. Geophys. Un.)*, **54**, 232.

Kolomiytseva, G. I. (1972) Distribution of electric conductivity in the mantle of the Earth, according to data on secular geomagnetic field variations. *Geomagnetism and Aeronomy*, **12**, 938.

Kresák, L., and Millman, P. M. (eds.) (1968) *Physics and dynamics of meteors. I.A.U. Symposium 33*. Dordrecht: Reidel.

Kuroda, P. K., Beck, J. N., Efurd, D. W., and Miller, D. K. (1974) Xenon isotope anomalies in the carbonaceous chondrite Murray. *J. Geophys. Res.*, **79**, 3981.

Lahiri, B. N., and Price, A. T. (1939) Electromagnetic induction in non-uniform conductors, and the determination of the conductivity of the Earth from terrestrial magnetic variations. *Phil. Trans. Roy. Soc. A*, **237**, 509.

Lamb, H. (1963) *Hydrodynamics* (6th ed., reprinted). Cambridge: Cambridge University Press. (1st ed., 1879.)

Lamb, H. H. (1970) Volcanic dust in the atmosphere; with a chronology and assessment of its meteorological significance. *Phil. Trans. Roy. Soc. A*, **266**, 425.

Lambeck, K. (1975) Effects of tidal dissipation in the oceans on the Moon's orbit and the Earth's rotation. *J. Geophys. Res.*, **80**, 2917.

Lambeck, K., and Cazenave, A. (1973) The Earth's rotation and atmospheric circulation—I Seasonal variations. *Geophys. J.*, **32**, 79.

Lambeck, K., and Cazenave, A. (1974) The Earth's rotation and atmospheric circulation—II The continuum. *Geophys. J.*, **38**, 49.

Lambeck, K., Cazenave, A., and Balmino, G. (1974) Solid earth and ocean tides estimated from satellite orbit analyses. *Revs. Geophys. Space Sci.*, **12**, 421.

Lammlein, D. R., Latham, G. V., Dorman, J., Nakamura, Y., and Ewing, M. (1974) Lunar seismicity, structure and tectonics. *Revs. Geophys. Space Phys.*, **12**, 1.

Lanphere, M. A., Wasserburg, G. J., Albee, A. L., and Tilton, G. R. (1964) Redistribution of strontium and rubidium isotopes during metamorphism, World Beater complex, Ponamint Range, California. In Craig et al. (1964), p. 269.

Larimer, J. W. (1973) Chemistry of the solar nebula. *Space Sci. Revs.*, **15**, 103.

Larson, E. E., and Strangway, D. W. (1966) Magnetic polarity and igneous petrology. *Nature*, **212**, 756.

Latham, G., Ewing, M., Dorman, J., Lammlein, D., Press, F., Toksöz, N., Sutton, G., Duennebier, F., and Nakamura, Y. (1971) Moonquakes. *Science*, **174**, 687.

Lee, W. H. K. (ed.) (1965) *Terrestrial heat flow*. Geophysical Monograph No. 8. Washington: American Geophysical Union.

Lee, W. H. K. (1970) On the global variations of terrestrial heat flow. *Phys. Earth Plan. Int.*, **2**, 332.

Leeds, A. R., Knopoff, L., and Kausel, E. G. (1974) Variations of upper mantle structure under the Pacific Ocean. *Science*, **186**, 141.

Le Pichon, X., Francheteau, J., and Bonnin, J. (1973) *Plate tectonics*. Amsterdam: Elsevier.

Leppaluoto, D. A. (1972) Melting of iron by significant structure theory. *Phys. Earth Plan. Int.*, **6**, 175.

Lewis, J. S. (1971) Consequences of the presence of sulfur in the core of the Earth. *Earth Plan. Sci. Letters*, **11**, 130.

Liebermann, R. C., and Ringwood, A. E. (1973) Birch's law and polymorphic phase transformations. *J. Geophys. Res.*, **78**, 6926.

Liebermann, R. C., and Schreiber, E. (1969) Critical thermal gradients in the mantle. *Earth Plan. Sci. Letters*, **7**, 77.

Linde, A. T., and Sacks, I. S. (1972) Dimensions, energy, and stress release for South American deep earthquakes. *J. Geophys. Res.*, **77**, 1439.

Lindemann, F. A. (1910) Uber die Berechnung molecularer Eigenfrequenzen. *Phys. Z.*, **11**, 609.

Liu, L.-g. (1974a) Birch's diagram: some new observations. *Phys. Earth Plan. Int.*, **8**, 56.

Liu, L.-g. (1974b) Silicate perovskite from phase transformations of pyropegarnet at high pressure and temperature. *Geophys. Res. Letters*, **1**, 277.

Longuet-Higgins, M. S. (1950) A theory of the origin of microseisms. *Phil. Trans. Roy. Soc. A*, **243**, 1.

Longuet-Higgins, M. S., and Ursell, F. (1948) Sea waves and microseisms. *Nature*, **162**, 700.

Loper, D. E. (1975) Torque balance and energy budget for the precessionally driven dynamo. *Phys. Earth Plan. Int.*, **11**, 43.

Lorell, J., and Shapiro, I. I. (1973) Mariner 9 celestial mechanics experiment: a status report. *J. Geophys. Res.*, **78**, 4327.

Lovell, A. C. B. (1954) *Meteor astronomy*. Oxford: Clarendon Press.

Lovering, J. F. (1962) Evolution of the meteorites—evidence for the coexistence of chondritic, achondritic and iron meteorites in a typical parent meteorite body. In Moore (1962).

Lowes, F. J. (1966) Mean square value on sphere of spherical harmonic vector fields. *J. Geophys. Res.*, **71**, 2179.

Lowes, F. J. (1974) Spatial power spectrum of the main geomagnetic field, and extrapolation to the core. *Geophys. J.*, **36**, 717.

Lowes, F. J., and Runcorn, S. K. (1951) The analysis of the geomagnetic secular variation. *Phil. Trans. Roy. Soc. A.*, **243**, 525.

Lowes, F. J., and Wilkinson, I. (1963) Geomagnetic dynamo: a laboratory model. *Nature*, **198**, 1158.

Lowes, F. J., and Wilkinson, I. (1967) Laboratory self-exciting dynamo. In Hindmarsh et al. (1967), p. 121.

Lowes, F. J., and Wilkinson, I. (1968) Geomagnetic dynamo: an improved laboratory model. *Nature*, **219**, 717.

MacGregor, I. D. (1974) The system $MgO$-$Al_2O_3$-$SiO_2$: solubility of $Al_2O_3$ in enstatite for spinel and garnet periodotite compositions. *Am. Mineralogist*, **59**, 110.

MacMillan, W. D. (1958) *The theory of the potential.* New York: Dover (reprinted from 1930 ed.).

Malin, S. R. C., and Clark, A. D. (1974) Geomagnetic secular variation, 1962.5 to 1967.5. *Geophys. J.*, **36**, 11.

Malkus, W. V. R. (1963) Precessional torques as the cause of geomagnetism. *J. Geophys. Res.*, **68**, 2871.

Malkus, W. V. R. (1968) Precession of the Earth as the cause of geomagnetism. *Science*, **160**, 259.

Malloy, R. J. (1964) Crustal uplift southwest of Montagu Island, Alaska. *Science*, **146**, 1048.

Mansinha, L., and Smylie, D. E. (1967) Effect of earthquakes on the Chandler wobble and the secular polar shift. *J. Geophys. Res.*, **72**, 4731.

Mansinha, L., and Smylie, D. E. (1968) Earthquakes and the Earth's wobble. *Science*, **161**, 1127.

Mansinha, L., and Smylie, D. E. (1973) Dislocation theory in real Earth models: a comparison of results (Abstract). *EOS (Trans. Am. Geophys. Un.)*, **54**, 1131.

Mansinha, L., Smylie, D. E., and Beck, A. E. (eds.) (1970) *Earthquake displacement fields and the rotation of the Earth*. Dordrecht: Reidel.

Manuel, O. K., Alexander, E. C., Roach, D. V. and Ganapathy, R. (1968) $^{129}I$-$^{129}Xe$ dating of chondrites. *Icarus*, **9**, 291.

Mao, H. K., and Bell, P. M. (1972) Electrical conductivity and the red shift of absorption in olivine and spinel at high pressure. *Science*, **176**, 403.

Mao, N.-h. (1974) Velocity-density systematics and its implication for the iron content of the mantle. *J. Geophys. Res.*, **79**, 5447.

Mark, H., and Fernbech, S. (1968) *Properties of matter under unusual conditions.* New York: Wiley-Interscience.

Marsden, B. G., and Cameron, A. G. W. (eds.) (1966) *The Earth-Moon system*. New York: Plenum Press.

Maruyama, T. (1964) Statical elastic dislocations in an infinite and semi-infinite medium. *Bull. Earthquake Res. Inst., Tokyo*, **42**, 289.

Mason, B. (1962) *Meteorites*. New York: Wiley.

Mason, B. (1966) Composition of the Earth. *Nature*, **211**, 616.

Mason, B. (1971) The lunar rocks. *Scientific American*, **225** (No. 4), 48.

Mason, W. P. (1969) Internal friction mechanism that produces an attenuation in the Earth's crust proportional to the frequency. *J. Geophys. Res.*, **74**, 4963.

Maxwell, A. E. (ed.) (1970) *The Sea, Vol. 4* (2 parts). New York: Wiley-Interscience.

Maxwell, A. E., Von Herzen, R. P., Hsü, K. J., Andrews, J. E., Saito, T., Percival, S. F., Milow, E. D., and Boyce, R. E. (1970) Deep sea drilling in the South Atlantic. *Science*, **168**, 1047.

McConnell, R. K. (1968) Viscosity of the mantle from relaxation time spectra of isostatic adjustment. *J. Geophys. Res.*, **73**, 7089.

McCord, T. B., and Gaffey, M. J. (1974) Asteroids: surface composition from reflection spectroscopy. *Science*, **186**, 352.

McDonald, K. L. (1957) Penetration of the geomagnetic secular field through a mantle with variable conductivity. *J. Geophys. Res.*, **62**, 117.

McDougall, I. (1964) Potassium-argon ages from lavas of the Hawaiian islands. *Bull. Geol. Soc. Am.*, **75**, 107.

McElhinny, M. W. (1971) Geomagnetic reversals during the Phanerozoic. *Science*, **172**, 157.

McElhinny, M. W. (1973) *Palaeomagnetism and plate tectonics*. Cambridge: Cambridge University Press.

McElhinny, M. W., and Briden, J. C. (1971) Continental drift during the palaeozoic. *Earth Plan. Sci. Letters*, **10**, 407.

McElhinny, M. W., and Evans, M. E. (1968) An investigation of the strength of the geomagnetic field in the early preCambrian. *Phys. Earth Planet. Int.*, **1**, 485.

McElhinny, M. W., and Merrill, R. T. (1975) Geomagnetic secular variation over the past 5 million years. *Revs. Geophys. Space Sci.*, **13**, 687.

McElhinny, M. W., and Opdyke, N. D. (1973) Remagnetization hypothesis discounted: a paleomagnetic study of the Trenton Limestone, New York State. *Bull. Geol. Soc. Am.*, **84**, 3697.

McEvilly, T. V., and Johnson, L. R. (1974) Stability of P and S velocities from central California quarry blasts. *Bull. Seism. Soc. Am.*, **64**, 343.

McGarr, A. (1974) Earthquake prediction: absence of a precursive change in seismic velocities before a tremor of magnitude 3-3/4. *Science*, **185**, 1047.

McKavanagh, B., and Stacey, F. D. (1974) Mechanical hysteresis of rocks at low strain amplitudes and seismic frequencies. *Phys. Earth Plan. Int.*, **8**, 246.

McKenzie, D. P. (1970) Temperature and potential temperature beneath island arcs. *Tectonophysics*, **10**, 357.

McKinley, D. W. R. (1961) *Meteor science and engineering*. New York: McGraw-Hill.

McQueen, H. W. S. and Stacey, F. D. (1976) Interpretation of low degree components of gravitational potential in terms of undulations of mantle phase boundaries. *Tectonophysics*, 34, T1.

McQueen, R. G., and Marsh, S. P. (1966) Shock wave compression of iron-nickel alloys and the Earth's core. *J. Geophys. Res.*, **71**, 1751.

McQueen, R. G., Marsh, S. P., and Fritz, J. N. (1967) Hugoniot equation of state of twelve rocks. *J. Geophys. Res.*, **72**, 4999.

Melchior, P. (1966) *The Earth tides*. Oxford: Pergamon.

Melchior, P., and Yumi, S. (eds.) (1972) *Rotation of the Earth*. Dordrecht: Reidel.

Menard, H. W., and Smith, S. M. (1966) Hypsometry of ocean basins. *J. Geophys. Res.*, **71**, 4305.

Mercer, J. H. (1972) Chilean glacial chronology 20,000 to 11,000 carbon-14 years ago: some global comparisons. *Science*, **176**, 1118.

Mercier, J.-C., and Carter, N. L. (1975) Pyroxene geotherms. *J. Geophys. Res.*, **80**, 3349.

Metchnik, V. I., Gladwin, M. T., and Stacey, F. D. (1974) Core convection as a power source for the geomagnetic dynamo—a thermodynamic argument. *J. Geomag. Geoelect.*, **26**, 405.

Metz, W. D. (1974) Plate tectonics: do hot spots really stand still? *Science*, **185**, 340.

Middlehurst, B. M., and Kuiper, G. P. (eds.) (1963) *The moon, meteorites and comets*. Vol. 4, *The solar system*. Chicago: University of Chicago Press.

Miller, G. R. (1966) The flux of tidal energy out of the deep oceans. *J. Geophys. Res.*, **71**, 2485.

Millman, P. M. (ed.) (1969) *Meteorite research*. Dordrecht: Reidel.

Mindlin, R. D., and Cheng, D. H. (1950) Nucleii of strain in the semi-infinite solid. *J. Appl. Phys.*, **21**, 926.

Minear, J. W., and Toksöz, M. N. (1970) Thermal regime of a downgoing slab. *Tectonophysics*, **10**, 367.

Ming, Li-C. and Bassett, W. A. (1975) The postspinel phases in the $Mg_2SiO_4$-$Fe_2SiO_4$ system. *Science*, **187**, 66.

Minster, J. B., and Archambeau, C. B. (1974) Elastodynamic radiation fields from stress relaxation source models (Abstract). *EOS (Trans. Am. Geophys. Un.)*, **55**, 683.

Minster, J. B., Jordan, T. H., Molnar, P., and Haines, E. (1974) Numerical modelling of instantaneous plate tectonics. *Geophys. J.*, **36**, 541.

Mitchell, J. G. (1968) The argon 40/argon 39 method for potassium-argon age determination. *Geochim. et Cosmochim. Acta*, **32**, 781.

Moorbath, S., O'nions, P. K., Pankhurst, R. J., Gale, N. H., and McGregor, V. R., (1972) Further rubidium-strontium age determinations on the very early precambrian rocks of the Godthaab district, West Greenland. *Nature, Phys. Sci.*, **240**, 78.

Moore, C. B. (ed.) (1962) *Researches on meteorites*. New York: Wiley.

Morgan, W. J. (1971) Convection plumes in the lower mantle. *Nature*, **230**, 42.

Morgan, W. J. (1972) Plate motions and deep mantle convection. *Geol. Soc. Am. Mem.*, **132**, 7.

Morse, P. M. (1969) *Thermal Physics, second edition*. New York: Benjamin.

Muelberger, W. R., Denison, R. E., and Lidiak, E. G. (1967) Basement rocks in continental interior of United States. *Bull. Am. Assoc. Petroleum Geologists*, **51**, 2351.

Munk, W. H., and Hassan, E. S. M. (1961) Atmospheric excitation of the Earth's wobble. *Geophys. J.*, **4**, 339.

Munk, W. H., and MacDonald, G. J. F. (1960a) Continentality and the gravitational field of the earth. *J. Geophys. Res.*, **65**, 2169.

Munk, W. H., and MacDonald, G. J. F. (1960b) *The Rotation of the Earth, a geophysical discussion*. Cambridge: Cambridge University Press.

Murnaghan, F. D. (1951) *Finite deformation of an elastic solid*. New York: Wiley.

Murray, B. C., Belton, M. J. S., Danielson, G. E., Davies, M. E., Gault, D., Hapke, B., O'Leary, B., Strom, R. G., Suonic, V., and Trask, N. (1974) Mariner 10 pictures of Mercury: first results. *Science*, **184**, 459.

Murthy, V. R., and Hall, H. T. (1970) The chemical composition of the earth's core: possibility of sulphur in the core. *Phys. Earth Plan. Int.*, **2**, 276.

Myerson, R. J. (1970) Long-term evidence for the association of earthquakes with the excitation of the Chandler wobble. *J. Geophys. Res.*, **75**, 6612.

Nagata, T. (1953) *Rock magnetism*. (1st ed.) Tokyo: Maruzen. (A second edition was published in 1961).

Nagata, T. (1965) Main characteristics of recent geomagnetic secular variation. *J. Geomag. Geoelect.*, **17**, 263.

Nagata, T. (1969) Tectonomagnetism. *Int. Assoc. Geomagnetism and Aeronomy Bull.*, **27**, 12.

Nagata, T., Uyeda, S., and Akimoto, S. (1952) Self-reversed thermoremanent magnetism of igneous rocks. *J. Geomag Geolect.*, **4**, 22.

Nairn, A. E. M. (ed.) (1964) *Problems in palaeoclimatology*. New York: Wiley-Interscience.

Néel, L. (1955) Some theoretical aspects of rock magnetism. *Adv. Phys.*, **4**, 191.

Nersesov, I. L., Semenov, A. N., and Simbireva, I. G. (1969) Space-time distribution of travel times of transverse and longitudinal waves in the Garm region. In Sadovsky (1969), p. 334.

Ness, N. F., Behannon, K. W., Lepping, R. P., Whang, Y. C., and Schatten, K. H. (1974a) Magnetic field observations near Venus: preliminary results from Mariner 10. *Science*, **183**, 1301.

Ness, N. F., Behannon, K. W., Lepping, R. P., Whang, Y. C., and Schatten, K. H. (1974b) Magnetic field observations near Mercury: preliminary results from Mariner 10. *Science*, **185**, 151.

Ness, N. F., Behannon, K. W., Lepping, R. P., and Whang, Y. C. (1975) Magnetic field of Mercury confirmed. *Nature*, **255**, 204.

Newton, R. R. (1969) Secular accelerations of the Earth and Moon. *Science*, **166**, 825.

Nicholls, G. D. (1955) The mineralogy of rock magnetism. *Adv. in Phys.*, **4**, 113.

Nicolaysen, L. O. (1961) Graphic interpretation of discordant age measurements on metamorphic rocks. *Annals of N.Y. Acad. Sci.*, **91**, 198.

Ninkovitch, D., Opdyke, N., Heezen, B. C., and Foster, J. H. (1966) Paleomagnetic stratigraphy, rates of deposition and tephrachronology in North Pacific deep sea sediments. *Earth and Plan. Sci. Letters*, **1**, 476.

Nur, A. (1972) Dilatancy, pore fluids, and premonitory variations of $t_s/t_p$ travel times. *Bull. Seism. Soc. Am.*, **62**, 1217.

Nur, A., and Booker, J. R. (1972) Aftershocks caused by pore fluid flow? *Science*, **175**, 885.

O'Connell, R. J. (1971) Pleistocene glaciation and the viscosity of the lower mantle. *Geophys. J.*, **23**, 299.

Oesterwinter, C., and Cohen, C. J. (1972) New orbital elements for Moon and planets. *Celest. Mech.*, **5**, 317.

Officer, C. B. (1974) *Introduction to theoretical Geophysics*. Berlin: Springer.

O'Keefe, J. A. (1970) Apollo 11: implications for the early history of the solar system. *EOS (Trans. Am. Geophys. Un.)*, **51**, 633.

Oliver, J. (1962) A summary of observed seismic surface wave dispersion. *Bull. Seism. Soc. Am.*, **52**, 81.

Opdyke, N. D. (1970) Paleomagnetism. In Maxwell (1970), Part 1, p. 157.

Opdyke, N. D. (1972) Paleomagnetism of deep-sea cores. *Revs. Geophys. Space Phys.*, **10**, 213.

Opdyke, N. D., Glass, B., Hays, J. D., and Foster, J. H. (1966) Paleomagnetic study of Antarctic deep sea cores. *Science*, **154**, 349.

Öpik, E. J. (1967) Climatic changes. In Runcorn et al. (1967), p. 139.

Oversby, V. M., and Ringwood, A. E. (1971) Time of formation of the Earth's core. *Nature*, **234**, 463.

Oversby, V. M., and Ringwood, A. E. (1972) Potassium distribution between metal and silicate and its bearing on the occurrence of potassium in the Earth's core. *Earth Plan. Sci. Letters*, **14**, 345.

Ozima, M. (1975) Ar isotopes and Earth-atmosphere evolution models. *Geochim. et Cosmochim. Acta*, **39**, 1127.

Schaeffer, O. A., and Zähringer, J. (1966) *Potassium-argon dating*. Berlin: Springer.

Schatz, J. F., and Simmons, G. (1972) Thermal conductivity of earth materials at high temperatures. *J. Geophys. Res.*, **77**, 6966.

Schilling, J.-G. (1975) Azores mantle blob: rare earth evidence. *Earth Plan. Sci. Letters*, **25**, 103.

Schofield, J. C. (1970) Correlation between sea level and volcanic periodicities of the last millenium. *New Zealand J. Geol. Geophys.*, **13**, 737.

Scholz, C., Sykes, L. R., and Aggarwal, Y. P. (1973) Earthquake prediction: a physical basis. *Science*, **181**, 803.

Schramm, D. N. (1973) Nucleo-cosmochronology. *Space Sci. Revs.*, **15**, 51.

Schwarz, E. J., and Symons, D. T. A. (1969) Geomagnetic intensity between 100 million and 2500 million years ago. *Phys. Earth Plan. Int.*, **2**, 11.

Sclater, J. G., and Francheteau, J. (1970) The implications of terrestrial heat flow observations on current tectonic and geochemical models of the crust and upper mantle of the Earth. *Geophys. J.*, **20**, 509.

Seitz, M. G., and Kushiro, I. (1974) Melting relations of the Allende meteorite. *Science*, **183**, 954.

Semenov, A. M. (1969) Variations in the travel-time of transverse and longitudinal waves before violent earthquakes. *Izv. Acad. Sci. USSR., Physics of the Solid Earth*, **1969**(4), 245.

Shamsi, S., and Stacey, F. D. (1969) Dislocation models and seismomagnetic calculations for California 1906 and Alaska 1964 earthquakes. *Bull. Seism. Soc. Am.*, **59**, 1435.

Shankland, T. J. (1967) Transport properties of olivines. In Runcorn (1967), p. 175.

Shankland, T. J. (1968) Band gap in forsterite. *Science*, **161**, 51.

Shankland, T. J. (1970) Pressure shift of infrared absorption bands in minerals and the effect on radiative heat transport. *J. Geophys. Res.*, **75**, 409.

Shankland, T. J. (1972) Simple sound speed systematics (Abstract). *EOS (Trans. Am. Geophys. Un.)*, **53**, 1120.

Shankland, T. J. (1975) Electrical conduction in rocks and minerals: parameters for interpretation. *Phys. Earth Plan. Int.*, **10**, 209.

Shankland, T. J., and Chung, D. H. (1974) General relationships among sound speeds II. Theory and discussion. *Phys. Earth Plan. Int.*, **8**, 121.

Sherby, O. D., and Simnad, M. T. (1961) Prediction of mobility in metallic systems. *Trans. Am. Met. Soc.*, **54**, 227.

Shillibeer, H. A., and Russell, R. D. (1955) The argon-40 content of the atmosphere and the age of the Earth. *Geochim. et Cosmochim. Acta*, **8**, 16.

Short, J. M., and Anderson, C. A. (1965) Electron microprobe analyses of the Widmanstätten structure of nine iron meteorites. *J. Geophys. Res.*, **70**, 3745.

Siebert, M. (1961) Atmospheric tides. *Adv. in Geophys.*, **7**, 105.

Simmons, G. (1964) Velocity of compressional waves in various minerals at pressures to 10 kilobars. *J. Geophys. Res.*, **69**, 1117.

Simpson, J. F. (1967) Earth tides as a triggering mechanism for earthquakes. *Earth Plan. Sci. Letters*, **2**, 473.

Singer, S. F. (1970a) How did Venus lose its angular momentum? *Science*, **170**, 1196.

Singer, S. F. (1970b) Origin of the Moon by capture and its consequences. *EOS (Trans. Am. Geophys. Un.)*, **51**, 637.

Sipkin, S. A., and Jordan, T. H. (1975) Lateral heterogeneity of the upper mantle determined from travel times of ScS. *J. Geophys. Res.*, **80**, 1474.

Slater, J. C. (1939) *Introduction to chemical physics*. New York: McGraw-Hill.

Slichter, L. B. (1967) Free oscillations of the Earth. In Runcorn et al. (1967), p. 331.

Slichter, L. B., MacDonald, G. J. F., Caputo, M., and Hager, C. L. (1966) Comparison of spectra for spheroidal modes excited by the Chilean and Alaskan quakes. (Abstract) *Geophys. J.*, **11**, 256.

Small, J. B., and Parkin, E. J. (1967) Horizontal and vertical crustal movement in the Prince William Sound, Alaska, earthquake of 1964. Paper presented to the fifth United Nations Regional Cartographic Conference for Asia and the Far East, Canberra, March 1967. Washington: E.S.S.A., U.S. Department of Commerce.

Smith, A. G., and Hallam, A. (1970) The fit of the southern continents. *Nature*, **225**, 139.

Smith, E. J., Davis, L., Jones, D. E., Coleman, P. J., Colburn, D. S., Dyal, P., Sonett, C. P., and Frandsen, A. M. A. (1974) The planetary magnetic field and magnetosphere of Jupiter: Pioneer 10. *J. Geophys. Res.*, **79**, 3501.

Smith, E. J., Davis, L., Jones, D. E., Coleman, P. J., Colburn, D. S., Dyal, P., and Sonett, C. P. (1975) Jupiter's magnetic field, magnetosphere, and interaction with the solar wind: Pioneer 11. *Science*, **188**, 451.

Smith, P. J. (1967) The intensity of the ancient geomagnetic field: a review and analysis. *Geophys. J.*, **12**, 321.

Smith, S. W. (1967) Free vibrations of the Earth. In Runcorn et al. (1967), p. 344.

Smith, S. W., and Jungels, P. (1970) Phase delay of the solid earth tide. *Phys. Earth Plan. Int.*, **2**, 233.

Smylie, D. E., and Mansinha, L. (1971) The elasticity theory of dislocations in real Earth models and changes in the rotation of the Earth. *Geophys. J.*, **23**, 329.

Sneddon, I. N. (1961) *Special functions of mathematical physics and chemistry*. Edinburgh: Oliver and Boyd.

Sonett, C. P., Colburn, D. S., Schwartz, K., and Keil, K. (1970) The melting of asteroidal-sized bodies by unipolar induction from a primordial T-Tauri sun. *Astrophys. and Space Sci.*, **7**, 446.

Spall, H. (1972) Paleomagnetism and precambrian continental drift. Proc. 24th International Geological Congress, Montreal, Section 3, p. 172.

Stacey, F. D. (1963) The theory of creep in rocks and the problem of convection in the Earth's mantle. *Icarus*, **1**, 304.

Stacey, F. D. (1964) The seismomagnetic effect. *Pure. and Appl. Geophys.*, **58**, 5.

Stacey, F. D. (1967) Convecting mantle as a thermodynamic engine. *Nature*, **214**, 476.

Stacey, F. D. (1972) Physical properties of the Earth's core. *Geophys. Surveys*, **1**, 99.

Stacey, F. D. (1975) Thermal regime of the Earth's interior. *Nature*, **255**, 44.

Stacey, F. D. (1976) Paleomagnetism of meteorites. *Rev. Earth Plan. Sci.*, **4**, 147.

Stacey, F. D., and Banerjee, S. K. (1974) *The physical principles of rock magnetism*. Amsterdam: Elsevier.

Stacey, F. D., Gladwin, M. T., McKavanagh, B., Linde, A. T., and Hastie, L. M. (1975) Anelastic damping of acoustic and seismic pulses. *Geophys. Surveys*, **2**, 133.

Stacey, F. D., and Johnston, M. J. S. (1972) Theory of the piezomagnetic effect in titanomagnetite-bearing rocks. *Pure Appl. Geophys.*, **97**, 146.

Stauder, W. (1962) The focal mechanism of earthquakes. *Advances in Geophysics*, **9**, 1.

Stauder, W., and Bollinger, G. A. (1966) The focal mechanism of the Alaska earthquake of March 28, 1964, and of its aftershock sequence. *J. Geophys. Res.*, **71**, 5283.

Steinbrugge, K. V., Zacher, E. G., Tocher, D., Whitten, C. A., and Claire, C. N. (1960) Creep on the San Andreas fault. *Bull. Seism. Soc. Am.*, **50**, 389.

Sternheimer, R. (1950) On the compressibility of metallic cesium. *Phys. Rev.*, **78**, 235.

Stocker, R. L., and Ashby, M. F. (1973) On the rheology of the upper mantle. *Revs. Geophys. Space Phys.*, **11**, 391.

Stoneley, R. (1961) The oscillations of the Earth. *Phys. and Chem. of Earth*, **4**, 239.

Strangway, D. W. (1970) *History of the Earth's magnetic field*. New York: McGraw-Hill.

Strutt, R. J. (1906) On the distribution of radium in the Earth's crust and on the Earth's internal heat. *Proc. Roy. Soc. A*, **77**, 472.

Stuart, W. D., and Johnston, M. J. S. (1974) Tectonic implications of anomalous tilt before central California earthquakes (Abstract). *EOS (Trans. Am. Geophys. Un.)*, **56**, 1196.

Subbarao, K. V., and Hedge, C. E. (1973) K, Rb, Sr and $^{87}Sr/^{86}Sr$ in rocks from the mid-Indian Ocean ridge. *Earth Plan. Sci. Letters*, **18**, 223.

Sverdrup, H. U., Johnson, M. W., and Fleming, R. H. (1942) *The oceans—their physics, chemistry and general biology*. Englewood Cliffs, N.J.: Prentice-Hall.

Sweetser, E. I., and Cohen, T. J. (1974) Evidence consistent with a multiple-event mechanism for large earthquakes. *Geophys. Res. Letters*, **1**, 363.

Sykes, L. R. (1967) Mechanism of earthquakes and nature of faulting in the mid-ocean ridges. *J. Geophys. Res.*, **72**, 2131.

Sykes, L. R., Isacks, B. L., and Oliver, J. (1969) Spatial distribution of deep and shallow earthquakes of small magnitudes in the Fiji-Tonga region. *Bull. Seism. Soc. Am.*, **59**, 1093.

Sykes, L. R., and Sbar, M. L. (1973) Intraplate earthquakes, lithospheric stresses and the driving mechanism of plate tectonics. *Nature*, **245**, 298.

Takahashi, T., and Bassett, W. A. (1964) High pressure polymorph of iron. *Science*, **145**, 483.

Takeuchi, H. (1950) On the Earth tide of the compressible Earth of variable density and elasticity. *Trans. Am. Geophys. Un.*, **31**, 651.

Tarling, D. H. (1967) On estimating secular variation from paleomagnetic data. In Collinson et al. (1967), p. 347.

Tarling, D. H. (1971) *Principles and applications of palaeomagnetism*. London: Chapman and Hall.

Tatsumoto, M. (1966) Genetic relations of oceanic basalts as indicated by lead isotopes. *Science*, **153**, 1094.

Tatsumoto, M., Knight, R. J., and Allegre, C. J. (1973) Time differences in the formation of meteorites as determined from the ratio of lead-207 to lead-206. *Science*, **180**, 1279.

Taylor, S. R. (1973) Tektites: a post Apollo view. *Earth Sci. Revs.*, **9**, 101.

Ter Haar, D., and Cameron, A. G. W. (1963) Historical review of theories of the origin of the solar system. In Jastrow and Cameron (1963), p. 1.

Thatcher, W. (1974) Strain release mechanism of the 1906 San Francisco earthquake. *Science*, **184**, 1283.

Thellier, E., and Thellier, O. (1959) Sur l'intensité du champ magnétique terrestre dans le passé historique e geologique. *Ann. de Geophys.*, **15**, 285.

Tilton, G. R., and Steiger, R. H. (1965) Lead isotopes and the age of the Earth. *Science*, **150**, 1805.

Tittman, B. R., Housley, R. M., and Cirlin, E. H. (1973) Internal friction of rocks and volatiles on the Moon. *Geochim. Cosmochim. Acta Suppl.* **4** (Proc. 4th Lunar Science Conference) Vol. 3, p. 2631.

Tocher, D. (1960) Creep on the San Andreas fault: creep rate and related measurements at Vinyard, California. *Bull. Seism. Soc. Am.*, **50**, 396.

Toksöz, M. N., Dainty, A. M., Solomon, S. C., and Anderson, K. R. (1974) Structure of the Moon. *Revs. Geophys. Space Phys.*, **12**, 539.

Tolland, H. G. (1974) Thermal regime in the Earth's core and lower mantle. *Phys. Earth Plan. Int.*, **8**, 282.

Toomre, A. (1966) On the coupling of the Earth's core and mantle during the 26,000-year precession. In Marsden and Cameron (1966), p. 33.

Torreson, O. W., Murphy, T., and Graham, J. W., (1949) Magnetic polarization of sedimentary rocks and the Earth's magnetic history. *J. Geophys. Res.*, **54**, 111.

Tozer, D. C. (1959) The electrical properties of the Earth's interior. *Phys. and Chem. of Earth*, **3**, 414.

Tozer, D. C. (1967) Towards a theory of thermal convection in the mantle. In Gaskell (1967), p. 325.

Tozer, D. C. (1970) Temperature, conductivity, composition and heat flow. *J. Geomag. Geoelect.*, **22**, 35.

Tozer, D. C. (1972) The present thermal state of the terrestrial planets. *Phys. Earth Planet. Int.*, **6**, 182.

Truran, J. W. (1973) Theories of nucleosynthesis. *Space Sci. Revs.*, **15**, 23.

Tsuboi, C., Wadati, K., and Hagiwara, T. (1962) *Prediction of earthquakes—progress to date and plans for further development.* Tokyo: Earthquake Research Institute.

Turcotte, D. L., and Schubert, G. (1971) Structure of the olivine-spinel phase boundary in the descending lithosphere. *J. Geophys. Res.*, **76**, 7980.

Usselman, T. M. (1975) Experimental approach to the state of the core, Parts I and II. *Am. J. Sci.*, **275**, 278 & 291.

Vacquier, V. (1972) *Geomagnetism in marine geology.* Amsterdam: Elsevier.

Valle, P. E. (1954) Adiabatic temperature gradients in the earth's interior. *Annali di Geofisica*, **5**, 41.

Van Dorn, W. G. (1965) Tsunamis, *Advances in Hydroscience*, **2**, 1. *(New York: Academic Press).*

Van Flandern, T. C. (1975) A determination of the rate of change of G. *Mon. Notices R. Astr. Soc.*, **170**, 333.

Van Zijl, J. S. V., Graham, K. W. T., and Hales, A. L. (1962) The palaeomagnetism of the Stormberg lavas of South Africa. *Geophys. J.*, **7**, 23 and 169.

Vashchenko, V. Ya., and Zubarev, V. N. (1963) Concerning the Grüneisen constant. *Soviet Physics—Solid State*, **5**(3), 653.

Verhoogen, J. (1961) Heat balance of the Earth's core. *Geophys. J.*, **4**, 276.

Verhoogen, J. (1965) Phase changes and convection in the Earth's mantle. *Phil. Trans. Roy. Soc. A*, **258**, 276.

Vestine, E. H. (1953) On variations of the geomagnetic field, fluid motions and the rate of the Earth's rotation. *J. Geophys. Res.*, **58**, 127.

Vestine, E. H., and Kahle, A. (1968) The westward drift and geomagnetic secular change. *Geophys. J.*, **15**, 29.

Vestine, E. H., LaPorte, L., Lange, I., Cooper, C., and Hendrix, W. C. (1947a) Description of the Earth's main magnetic field and its secular change, 1905–1945. *Carnegie Institution of Washington*; *Publication* **578**.

Vestine, E. H., Lange, I., LaPorte, L., and Scott, W. E. (1947b) The geomagnetic field, its description and analysis. *Carnegie Institution of Washington; Publication* **580**.

Vine, F. J., and Mathews, D. H. (1963) Magnetic anomalies over ocean ridges. *Nature*, **199**, 947.

Vinogradov, A. P., Surkov, Yu. A., Kirnozov, F. F., and Glazov, V. N. (1973) The content of natural radioactive elements in Venusian rock according to data obtained by the automatic probe 'Venera-8'. *Geochemistry International* (Translation of *Geokhimiya Akad. Nauk SSSR*), **1973**(1), 1.

Von Herzen, R. P., and Maxwell, A. E. (1964) Measurement of heat flow at the preliminary Mohole site off Mexico. *J. Geophys. Res.*, **69**, 741.

Vostryakov, A. A., Vatolin, N. A., and Yesin, O. A. (1964) Viscosity and electrical resistivity of molten alloys of iron with phosphorus and sulphur. *Phys. Metals Metaltography*, **18**(3), 167.

Waddington, C. J. (1967) Paleomagnetic field reversals and cosmic radiation. *Science*, **158**, 913.

Walter, M. R. (1972) Stromatolites and the biostratigraphy of the Australian pre-Cambrian and Cambrian. *Special Papers in Palaeontology*, **11**.

Wänke, H. (1968) Radiogenic and cosmic ray exposure ages of meteorites, their orbits and parent bodies. In Ahrens (1968) p. 411.

Wasserburg, G. J., and Burnett, D. S. (1969) The status of isotopic age determinations on iron and stone meteorites. In Millman (1969), p. 467.

Wasserburg, G. J., MacDonald, G. J. F., Hoyle, F., and Fowler, W. A. (1964) The relative contributions of uranium, thorium and potassium to heat production in the Earth. *Science*, **143**, 465.

Wasson, J. T. (1974) *Meteorites; classification and properties*. Berlin and New York: Springer.

Watkins, N. D., and Goodell, H. G. (1967) Geomagnetic polarity change and faunal extinction in the Southern Ocean. *Science*, **156**, 1083.

Watt, J. P., Shankland, T. J., and Mao, N. H. (1975) Uniformity of mantle composition. *Geology*, **3**, 91.

Weertman, J. (1968) Dislocation climb theory of steady state creep. *Trans. Am. Soc. Met.*, **61**, 680.

Weertman, J. (1970) The creep strength of the Earth's mantle. *Revs. Geophys. Space Phys.*, **8**, 145.

Wegener, A. (1966) *The origin of continents and oceans*. New York: Dover. (This is a translation by J. Biram from the fourth German edition (1929). A translation of the third edition (1922) is also available—Methuen 1924).

Weinstein, D. H., and Keeney, J. (1973) Apparent loss of angular momentum in the Earth-Moon system. *Nature*, **244**, 83.

Weizsäcker, C. F. von. (1943) Uber die Entstehung des Planetensystems. *Z. Astrophysik*, **22**, 319.

Wetherill, G. W. (1968) Stone meteorites: time of fall and origin. *Science*, **159**, 79.

Wetherill, G. W., Mark, R., and Lee-Hu, C. (1973) Chondrites: initial strontium$^{87}$/strontium$^{86}$ ratios and the early history of the solar system. *Science*, **182**, 281.

Whitcomb, J. H., Garmany, J. D., and Anderson, D. L. (1973) Earthquake prediction: variation of seismic velocities before the San Fernando earthquake. *Science*, **180**, 632.

Wier, S. (1975) Martian density models and the moment of inertia. *EOS (Trans. Am. Geophys. Un.)*, **56**, 387.

Wilson, C. R., and Haubrich, R. A. (1974) Atmospheric excitation of the Earth's wobble (Abstract). *EOS (Trans. Am. Geophys. Un.)*, **55**, 220.

Wilson, J. T. (1963) A possible origin of the Hawaiian Islands. *Can. J. Phys.*, **41**, 863.

Wilson, J. T. (1965a) A new class of faults and their bearing on continental drift. *Nature*, **207**, 343.

Wilson, J. T. (1965b) Convection currents and continental drift: evidence from ocean islands suggesting movement in the Earth. *Phil. Trans. Roy. Soc. A*, **258**, 145.

Wilson, R. L. (1962) The paleomagnetism of baked contact rocks and reversals of the Earth's magnetic field. *Geophys. J.*, **7**, 194.

Wilson, R. L. (1972) Paleomagnetic differences between normal and reversed field sources, and the problem of far-sided and right-handed pole positions. *Geophys. J.*, **28**, 295.

Wilson, R. L., and Watkins, N. D. (1967) Correlation of petrology and natural magnetic polarity in Columbia Plateau basalts. *Geophys. J.*, **12**, 405.

Wood, J. A. (1964) The cooling rates and parent planets of several iron meteorites. *Icarus*, **3**, 429.

Wood, J. A. (1967) Chondrites: their metallic minerals, thermal histories and parent planets. *Icarus*, **6**, 1.

Wood, J. A. (1968) *Meteorites and the origin of planets*. New York: McGraw-Hill.

Woolfson, M. M. (1969) The evolution of the solar system. *Reports on Progress in Physics*, **32**(1), 135.

Worrall, G., and Wilson, A. M. (1972) Can astrophysical abundances be taken seriously? *Nature*, **236**, 15.

Wunsch, C. (1974) Dynamics of the pole tide and the damping of the Chandler wobble. *Geophys. J.*, **39**, 539.

Wyllie, P. J. (1971) *The dynamic Earth*. New York: Wiley.

York, D. and Farquhar, R. M. (1972) The Earth's age and geochronology. Oxford: Pergamon.

Yukutake, T. (1972) The effect of change in the geomagnetic dipole moment on the rate of the earth's rotation. *J. Geomag. Geoelect.*, **24**, 19.

Yukutake, T. (1973) Fluctuations in the earth's rate of rotation related to changes in the geomagnetic dipole field. *J. Geomag. Geoelect.*, **25**, 195.

Yukutake, T., and Tachinaka, H. (1969) Separation of the Earth's magnetic field into the drifting and the standing parts. *Bull. Earth. Res. Inst., Tokyo*, **47**, 65.

Zharkov, V. N., and Kalinin, V. A. (1971) Equations of state for solids at high pressures and temperatures (Translated from Russian) New York and London: Consultants Bureau.

# Name Index

# Subject Index

LADY MARGARET HALL LIBRARY
OXFORD

Pandit, B. I., and Tozer, D. C. (1970) Anomalous propagation of elastic energy within the Moon. *Nature*, **226**, 335.

Panella, G. (1972) Paleontological evidence on the earth's rotational history since the early pre-Cambrian. *Astrophys. Space Sci.*, **16**, 212.

Papanastassiou, D. A., and Wasserburg, G. J. (1969) Initial strontium isotopic abundances and the resolution of small time differences in the formation of planetary objects. *Earth Plan. Sci. Letters*, **5**, 361.

Parker, R. L. (1971) The inverse problem of electrical conductivity in the mantle. *Geophys. J.*, **22**, 121.

Parker, R. L. (1972) Inverse theory with grossly inadequate data. *Geophys. J.*, **29**, 123.

Parker, R. L., and Oldenburg, D. W. (1973) Thermal model of ocean ridges. *Nature Physical Science*, **242**, 137.

Pastine, D. J. (1965) Formulation of the Grüneisen parameter for monatomic cubic crystals. *Phys. Rev.*, **138A**, 767.

Paterson, M. S., and Weiss, L. E. (1961) Symmetry concepts in the structural analysis of deformed rocks. *Bull. Geol. Soc. Am.*, **72**, 841.

Patterson, C. (1956) Age of meteorites and the Earth. *Geochim. et Cosmochim. Acta*, **10**, 230.

Pearson, C. E. (1959) *Theoretical elasticity*. Cambridge, Mass.: Harvard University Press.

Pedersen, G. P. H., and Rochester, M. G. (1972) Spectral analyses of the Chandler wobble. In Melchior and Yumi (1972), p. 33.

Pekeris, C. L. (1935) Thermal convection in the interior of the Earth. *Mon. Not. Roy. Astr. Soc. Geophys. Suppl.*, **3**, 343.

Pekeris, C. L., Alterman, Z., and Jarosch, H. (1961) Terrestrial spectroscopy. *Nature*, **190**, 498.

Plafker, G. (1965) Tectonic deformation associated with the 1964 Alaska earthquake. *Science*, **148**, 1675.

Podosek, F. A. (1970) Dating of meteorites by the high-temperature release of iodine-correlated $Xe^{129}$. *Geochim. Cosmochim. Acta*, **34**, 341.

Post, R. L., and Griggs, D. T. (1973) The Earth's mantle: evidence for non-Newtonian flow. *Science*, **181**, 1242.

Press, F. (1965) Displacements, strains and tilts at teleseismic distances. *J. Geophys. Res.*, **70**, 2395.

Press, F. (1970) Earth models consistent with geophysical data. *Phys. Earth Plan. Int.*, **3**, 3.

Press, F., Ben-Menahem, A., and Toksöz, M. N. (1961) Experimental determination of earthquake fault length and rupture velocity. *J. Geophys. Res.*, **66**, 3471.

Press, F., and Briggs, P. (1975) Chandler wobble, earthquakes, rotation and geomagnetic changes. *Nature*, **256**, 270.

Press, F., and Jackson, D. (1965) Alaskan earthquake, 27 March, 1964: vertical extent of faulting and elastic strain release. *Science*, **147**, 867.

Proudman, J. (1953) *Dynamical oceanography*. London: Methuen.

Qamar, A., and Eisenberg, A. (1974) The damping of core waves. *J. Geophys. Res.*, **79**, 758.

Radbruch, D. H., and others (1966) Tectonic creep in the Hayward fault zone, California. *U.S. Geol. Survey Circular*, **525**.

Raitt, R. W. (1969) Anisotropy of the upper mantle. In Hart (1969), p. 250.

Raitt, R. W., Shor, G. G., Morris, G. B., and Kirk, H. K. (1971) Mantle anisotropy in the Pacific Ocean. *Tectonophysics*, **12**, 173.

Randall, M. J. (1973) The spectral theory of seismic sources. *Bull. Seism. Soc. Am.*, **63**, 1133.

Rao, M. N., and Gopalan, K. (1973) Curium-248 in the early solar system. *Nature*, **245**, 304.

Rapp, R. H. (1968) Comparison of two methods for the combination of satellite and gravimetric data. Report No. 113, Dept. of Geodetic Science, Ohio State University, Columbus.

Rapp, R. H. (1973) Numerical results from the combination of gravimetric and satellite data using the principles of least squares collocation. *Report No. 200*, Department of Geodetic Science, Ohio State University, Columbus.

Rapp, R. H. (1974) Current estimates of mean earth ellipsoid parameters. *Geophys. Res. Letters*, **1**, 35.

Reid, H. F. (1911) The elastic rebound theory of earthquakes. *Bull. Dept. Geology. Univ. California*, **6**, 413.

Reynolds, J. H. (1960) The age of the elements in the solar system. *Scientific American*, **203**(5), 171.

Reynolds, M. A., Rao, M. N., Meason, J. L., and Kuroda, P. K. (1969) Fissiogenic and radiogenic xenon in the chondrites Beardsley and Holbrook. *J. Geophys. Res.*, **74**, 2711.

Reynolds, R. T., and Summers, A. L. (1969) Calculations on the composition of the terrestrial planets. *J. Geophys. Res.*, **74**, 2494.

Rice, M. H., McQueen, R. G., and Walsh, J. M. (1958) Compression of solids by strong shock waves. *Solid State Phys.*, **6**, 1.

Richter, C. F. (1958) *Elementary seismology*. San Francisco: Freeman.

Rikitake, T. (1958) Oscillations of a system of disk dynamos. *Proc. Camb. Phil. Soc.*, **54**, 89.

Rikitake, T. (1966a) Westward drift of the equatorial component of the Earth's magnetic dipole. *J. Geomag, Geoelect.*, **18**, 383.

Rikitake, T. (1966b) *Electromagnetism and the Earth's interior*. Amsterdam: Elsevier.

Ringwood, A. E. (1966a) Chemical evolution of the terrestrial planets. *Geochim. et Cosmochim. Acta*, **30**, 41.

Ringwood, A. E. (1966b) Genesis of chondritic meteorites. *Revs. Geophys.*, **4**, 113.

Ringwood, A. E. (1966c) The chemical composition and origin of the Earth. In Hurley (1966), p. 287.

Ringwood, A. E. (1969) Composition and evolution of the upper mantle. In Hart (1969), p. 1.

Ringwood, A. E. (1970a) Origin of the Moon: the precipitation hypothesis. *Earth and Plan. Sci. Letters*, **8**, 131.

Ringwood, A. E. (1970b) Phase transformations and the constitution of the mantle. *Phys. Earth Plan. Int.*, **3**, 109.

Ringwood, A. E. (1971) Internal constitution of Mars. *Nature*, **234**, 89.

Ringwood, A. E., and Green, D. H. (1966) An experimental investigation of the gabbro-ecolgite transformation and some geophysical implications. *Tectonophysics*, **3**, 383.

Ringwood, A. E., and Major, A. (1970) The system $Mg_2SiO_4$-$Fe_2SiO_4$ at high pressures and temperatures. *Phys. Earth Plan. Int.*, **3**, 89.

Roberts, P. H. (1967) *An introduction to magnetohydrodynamics*. London: Longmans.

Roberts, P. H., (1972) Kinematic dynamo models. *Phil. Trans. Roy. Soc. A*, **272**, 663.

Robertson, E. C. (ed.) (1972) *The Nature of the solid Earth*. New York: McGraw-Hill.

Robinson, R., Wesson, R. L., and Ellsworth, W. L. (1974). Variation of P-wave velocity before the Bear Valley, California, earthquake of 24 February, 1972. *Science*, **184**, 1281.

Rochester, M. G. (1960) Geomagnetic westward drift and irregularities in the Earth's rotation. *Phil. Trans. Roy. Soc. A*, **252**, 531.

Rochester, M. G. (1973) The Earth's rotation. *EOS (Trans. Am. Geophys. Un.)*, **54**, 769.

Rochester, M. G. (1974) Can precession power the dynamo? *EOS (Trans. Am. Geophys. Un.)*, **56**, 1108.

Rochester, M. G. (1976) The secular decrease of obliquity due to dissipative core-mantle coupling. *Geophys. J.*, **46**, 109.

Rochester, M. G., Jacobs, J. A., Smylie, D. E., and Chong, K. F. (1975) Can precession power the geomagnetic dynamo? *Geophys. J.*, **43**, 661.

Rochester, M. G., and Smylie, D. E. (1965) Geomagnetic core-mantle coupling and the Chandler wobble. *Geophys. J.*, **10**, 289.

Roden, R. B. (1963) Electromagnetic core-mantle coupling. *Geophys. J.*, **7**, 361.

Ross, J. E. and Aller, L. H. (1976) The chemical composition of the Sun. *Science*, **191**, 1223.

Roufosse, M. C., and Klemens, P. G. (1974) Lattice thermal conductivity of minerals at high temperatures. *J. Geophys. Res.*, **79**, 703.

Roy, A. E. (1967) Bode's law. In Runcorn et al. (1967), p. 146.

Roy, A. E., and Ovenden, M. W. (1954) On the occurrence of commensurable mean motions in the solar system. *Mon. Not. R. Astr. Soc.*, **114**, 234.

Roy, R. F., Blackwell, D. D., and Birch, F. (1968) Heat generation of plutonic rocks and continental heat flow provinces. *Earth Plan. Sci. Letters*, **5**, 1.

Rubey, W. W. (1951) Geologic history of sea water. *Bull. Geol. Soc. Am.*, **62**, 1111.

Runcorn, S. K. (1955) The electrical conductivity of the Earth's mantle. *Trans. Am. Geophys. Un.*, **36**, 191.

Runcorn, S. K. (ed.) (1962) *Continental drift*. New York: Academic Press.

Runcorn, S. K. (1964) Changes in the Earth's moment of inertia. *Nature*, **204**, 823.

Runcorn, S. K. (ed.) (1967) *The application of modern physics to the Earth and planetary interiors*. New York: Wiley.

Runcorn, S. K. (1970) A possible cause of the correlation between earthquakes and polar motion. In Mansinha et al. (1970), p. 181.

Runcorn, S. K., and Urey, H. C. (eds.) (1972) *The Moon* (I.A.U. Symposium 47). Dordrecht: Reidel.

Runcorn, S. K. et al. (eds.) (1967) *International Dictionary of Geophysics*. (2 vols.) Oxford: Pergamon.

Russell, R. D. (1972) Evolutionary model for lead isotopes in conformable ores and in ocean volcanics. *Revs. Geophys. Space Phys.*, **10**, 529.

Sacks, I. S. (1970a) Anelasticity of the outer core. Carnegie Institution, Department of Terrestrial Magnetism Annual Report, 1969–70, p. 414.

Sacks, I. S. (1970b) Anelasticity of the inner core. Carnegie Institution, Department of Terrestrial Magnetism Annual Report, 1969–70, p. 416.

Sacks, I. S. (1971) The Q structure of South America. Carnegie Institution, Department of Terrestrial Magnetism Annual Report, 1970–71, p. 340.

Sadovsky, M. A. (ed.) (1969) *Experimental seismology*. Moscow: Science Publishing House.

Sanz, H. G., Burnett, D. S., and Wasserburg, G. J. (1970) A precise $^{87}$Rb/$^{87}$Sr age and initial $^{87}$Sr/$^{86}$Sr for the Colomera iron meteorite. *Geochim. et Cosmochim. Acta*, **34**, 1227.

Sasajima, S., (1965) Geomagnetic secular variation revealed in the baked earths in West Japan (Part 2). Change of the field intensity. *J. Geomag. Geoelect.*, **17**, 413.

Sassa, K., and Nishimura, E. (1956) On phenomena forerunning earthquakes. *Bulletin of Disaster Prevention Research Institute, Kyoto University*, **13**, 1. (Also *Trans. Am. Geophys. Un.*, **32**, (1951), 1.)

Savage, J. C., and Hastie, L. M. (1966) Surface deformation associated with dip slip faulting. *J. Geophys. Res.*, **71**, 4897.